OXFORD STATISTICAL SCIENCE SERIES

OXFORD STATISTICAL SCIENCE SERIES

A. C. Atkinson: *Plots, Transformations, and Regression*
M. Stone: *Coordinate-Free Multivariable Statistics*
W. J. Krzanowski: *Principles of Multivariate Analysis: A User's Perspective*
M. Aitkin, D. Anderson, B. Francis, and J. Hinde: *Statistical Modelling in GLIM*
P. J. Diggle: *Time Series: A Biostatistical Introduction*

Statistical Modelling in GLIM

MURRAY AITKIN

Department of Statistics, University of Tel Aviv, Israel

DOROTHY ANDERSON

Division of Mathematics and Statistics,
C.S.I.R.O., Melbourne, Australia

BRIAN FRANCIS

Centre for Applied Statistics, University of Lancaster,
Fylde College, Lancaster

JOHN HINDE

Department of Mathematical Statistics and Operational Research,
University of Exeter, Exeter

CLARENDON PRESS · OXFORD

Oxford University Press, Walton Street, Oxford OX2 6DP
Oxford New York Toronto
Delhi Bombay Calcutta Madras Karachi
Petaling Jaya Singapore Hong Kong Tokyo
Nairobi Dar es Salaam Cape Town
Melbourne Auckland
and associated companies in
Berlin Ibadan

Oxford is a trade mark of Oxford University Press

Published in the United States
by Oxford University Press, New York

First published 1989
Reprinted 1989, 1990

British Library Cataloguing in Publication Data
Statistical modelling in Glim.
1. Statistical analysis. Applications of computer systems.
Software packages: GLIM
1. Aitkin, Murray
519.5'028'553
ISBN 0–19–852204–5
ISBN 0–19–852203–7 (Pbk)

Library of Congress Cataloging in Publication Data
Statistical modelling in GLIM / Murray Aitkin . . . [et al.].
p. cm—(Oxford statistical science series)
Bibliography: p. Includes index.
1. GLIM (Computer program) 2. Linear models (Statistics)—Data process. I. Aitkin, Murray A. II. Series.
QA279.S725 1989 519.5—dc19 88–12567
ISBN 0–19–852204–5
ISBN 0–19–852203–7 (Pbk)

Printed in Great Britain
by Biddles Ltd,
Guildford & King's Lynn

Preface

The aim of this book is twofold: to give an exposition of the principles of statistical modelling with the necessary statistical theory, and to describe the application of these principles to the analysis of a wide range of practical examples using GLIM3, the statistical package for Generalized Linear Interactive Modelling developed by the Working Party on Statistical Computing of the Royal Statistical Society.

There are few published accounts of the principles and practice of statistical modelling. The recent book by McCullagh and Nelder (1983), to which we make frequent reference, provides an authoritative account of much of the theory but does not give details of computing systems used for model fitting. Our book gives a full description of the use of GLIM3 for model fitting, and detailed discussions of many examples. The theoretical orientation of our book is slightly different, as we place greater emphasis on the direct use of the likelihood function for inference.

This book was written over the period 1982–1987, as part of an Economic and Social Research Council research programme at the Centre for Applied Statistics in the analysis of complex social data, which supported Dorothy Anderson and John Hinde. Drafts of chapters were developed and revised from regular intensive courses in statistical modelling in GLIM3 given at the Centre and elsewhere over this period. We are grateful to many of the students on these courses, some of them experienced statisticians, whose comments and criticisms improved these early drafts and our own understanding of statistical modelling and GLIM3.

The book can be used in several ways. It is written in a sequence which we have found appropriate for intensive courses. Chapter 1 gives a gentle introduction to GLIM3 for novices, and Chapter 2 a general introduction to the principles of statistical modelling, with two simple examples. This chapter also develops the necessary theory of maximum likelihood estimation and likelihood ratio testing. Chapter 3 discusses the normal model, Chapter 4 binomial data, Chapter 5 multinomial and Poisson data, and Chapter 6 survival data.

The book is designed to be used for graduate and advanced undergraduate courses in statistics, in conjunction with interactive use of GLIM, either in lectures or in practical/tutorial sessions. We have included relevant parts of the actual display from GLIM sessions (mostly plots and parameter estimates) to make the exposition self-contained as far as possible, so that it can be used as a lecture text without access to GLIM if necessary. The book is equally suitable as a self-teaching manual for professional statisticians and research workers. The data sets and macros in Appendix 4 can be bought separately from the University of Lancaster in machine-readable form.

We have made some compromises in coverage in keeping the book to a reasonable length. Our aim has been to concentrate on the central issues of modelling and data analysis. Many GLIM macros for specialized problems have been published in the GLIM Newsletters, in the GLIM82 and GLIM85 conference proceedings, and in several journals in the field of statistical computing.

We are grateful for the generous support of the Economic and Social Research Council and the University of Lancaster. We particularly thank Doreen Hough who processed the manuscript through numerous drafts with great patience and prodigious speed and accuracy. We would like to thank Mick Green, whose macros for the piecewise exponential distribution are included in Chapter 6, and to other colleagues who commented on earlier drafts of this book. We are also grateful to Alan Airth and Ron McKay for their careful checking of the examples: any remaining errors are entirely our responsibility.

1988 MA
Tel Aviv, Melbourne, Lancaster, DA
and Exeter BF
JH

Acknowledgements

We are grateful to the following publishers of journals and books for the use of the datasets reproduced in this book:

American Statistical Association
Biometrika Trustees
Duxbury Press
International Statistical Institute
McGraw-Hill Ryerson Limited
Open University Educational Enterprises
PWS Publishers
Royal Statistical Society
The Biometric Society

Acknowledgements to individual data sources are given in detail in Appendix 4.

Contents

1
Introducing GLIM3

1.1 Statistical packages and modelling

We begin this book by providing an introduction to the main facilities of GLIM3. GLIM3 is an interactive statistical modelling package which provides a powerful set of commands for fitting a wide variety of models to different types of data. The package can be used by both the experienced data analyst and the computer novice, and the command language can be learnt quickly.

We have assumed that GLIM is being used interactively. In other words, we have assumed that the user can interact with the package either on a microcomputer or through a terminal connected to a mainframe or minicomputer. GLIM can also be run in batch mode, that is, by submitting a file of previously assembled commands; however, it is primarily designed as a compact package for interactive use. Batch use of the package may be appropriate for time-consuming tasks such as the fitting of complex models to large sets of data.

We have written this book around GLIM in preference to other packages for two reasons. First, GLIM provides powerful statistical modelling facilities which extend beyond the standard normal linear model and enable many different types of response variable to be modelled in a consistent way. Second, the command language sets GLIM apart from many other interactive statistical packages. Many of these are conversational or provide the user with menus of options. These packages are easy to use, but often restrict the user to analysing data in a fixed way, with little or no flexibility to enable the user to investigate and model the data in different ways.

Statistical modelling is an iterative process. An initial model is developed, fitted and examined. Residuals are plotted, model assumptions checked and parameter estimates examined. Based on this information, a second model may be proposed, possibly involving some transformation of the original data. Further models for the data may then be specified, with the form of the current model being based on the information provided by the previous models. An interactive statistical package is necessary to carry out this iterative process, but it should provide the user with the power to control the direction of the analysis. This power is provided in GLIM by a command language. The user needs to invest some time in learning such a language, but the advantages are numerous. GLIM uses this approach, and has a set of simple commands which are easily memorized. Commands for data manipulation, data transformation, data display and model fitting may be entered, with a few restrictions, in any order. In addition, sequences of frequently used commands (known in GLIM as *macros*) may be named and saved for later use, providing a basic facility for users to define their own procedures.

We have assumed that the reader has access to GLIM3.77, the most recent release of GLIM3. GLIM3.77 provides many extra features which are useful in the statistical modelling process. It also provides a library of useful statistical procedures called the GLIM macro library, and many of these macros are used in the text. We have generally indicated in the text where we have used features of GLIM3.77 which are not available in previous releases of GLIM3, and although it is possible to work through the GLIM examples using a previous release of the package, GLIM3.77 should be used if it is available.

1.2 Getting started in GLIM3

We shall assume that you have already loaded the GLIM program into your computer, and that you are using GLIM interactively. On many computers, the program is loaded simply by issuing a command such as:

```
GLIM
```

or

```
GLIM3
```

Documentation for your particular machine range or type should be consulted if you are unsure how to do this.

On loading GLIM, an introductory message appears:

```
GLIM 3.77 update 2 (copyright)1985 Royal Statistical Society, London

?
```

The question mark which follows the message is the GLIM prompt (in some versions of GLIM the prompt may be different). This prompt tells you that GLIM is waiting to receive instructions or commands from you. If you are using a previous release of GLIM3 a prompt may not appear. Note that we introduce here the convention of representing *input* to and *output* from GLIM in distinctive typefaces, to distinguish them from ordinary text.

Instructions are issued to the GLIM system by means of *directives*. All directives consist of a *directive name*, possibly followed by a set of *items*.

Directive names begin with the *directive symbol*, which is usually a dollar ($). (If the dollar symbol is not available on your keyboard, try another currency symbol such as £.) We assume here that the directive symbol is a dollar. The dollar sign provides the means by which GLIM recognizes directive names. We illustrate this using the $PRINT directive. Try issuing the following simple directive, which instructs GLIM to print the mathematical constant π on the screen.

```
$PRINT %PI $
```

If all is well, GLIM should respond

```
   3.142
?
```

displaying the value of π on the screen.

Following the directive name $PRINT, we have one item %PI which is the GLIM name or *identifier* for the mathematical constant π. The directive is terminated by another dollar symbol which indicates the end of the directive and tells GLIM that no more items are to be printed.

Most versions of GLIM3.77 accept input in lower case as well as upper case. The above directive could have been entered as

```
$print %pi $
```

or even

```
$pRInT %Pi $
```

We adopt the convention of representing GLIM directives in upper case only, which is consistent with previous releases of GLIM3.

All GLIM directive names can be abbreviated to the first four characters, *including the directive symbol.* Thus $PRINT can be abbreviated to $PRI

```
$PRI %PI $
```

Many directive names can also be abbreviated further. The $PRINT directive can be abbreviated to $PR, for example. This is useful for compact code, but can make it more difficult to understand.

More than one item can be printed in the $PRINT directive and text can be printed by enclosing it in primes ('). We can therefore extend our GLIM directive to:

```
$PRI 'THE VALUE OF PI IS ' %PI $
```

GLIM will now display

```
THE VALUE OF PI IS     3.142
```

What happens if we omit the terminating dollar?

```
$PRINT 'THE VALUE OF PI IS ' %PI

$PRI?
```

GLIM responds with a prompt, but preceding the prompt symbol is the abbreviated directive name $PRI. GLIM is indicating that the end of the directive has not been reached. Nothing has been printed as the list of items following the $PRINT directive has not been terminated by a dollar. In response to this prompt, we can either enter a dollar (to indicate the end of the list of items to be printed) or enter further items.

```
$PRI?    ' (TO THREE DECIMAL PLACES).'$

THE VALUE OF PI IS     3.142 (TO THREE DECIMAL PLACES).
```

This provides an illustration of the concept of free format directive input in GLIM. Items can be separated from each other, or from a directive name, by any number of spaces, limited only by the number of characters which can be entered on one line. This is usually set by GLIM installers to be 72 or 80 (this width or *Primary Input channel Length* for your particular installation can be discovered by issuing the GLIM directive $PRI %PIL$). As we have seen, the line width is not a practical restriction, as items can also be spread over many lines of input. Directives need not begin on a fresh line; indeed it is possible to have many directives on one line of input. GLIM recognizes the start of a new directive by finding a directive name, and will then search for items forming part of that directive, until a dollar symbol is encountered. The dollar symbol which begins one directive name may be used as the terminating dollar symbol for the previous directive. Thus, the GLIM directive

```
$PRI                'PI = '                        %PI              $PRI
   'PI WITH MORE DECIMAL PLACES = '       *6        0%PI    $
```

will therefore have the same effect as the more conventional

```
$PRI 'PI = ' %PI $
$PRI 'PI WITH MORE DECIMAL PLACES = ' *6 %PI $
```

and both will display

```
PI =    3.142
PI WITH MORE DECIMAL PLACES =    3.14159
```

We have introduced a new item for the $PRINT directive; the *6 indicates that all items following, within the same directive, are to be printed to six significant figures, and not the usual four. We discuss this further in Section 1.4. Note also that spaces *within* text are taken account of; the spaces within the text string 'PI = ' are printed, not ignored.

Spaces are *not* allowed within directive names, or within any other names defined or set up in GLIM. For instance, the GLIM directives

```
$PRINT % PI $
```

and

```
$ PRINT %PI $
```

will both fail. The first will fail because no spaces are allowed within names of identifiers like %PI. The second will fail because the directive symbol $ is considered to be part of a directive name, and no spaces are allowed within directive names.

Try entering the first of these incorrect directives. A GLIM error message will be produced.

```
** identifier invalid or of incorrect type, at [ $PRINT % ]

The $PRI directive expected an identifier of a different type to the one
entered.  Check the  syntax of the directive and/or the spelling of the
identifier.

?
```

The first part of the message indicates the *fault* which has occurred, and the second part is a *help* message which provides some guidance on the fault. If only the first part of this message was produced, your version of GLIM has been installed with the help messages suppressed by default. If this is so, the directive $HELP will print the help message, and instruct GLIM to provide help messages by default if further errors occur within the same job. In fact, this directive acts as a switch, switching the help facility on if it is off, and switching it off if it is on. $HELP is an example of a directive which consists of the directive name only; no items following the directive name are necessary. GLIM recognizes directives of this type and will not prompt the user for $HELP items if the terminating dollar is omitted.

We see that the prompt symbol is given after the error message. The error has not caused the program to terminate; the user has the opportunity to correct the error and re-enter the offending line, or to enter a completely different directive or directives.

We how examine the error message in more detail. The first line of the fault message reads

```
** identifier invalid or of incorrect type, at [ $PRINT % ]
```

The group of characters enclosed in square brackets is part of the line of input in which the fault was discovered; the last character is the point at which the error was detected by GLIM. The fault occurred at the space following the per cent symbol (%). The fault message on the same line states that "the identifier is invalid or of incorrect type". If we re-examine the input line, we see that the space separating the % character from the characters 'PI' means that % is interpreted by GLIM as an identifier. We are trying to print an identifier called %, and no identifier of this name is allowed in GLIM.

Now try the second of the invalid directives and analyse the error message produced.

There are many further uses of the $PRINT directive, and these are discussed in Section 1.4. From now on, we omit the GLIM prompts from any GLIM output reproduced in this book.

1.3 Reading data into GLIM

We start by introducing a simple set of data. Table 1.1 presents a small set of data, given in Erickson and Nosanchuk (1979), extracted from a larger set collected by Atkinson and Polivy (1976) to examine the effects of unprovoked verbal attack. Ten male and nine female subjects were asked to fill out a questionnaire which mixed innocuous questions with questions attempting to assess the subject's self-reported hostility. A hostility score for each individual was calculated from these responses. After completing the questionnaire, the subjects were then left waiting for a long time, and were subjected to insults and abuse by the experimenter when the questionnaire was eventually collected. All subjects were told that they had filled out the questionnaire incorrectly, and were instructed to fill it out again. A second hostility score was then calculated from these later responses.

For each of the 19 individuals involved in this psychological experiment we have three items of information, or variables:

(1) the hostility score of the individual before being insulted;

(2) the hostility score of the individual after being insulted;

(3) the sex of the individual.

One of the variables (the sex of the individual) is a categorical variable. In this case, the variable "the sex of the individual" has only two categories: male and female. The remaining two variables are continuous—that is, variables which are measured on an underlying continuous scale to a certain measurement accuracy (Section 2.2 contains a further discussion of types of variables). The data are arranged in a data matrix, with the rows of the matrix representing individuals and the columns representing the variables recorded for the individuals. There are no missing data; we have complete information on all the variables for each individual.

Table 1.1 Hostility scores before and after insult for 19 individuals

	Hostility score		
Individual	Before insult	After insult	Sex of individual
1	51	58	Female
2	54	65	Female
3	61	86	Female
4	54	77	Female
5	49	74	Female
6	54	59	Female
7	46	46	Female
8	47	50	Female
9	43	37	Female
10	86	82	Male
11	28	37	Male
12	45	51	Male
13	59	56	Male
14	49	53	Male
15	56	90	Male
16	69	80	Male
17	51	71	Male
18	74	88	Male
19	42	43	Male

How do we define and read this set of data into GLIM? We encountered in the previous section the two identifiers %PI and %PIL. Both of these contain single numbers or *scalars* and both are pre-defined by the GLIM system. Because of this, they are formally known as *system scalars*. A system scalar is only one type of GLIM *structure*; we now introduce a second type of structure called a *vector*. Any vector, once defined, has a fixed length (that is, it can hold a fixed number of data values). A vector also has a name or *identifier*, and this name may be chosen by the user. The data values stored in a vector must be numerical; GLIM cannot read non-numeric characters as data. We can see that vectors are appropriate structures to store the data values for our three variables. The two continuous variables will cause no problem as their values are numeric, but the categorical variable will need to be *coded*, associating a different number with each category.

There are many ways of coding categorical values. For our categorical variable SEX, we could code males as 0 and females as 1, or females as -1 and males as 1: any two different numbers will do. Here, we code females as 1 and males as 2; the reason for this choice will become clear later in this section.

We need to issue *directives* to GLIM to define the length and names of the three vectors that we wish to define. Once this is done, we need to instruct GLIM to read the data values into these vectors.

We first define the length of the data. There are 19 individuals (subjects, cases) in the data matrix.

```
$UNITS 19$
```

GLIM uses the terminology "units" to refer to cases or individuals. The $UNITS directive has one item following the directive name: the "number of units". This directive sets the standard length for all vectors which are to be defined in GLIM. As we shall see, this does not prohibit other vectors being defined with lengths other than the standard length.

The simplest method of defining vectors to be read in, is to use the $DATA directive.

```
$DATA HBEFORE HAFTER SEX$
```

This directive has two functions. First, if the vectors specified as items have not been defined, it will define them with the current standard length set by $UNITS: so here we define three vectors called HBEFORE, HAFTER and SEX, all with standard length 19. The names or identifiers of the vectors have been chosen to be meaningful and can easily be remembered. In general, vector identifiers such as X1 or X2 should be avoided as these provide no information about what each of the vectors contain as data values.

User-defined identifiers in GLIM must start with a letter, which can be optionally followed by any sequence of numbers and letters. Thus, K, B1X, ABCDE and Z123456789 are all valid identifiers. However, only the first four characters are stored by GLIM; the remaining characters in an identifier are simply ignored, and do not cause an error. HBEFORE is therefore stored by GLIM as HBEF. Note that if instead we had chosen the identifiers HOSTB and HOSTA for the hostility scores, GLIM would be unable to distinguish between them. Only one variate would have been defined, and this would be called HOST. Care must therefore be taken in choosing suitable identifiers for user-defined structures. If lower case input is being used, the names will be converted to upper case by GLIM and stored in upper case; GLIM therefore cannot distinguish between the identifiers HBEFORE, hbefore and HBefore. GLIM3.77 also allows the underline symbol (_) to be used in identifiers. The GLIM macro library, however, uses vector identifiers ending with the underline symbol (for instance, RES_, YV_) and it is best to avoid using identifiers of this form if you are using any of the library macros.

The second function of the $DATA directive is to inform GLIM of the names of the vectors which are to be read in. The list of identifiers following the $DATA directive name is known as a *data list*. The reading of the data values is initiated by the $READ directive. In our example, the $READ directive name will be followed by 57 numbers, as follows:

```
$DATA HBEF HAFT SEX $
$READ
 51 58 1
 54 65 1
 61 86 1
 54 77 1
 49 74 1
 54 59 1
 46 46 1
 47 50 1
 43 37 1
 86 82 2
 28 37 2
 45 51 2
 59 56 2
 49 53 2
 56 90 2
 69 80 2
 51 71 2
 74 88 2
 42 43 2  $
```

When the $READ directive is encountered, the data values are read in *parallel*, one unit at a time. For each unit, the data values for the vectors in the data list are read in order. Thus, the first data value of HBEFORE is followed by the first data value of HAFTER, the first value of SEX, the second value of HBEFORE, and so on.

Spaces separate the data values. We are reading the numbers in *free format*; data values may be separated by one or more spaces or new lines. In our example, we have entered the data tidily, with each case or unit starting on a new line and the data values neatly arranged in columns, but this is unnecessary. We could instead have entered

```
$DATA HBEF HAFT SEX $READ   51 58 1   54   65 1
  61 86  1  54
    77 1 49 74  1   54 59 1 46
46  1 47 50 1 43 37  1 86  82   2  28 37 2   45 51 2  59
  56  2 49 53
2 56 90 2 69  80  2 51 71 2  74 88 2
42  43  2    $
```

Note that we do not enter a dollar immediately after the $READ directive name, as this directive takes as items the data values of the vector or vectors being read in. The terminating dollar instead follows the 57th data value.

It is also possible to define and read the data serially, with the data values for each vector being read with separate $READ and $DATA directives.

```
$DATA  HBEFORE$
$READ  51 54 61 54 49 54 46 47 43
         86 28 45 59 49 56 69 51 74 42 $
$DATA  HAFTER $
$READ  58 65 86 77 74 59 46 50 37
         82 37 51 56 53 90 80 71 88 43 $
$DATA  SEX$
$READ   1 1 1 1 1 1 1 1 1 2 2 2 2
         2 2 2 2 2 2 $
```

The first $DATA directive now defines one vector HBEFORE whose values are to be read in. The data list now consists of the single identifier HBEFORE, and the 19 data values following the $READ directive are therefore stored in the vector HBEFORE. As before, the data values are ordered, with the first data value corresponding to unit 1 of HBEFORE, the second to unit 2, and so on.

The second and third $DATA directives redefine the data list twice—first to be the single vector identifier HAFTER, then finally to be the vector identifier SEX. The $READ directive following each $DATA directive will read data values into the vector specified by the currently defined data list.

We have thus defined three vectors, HAFTER, HBEFORE and SEX, and stored data values in them. For the fitting of statistical models and certain other applications, GLIM makes a distinction between two kinds of vectors: *variates* and *factors*. Vectors representing categorical variables may be explicitly defined as factors by using the $FACTOR directive. For a categorical variable with k categories to be represented as a factor, the categories must be coded using the integers $1,2,\ldots,k$. We declare the vector SEX to be a factor:

```
$FACTOR  SEX  2$
```

The $FACTOR directive defines a factor SEX with two values (1 and 2) referred to in GLIM as *levels*. Here, we have coded female subjects as level 1 and male subjects as level 2. Of course, there is nothing to stop us adopting the alternative strategy of coding males as 1 and females as 2; the decision is entirely ours.

If vectors are not explicitly defined as factors, they are by default variates. In this study, we have two variates HBEFORE and HAFTER and one factor SEX. It is possible to define vectors explicitly as variates; the statement

```
$VARIATE HBEFORE HAFTER$
```

defines two variates of standard length called HBEFORE and HAFTER. We could have used the declarations

```
$VARIATE 19 HBEFORE HAFTER  $FACTOR 19 SEX 2$
```

to define the factor and the two variates. If the first item encountered after the directive name $VAR or $FAC is a number, then this is taken to be the length of the vectors which are being declared. As a previous $UNITS directive has been used to set the standard length to 19, the length is redundant as the two variates have standard length. Vectors may be explicitly declared as variates or factors either before or after they have been assigned values.

The above method of reading in variables is obviously suitable only for small data sets. In most applications, the data would be entered into a file outside the GLIM system and stored on disk or other permanent storage medium available on the computer. There are many advantages to this approach; the dataset needs to be typed in to the filestore only once, and if corrections to the data values are necessary, the dataset can be edited. GLIM has facilities for reading files, and we discuss these further in Section 1.9.

An alternative way of reading in variables is to use the $ASSIGN directive. This directive, which is available only in GLIM3.77, is used to assign data values to a vector. It differs from the directives which we have so far encountered, in that the length of the vector does not need to be specified before the values are assigned. If the vector is undefined, GLIM will declare it automatically as a variate and will take the length of the vector from the number of data values entered.

```
$ASSIGN HBEFORE=51,54,61,54,49,54,46,47,43,86,28,45,
                59,49,56,69,51,74,42 $
$ASSIGN HAFTER=58,65,86,77,74,59,46,50,37,82,37,51,
               56,53,90,80,71,88,43 $
$ASSIGN SEX=1,1,1,1,1,1,1,1,1,2,2,2,2,2,2,2,2,2,2 $
$FACTOR SEX 2$
```

Here, as before, we have declared the vector identifier SEX to be a factor *after* the values have been assigned. SEX is declared by default to be a variate by the $ASSIGN directive, then redeclared as a factor by the $FACTOR directive. This is allowed in GLIM; variates can be redeclared as factors and vice versa. This facility is useful in statistical modelling and will be used a great deal in the modelling of the datasets which we introduce in later chapters. The $ASSIGN directive is discussed further in Section 1.8.

1.4 Displaying data using $PRINT and $LOOK

We now return to the $PRINT directive, which was introduced in Section 1.2. This directive can also be used to display the contents of vectors.

```
$PRINT HBEFORE$
```

```
51.00   54.00   61.00   54.00   49.00   54.00   46.00   47.00   43.00
86.00   28.00   45.00   59.00   49.00   56.00   69.00   51.00   74.00
42.00
```

The data values are printed across the screen, to an accuracy of four significant figures and with nine data values on each line. The number of data values on each line is determined solely by the *output line width*; this is set in most installations of GLIM to 72 or 80 characters. Throughout this section, we assume that GLIM is installed with an output line width of 80 and an output line height of 24. Your particular installation of GLIM may have different default values, and your screen output may therefore differ slightly from that given in this book. The values used here may be temporarily reset for your installation of GLIM by using

```
$OUTPUT %POC 80 24 $
```

where %POC is a system scalar available in GLIM3.77 which contains the Primary Output Channel number. (We leave discussion of channel numbers to Section 1.9.)

The accuracy of four significant figures is the default accuracy and we have seen in Section 1.2 that this can be altered by introducing the * symbol followed by a positive number in the range 1–9. This item is known as a *phrase*: the *accuracy phrase* of the $PRINT directive. For example,

```
$PRINT *6 HBEFORE$
```

will produce

```
51.0000   54.0000   61.0000   54.0000   49.0000   54.0000   46.0000
47.0000   43.0000   86.0000   28.0000   45.0000   59.0000   49.0000
56.0000   69.0000   51.0000   74.0000   42.0000
```

It is also possible to use the phrase *−1, to specify that no decimal places are to be printed. Values are rounded to the nearest integer, and are printed with a decimal point.

```
$PRINT *-1 HBEFORE$

51.  54.  61.  54.  49.  54.  46.  47.  43.  86.  28.  45.  59.
49.  56.  69.  51.  74.  42.
```

Specifying the accuracy of printed numbers in this way will only hold for the duration of the $PRINT statement. Subsequent $PRINT statements will use the default accuracy, and if greater accuracy is required, a new accuracy phrase in the $PRINT directive will need to be specified. The $ACCURACY directive provides a way round this difficulty. It takes one item, a number, which should lie in the range from 0 to 9, and is used to reset the *default* accuracy to a higher or lower setting. The new setting remains in force until a new $ACCURACY directive is encountered, or until the end of the job. An accuracy of zero resets the accuracy setting to the GLIM default value of four significant figures.

```
$ACCURACY 6$
$ACC 0$
```

More than one vector may be specified in a $PRINT directive, but care must be taken, as the vectors are printed *serially* and *concatenated*: that is, all the values of the first vector in the list are printed first, followed immediately by the values of the next vector in the list, and so on. A new line is not started for each new vector.

```
$PRINT HBEFORE HAFTER SEX $
```

will produce

```
51.00  54.00  61.00  54.00  49.00  54.00  46.00  47.00  43.00
86.00  28.00  45.00  59.00  49.00  56.00  69.00  51.00  74.00
42.00  58.00  65.00  86.00  77.00  74.00  59.00  46.00  50.00
37.00  82.00  37.00  51.00  56.00  53.00  90.00  80.00  71.00
88.00  43.00  1.000  1.000  1.000  1.000  1.000  1.000  1.000
1.000  1.000  2.000  2.000  2.000  2.000  2.000  2.000  2.000
2.000  2.000  2.000
```

It is hard to tell from the output where one vector ends and another begins. We can circumvent this by using separate $PRINT statements, as the output from a $PRINT directive always starts on a new line.

```
$PRINT HBEFORE  $PRINT HAFTER  $PRINT SEX $
```

This is rather tedious to type; we can improve the above code by using the repetition symbol, which is usually a colon (:).

```
$PRINT HBEFORE : HAFTER : SEX$
```

The repetition symbol means "repeat the last directive name". There are still three separate calls to $PRINT. Both methods will display.

```
51.00   54.00   61.00   54.00   49.00   54.00   46.00   47.00   43.00
86.00   28.00   45.00   59.00   49.00   56.00   69.00   51.00   74.00
42.00
58.00   65.00   86.00   77.00   74.00   59.00   46.00   50.00   37.00
82.00   37.00   51.00   56.00   53.00   90.00   80.00   71.00   88.00
43.00
1.000   1.000   1.000   1.000   1.000   1.000   1.000   1.000   1.000
2.000   2.000   2.000   2.000   2.000   2.000   2.000   2.000   2.000
2.000
```

For clarity, when printing many vectors in serial, it is useful to produce a blank line to separate the values of one vector from that of the next. A blank line is printed by a $PRINT directive with no items. Thus, we could use

```
$PRINT HBEFORE : : HAFTER : : SEX $
```

We mention briefly some other facilities of the $PRINT directive available in GLIM3.77. Names of identifiers can be printed by using the phrase *NAME or *N before the identifier, and real numbers may be printed using the phrase *REAL to a specified field width and number of decimal places, with the number right or left justified within the field. The *INTEGER phrase may be used to print integer numbers, scalars and vectors with a high degree of control over the format. The left and right margins within which the output is to be constrained can also be set explicitly by using the *MARGIN phrase. All phrase names may be abbreviated to the first letter of the name. We give one example.

```
$PRINT *M 1,20 *I SEX, 2$
```

```
1 1 1 1 1 1 1 1 1
2 2 2 2 2 2 2 2 2
2
```

The *M phrase specifies that the left margin for the current print directive is to be set to 1, and the right margin to 20. The *I phrase specifies that the vector SEX is to be printed as integers, allowing two printing positions for each number. Further details on the use of the $PRINT phrases may be found in the GLIM3.77 user guide.

The $PRINT directive provides a flexible facility for displaying vectors serially but cannot be used to print vectors in *parallel*. The $LOOK directive is provided to do this.

```
$LOOK HBEFORE HAFTER SEX$
```

```
     HBEF    HAFT     SEX
 1   51.00   58.00   1.000
 2   54.00   65.00   1.000
 3   61.00   86.00   1.000
 4   54.00   77.00   1.000
 5   49.00   74.00   1.000
 6   54.00   59.00   1.000
 7   46.00   46.00   1.000
 8   47.00   50.00   1.000
 9   43.00   37.00   1.000
10   86.00   82.00   2.000
11   28.00   37.00   2.000
12   45.00   51.00   2.000
13   59.00   56.00   2.000
14   49.00   53.00   2.000
15   56.00   90.00   2.000
16   69.00   80.00   2.000
17   51.00   71.00   2.000
18   74.00   88.00   2.000
19   42.00   43.00   2.000
```

The values of the three vectors are printed in columns, with each row corresponding to one unit. The unit number is displayed in an additional column on the left of the screen. Each column is headed with the identifier of the vector. Subsets of the vector values may be displayed in parallel by specifying the lower and upper unit numbers of the range of units. These integers are positioned immediately following the directive name. For example,

```
$LOOK 5 8 HBEFORE HAFTER SEX$
```

```
    HBEF    HAFT     SEX
5   49.00   74.00   1.000
6   54.00   59.00   1.000
7   46.00   46.00   1.000
8   47.00   50.00   1.000
```

Units 5 to 8 inclusive of the vectors HBEFORE, HAFTER and SEX are displayed in parallel. If only a single unit is of interest, the $LOOK statement may be shortened by omitting the second integer.

```
$LOOK 15 HAFTER HBEFORE $
```

```
     HAFT    HBEF
15   90.00   56.00
```

The $LOOK directive also provides a number of printing styles (GLIM3.77 only). The default style (style 0) prints the vector identifiers as headings to each column and prints the unit number column on the left of the display. Style 1 provides a more attractive display with border lines arount the output and separating columns, while style -1 provides a basic display with no border lines, vector identifiers or unit numbers being printed. Styles are specified in an *option list* immediately following the directive name.

```
$LOOK (STYLE=1) 1 4 HAFTER$
```

or, more succinctly,

```
$LOO(S=1) 1 4 HAFT$
```

```
    +----------+
    |   HAFT   |
+---+----------+
| 1 |  58.00   |
| 2 |  65.00   |
| 3 |  86.00   |
| 4 |  77.00   |
+---+----------+
```

Option lists provide greater user control on the actions of some directives. They are currently used only for directives which print or display information. The *option name* and *option setting* are separated by an equals sign, and the whole is enclosed in brackets. The option name may be abbreviated to one character.

There is no limit to the number of vector identifiers which can follow the $LOOK directive name, but there is a limit to the number of characters which can be displayed on one line of a screen, or on one line of printed output. If the number of vector identifiers specified would cause GLIM to exceed the output line width, then GLIM truncates the list of identifiers until the width of the table is less than the output line width. This may be demonstrated by the following artificial example, which is intended to print eight identical columns of data, with each column containing the values of the vector HAFTER.

```
$LOOK(S=1) HAFTER HAFTER HAFTER HAFTER
             HAFTER HAFTER HAFTER HAFTER $
```

```
-- list truncated
```

A warning message is produced, followed by the truncated output from the $LOOK directive, which displays six columns of identical data values. The remaining two identifiers have been ignored. The warning message may be suppressed in such situations by using the $WARN directive, which acts in a similar fashion to the $HELP directive (Section 1.2). The directive reverses the setting which controls whether or not most warning messages are produced.

It is possible to use our knowledge of the data to print more columns of vectors. The hostility score HAFTER is given in the original table as a vector of integers. We can therefore temporarily reduce the accuracy of the display by using the $ACCURACY directive. As each data value now needs less space to be displayed, more vectors may be printed in parallel.

```
$ACC 1$
$LOOK(S=1) HAFTER HAFTER HAFTER HAFTER
             HAFTER HAFTER HAFTER HAFTER $
$ACC $
```

Using the default style of printing will also allow more columns of vectors to be printed.

1.5 Graphical facilities in GLIM

Statistical modelling of data is aided considerably by the graphical display of data. Two directives, $PLOT and $HISTOGRAM, are provided in GLIM to do this. Try entering the following GLIM statement:

```
$PLOT HAFTER HBEFORE$
```

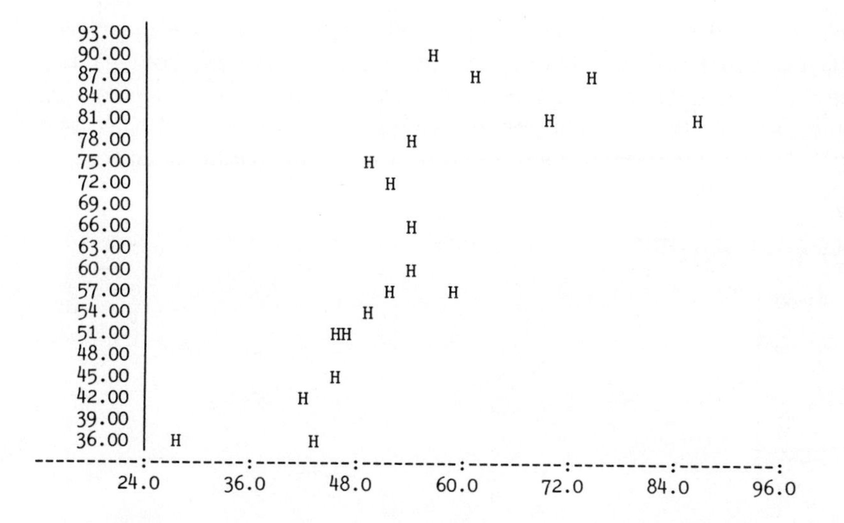

A scatterplot is produced. The $PLOT directive name is followed by two vector identifiers. The first gives the y-axis vector identifier and the second the x-axis vector identifier; both vectors need to have the same length. For each unit, the (x,y) pair of data values is plotted as a point on the scatterplot. In our example, the values of the variate HAFTER are taken as values on the y-axis, and are plotted against the values of the variate HBEFORE. The plot contains 19 points, and each is presented by the initial letter of the y-axis vector identifier, in this case the letter H. The ranges of the axes and the printed scale are automatically chosen by GLIM. The positions of the plotted points are approximate, as there is only a limited number of plotting positions available. The above plot has twenty plotting positions on the y-axis (rows) and sixty plotting positions on the x-axis (columns).

We can see from the plot that there is a strong relationship between the hostility scores before and after the insults. There is also some evidence of increasing spread in the plot as the HBEFORE score increases.

We can improve the look of the plot by using a plotting symbol (not available in previous releases of GLIM3) to replace the letter H. Popular choices for a plotting symbol are +, o, @, * and ×. Any of the characters in the GLIM character set may be chosen as a plotting symbol, with the exception of the dollar symbol, the quote symbol and space. We choose the plotting symbol +, which is placed in quotes following the *x*-axis vector identifier. Any spaces found within the two quote symbols are ignored; thus '+' and ' + ' are equivalent.

```
$PLOT HAFTER HBEFORE '+' $
```

The increasing spread noted in the above plot may be due to different behaviour of the two sexes. However, the plot does not distinguish between male and female subjects; both are plotted with the same symbol (+). If a factor has been defined with a number of levels, we can choose to identify each of these levels with a different plotting symbol. Here, the factor SEX has two levels, and we use the plotting symbol '+' for females (coded 1) and 'o' for males (coded 2).

```
$PLOT HAFTER HBEFORE '+o' SEX $
```

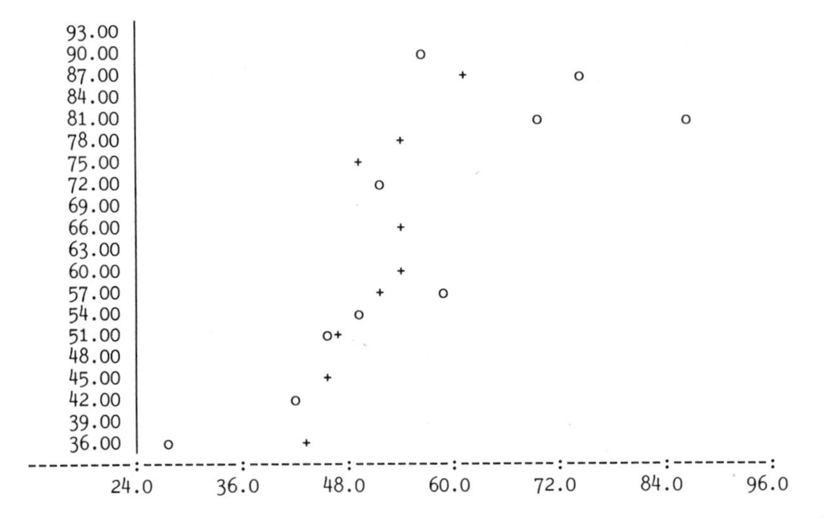

The two plotting symbols, enclosed in quotes, and the factor identifier now follow the *x*-axis vector identifier. The increase in variation with increasing hostility score is evident for males. Increasing variation is less evident for females, partly because the range of HBEFORE scores for females is small compared with that of the males.

It is difficult for the eye to concentrate on one group of points to the exclusion of other groups. Can the males and the females be plotted

separately? In GLIM, plots can be *restricted* so that only a subset of points is plotted. This facility will enable us to plot the before and after hostility scores for the female subjects only. We define a *system vector* %RE (the REstrict vector) which has non-zero values for the points to be included and zero values for the points to be excluded from the plot. This system vector needs to be defined with the same length as the other vectors specified as items to the $PLOT directive. The vector is a system vector because its name is known to GLIM and not user-defined. Again, we note that the system identifier %RE begins with the % symbol. This is a general rule in GLIM: identifiers for system vectors and scalars all begin with the per cent symbol.
We set up the vector %RE to have zero values for males, and non-zero values for females. To produce the desired contents of %RE, we use the $CALCULATE directive.

```
$CALCULATE %RE= 2 - SEX $
```

The expression to the right of the equals sign is evaluated for each element of SEX, and stored in the corresponding element of %RE. As both vectors have length 19, this causes no problem. The factor SEX contains data values which are 1 for female and 2 for male subjects. The expression 2 – SEX will therefore produce values of zero for males, and values of one for females. We leave further explanation of this directive to Section 1.7.

As an alternative to the $CALCULATE directive, we can use the $ASSIGN directive:

```
$ASSIGN %RE=1,1,1,1,1,1,1,1,0,0,0,0,0,0,0,0,0,0,0 $
$PLOT HAFTER HBEFORE '+'$
```

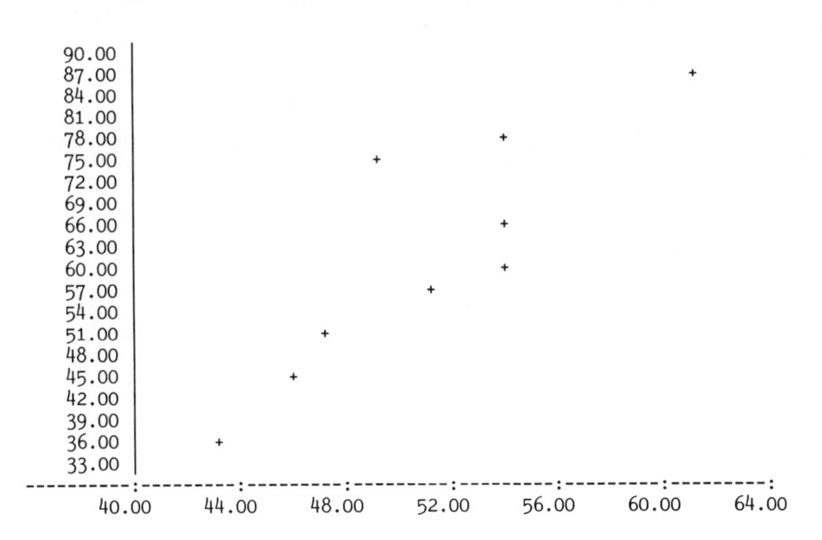

The first nine units are plotted, and a scatterplot is produced for female subjects only. Note that the scales of the restricted plot have changed. The ranges of the *x*- and *y*-axes are calculated from the restricted set of points.

There are some occasions when the original scaling from the unrestricted plot needs to be retained. This may be achieved by using the options XLIMITS and YLIMITS in an option list following the $PLOT directive name. The range of the *x*-axis for the unrestricted plot is from 24 to 96, while that for the *y*-axis is 36 to 93. We respecify the $PLOT directive using the relevant options:

```
$PLOT ( XLIMITS=24,96  YLIMITS = 36,93 ) HAFTER HBEFORE '+' $
```

The option names XLIMITS and YLIMITS may be abbreviated to X and Y respectively. The comma separating the minimum and maximum values of the axis range is obligatory. A scatterplot is produced, but this time the original scaling has been kept and the nine points are bunched in the centre of the *x*-axis. The plot is still restricted to display female subjects only as the contents of the system vector %RE have not been altered.

Further options are available for the $PLOT directive. The ROW and COLUMN options are used to alter the number of rows and the number of columns used for the plot, that is, the number of plotting positions on the *y*-axis and on the *x*-axis. The STYLE option produces a similar effect to the STYLE option on the $LOOK directive. A STYLE of 1 produces a more attractive display, with a box drawn around the plot. The default style is 0, and a style of −1 produces a scatterplot with no axes or scales, but which executes very quickly. All option names may be abbreviated to the first letter only and may be specified in any order. We give one example.

```
$PLOT(S=1 R=10 C=30) HAFTER HBEFORE '+' $
```

```
+----------:---------:---------:
|    90.00 |                   |
|    84.00 |         +         |
|    78.00 |     +             |
|    72.00 |   +               |
|    66.00 |     +             |
|    60.00 |     ++            |
|    54.00 |                   |
|    48.00 |  ++               |
|    42.00 |                   |
|    36.00 | +                 |
|----------:---------:---------:
        40.0      60.0      80.0
```

It is possible to plot more than one *y*-axis vector against the same *x*-axis vector. What happens if we plot both HBEFORE and HAFTER against HBEFORE? We can specify different plotting symbols for the two *y*-axis vectors and we choose / to represent HBEFORE and + to represent HAFTER. We illustrate this for the ten male subjects.

```
$CALCULATE %RE = SEX-1$
$PLOT HBEFORE HAFTER HBEFORE '/+' $
```

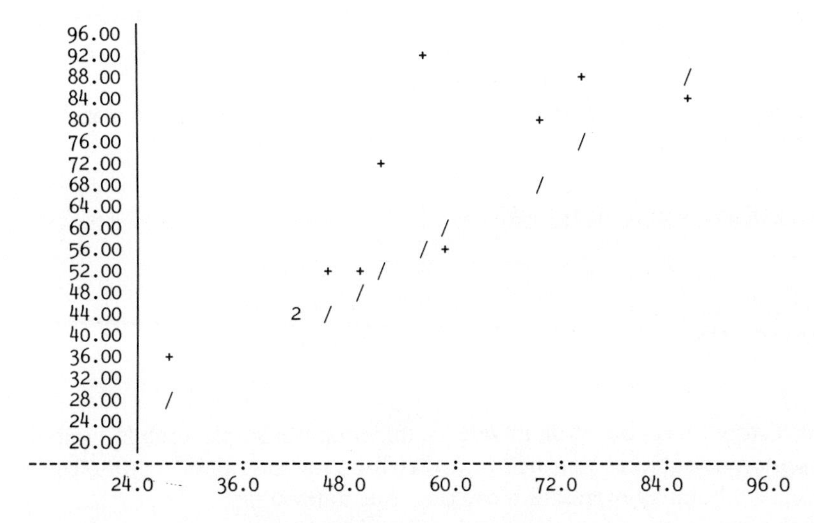

We use the $CALCULATE directive to restrict the plot to male subjects only. The plot of HBEFORE against HBEFORE is represented by the / symbol. The points lie on a straight line (to the accuracy of the plot) and this line represents the "no change" position. The plot of HAFTER against HBEFORE using the plotting symbol + is superimposed on this plot. Points falling above and to the left of the imaginary line formed by the / symbols have higher hostility scores after the experiment than before, and points lying below and to the right have lower hostility scores.

Although there are ten male subjects, only nine + symbols and nine / symbols have appeared in the plot. The two "missing" points are coincident and the figure 2, which also appears in the above plot, indicates two coincident points at that plotting position. These points are in fact not coincident; they correspond to one male with an HBEFORE value of 42 and a HAFTER value of 43. This difference is too small for the coarse resolution of the plot, and they are plotted as coincident values. In general, GLIM represents coincident points by using the plotting symbols 2, 3 up to 8 to indicate the number of points at that plotting position. If nine or more points are coincident at a particular plotting position, then GLIM uses the plotting symbol 9.

We use the $DELETE directive to remove the vector %RE from the workspace, and its name from the list of defined identifiers. The above plot may now be repeated for all nineteen subjects.

```
$DELETE %RE $
$PLOT HBEFORE HAFTER HBEFORE '/+' $
```

The plot is more crowded, with many coincident points.

Histograms

These may be plotted by using the $HISTOGRAM directive. We do not describe this directive in detail, but instead illustrate typical uses of it.

The simplest use of $HISTOGRAM is in the production of a simple histogram of one vector.

```
$HISTOGRAM HBEFORE $
```

```
[27.,42.) 1  H
[42.,57.) 13 HHHHHHHHHHHHH
[57.,72.) 3  HHH
[72.,87.] 2  HH
```

The histogram is printed on its side. In this simple example, each H symbol represents one subject. You will find that for very long data vectors, each plotting symbol may represent more than one data value.

To the left of the histogram, enclosed in brackets, is the range of each interval followed by the count of the number of data values found in that interval. The number of intervals, the width of the intervals used and the total range covered by the histogram are all automatically calculated by GLIM. In this case, we have four intervals each with width 15 hostility score units, and the total range covered is from 27 to 87. What happens to a data value like 42, which falls on the boundary between two adjacent intervals? The convention used for the interval range provides the answer. A square bracket at the beginning or end of an interval range indicates that the end-point of the range is included in the interval, whereas a curved bracket starting or finishing an interval indicates that the corresponding end-point is *not* included in the interval. An interval range of [27.,42.) will include potential data values of 27, 30, 41, 41.9 and 41.9999, but not the end-point 42.0, which is included in the adjacent interval [42.,57.).

It is possible, as in the $PLOT directive, to specify a plotting symbol; this follows the vector identifier.

```
$HIST HBEFORE '*' $
```

A wide range of options is provided for this directive, and the user has a great deal of control over the form of the display produced. The number of

ROWS, the total interval range or YLIMITS, the STYLE and the maximum number of COLUMNS to be plotted may all be specified using option lists. If the option YLIMITS is used, the option TAILS may be used to determine whether values falling outside the interval range are to be plotted. We redisplay the histogram using style 1, specifying 11 rows, with a total interval range of [10,120] and with both tail intervals present. As usual, all option names may be abbreviated to one character.

```
$HIST (S=1 R=11 Y=10,120 T=1,1) HBEFORE '*' $

+-----------+---:-------+
|( -o0, 10.)| 0 |       |
|[ 10., 20.)| 0 |       |
|[ 20., 30.)| 1 |*      |
|[ 30., 40.)| 0 |       |
|[ 40., 50.)| 7 |*******|
|[ 50., 60.)| 7 |*******|
|[ 60., 70.)| 2 |**     |
|[ 70., 80.)| 1 |*      |
|[ 80., 90.)| 1 |*      |
|[ 90.,100.)| 0 |       |
|[100.,110.)| 0 |       |
|[110.,120.]| 0 |       |
|(120., +o0)| 0 |       |
+-----------+---:-------+
                0
```

1.6 Vectors, scalars and the workspace

We begin by reviewing the various types of GLIM structure which are used for storing data values. Data values may be stored in *scalars* or *vectors*. Scalars hold single real numbers, and there are two types of scalars available in GLIM. *System scalars* contain a single number which is pre-defined by GLIM; for instance, the system scalar %POC was introduced in Section 1.4 and it contains the primary output channel. There are also 26 *ordinary scalars* available and these may be assigned values by the user. We saw in Section 1.3 that vectors may be subdivided into variates and factors and in Section 1.5 that vectors may also be subdivided into user-defined vectors and system vectors.

At this point, we also need to mention macros. A macro is an additional type of GLIM structure which is used for storing text or for storing sequences of GLIM directives. We mention them only briefly in this section as they are not normally used to store data values; they are treated in more detail in Section 1.10. We may now summarize the various types of GLIM structure in the following diagram:

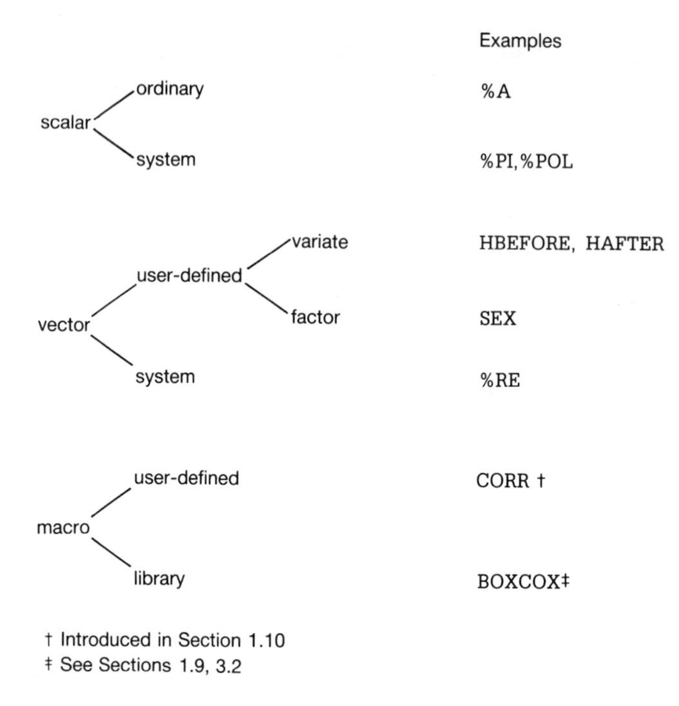

Twenty-six ordinary scalars are available with identifiers %A, %B, through to %Z. The ordinary scalars are available to users for the storage of single numbers. At the start of a GLIM session or job, the contents of all 26 ordinary scalars are set to zero. Remember that GLIM does not distinguish between upper and lower case letters in identifiers, so the identifiers %k and %K will refer to the same ordinary scalar. The identifiers of the ordinary scalars start with a % symbol as their names are all predefined by GLIM.

Scalars are assigned values by means of the $CALCULATE directive. As an example, the directive

```
$CALCULATE %A=4$
```

will assign the real value 4.000 to the ordinary scalar %A. A scalar may be overwritten with a new value at any time during a GLIM session. The above $CALCULATE directive has a simple syntax. Following the $CALCULATE directive name is an *expression*. The equals sign within the expression is used in this case to assign a value to a scalar; on the left of the equals sign is the scalar identifier, on the right is the value which we wish to assign to the scalar.

Many system scalars are defined in GLIM3.77. We have already encountered the following system scalars in earlier sections:

%PI the value of the mathematical constant π;

%POC the primary output channel number;

%PIL the primary input channel length.

We now introduce the following new system scalars:

%NU contains the current Number of Units set by the last $UNITS directive;

%HEL contains 1 if the $HELP facility is switched on, and zero otherwise;

%WAR contains 1 if the $WARN facility is switched on, and zero otherwise;

%ACC contains the current setting of the $ACCURACY directive.

The scalar %NU therefore contains 19, both %HEL and %WAR contain 1 and %ACC contains 4 (the default accuracy setting). We describe many of the remaining system scalars in subsequent chapters of this book. A full list may be found in Appendix 3.

All system scalars in GLIM provide details on the current setting of various features of GLIM. Assigning values to a system scalar does *not* affect that setting. Consider the following piece of GLIM code:

```
$ACC 4 $PRINT %ACC $
$CAL %ACC=9 $PRINT %ACC : HAFTER $
```

The system scalar %ACC has been set to 4 by the $ACCURACY directive. It has then been overwritten with the value 9 by the $CALCULATE statement. When the second $PRINT directive is executed, we see that the vector HAFTER is printed to four significant figures, and not nine. Why is this? The stored accuracy setting within GLIM may only be altered by using the $ACCURACY directive. The system scalar %ACC is a *copy* of the accuracy setting within GLIM. Altering the contents of the copy by overwriting its contents does *not* affect the accuracy setting and should be avoided.

Scalars may be used anywhere in GLIM where a number or an integer is expected. For instance, the $PLOT directive

```
$PLOT(STYLE=1 COLS=50) HAFTER HBEFORE$
```

could be rewritten

```
$CALC %B=1 : %C=50 $
$PLOT (S=%B C=%C ) HAFTER HBEFORE $
```

We emphasize that scalars, in common with all numeric structures in GLIM, store numbers as real values. The scalar %C will therefore not contain the integer 50 but the real value 50. which is stored to the accuracy of the computer. The $PLOT directive, however, expects a columns setting which is a whole number or integer. In such circumstances, GLIM will round the scalar to the nearest whole number.

We have seen that the contents of scalars may be examined by using the $PRINT directive, but we may also use the $LOOK directive.

```
$LOOK %B %A %NU $

1.000   4.000   19.00
```

We see that the output from the $LOOK directive is similar to that obtained when the $LOOK directive is used to display vectors. The main difference is that the scalar names are not displayed above the appropriate scalar value. The STYLE option will also have no effect. Note the difference between the above directive and the very similar

```
$LOOK %B %A HAFTER $
```

where the last item is a vector and not a scalar. Here, %B and %A determine which elements of the vector HAFTER are to be printed.

We now turn our attention to vectors. Unlike scalars, user-defined vectors do not have reserved identifiers. Additionally, GLIM does not reserve a fixed amount of space either for user-defined vectors or for system vectors. User-defined vectors are created by the user as they are needed, and system vectors may either be created by the user (e.g. %RE) or may be initialized and assigned values by the system following certain directives such as $FIT. How may we find out which vectors are currently defined? The $ENVIRONMENT directive is used to do this.

```
$ENVIRONMENT D$
```

produces a *Directory* of the identifiers known to the system.

```
Directory:   levels length  space
   HBEF           0     19     10
   HAFT           0     19     10
   SEX            2     19     10
```

The identifiers listed include system and user-defined vectors and also macro identifiers (Section 1.10). We see that we have three identifiers defined, called HBEF, HAFT and SEX. The "levels" column provides information on which type of structure the identifier represents. Identifiers with zero levels have been defined as variates, whereas identifiers with a positive number of levels have been defined as factors. Although we have no macros defined, we shall see in a later section that they may be identified by a -1 in the "levels" column.

The "length" column gives the defined length of each vector, and a further column, which is labelled by the heading "space", gives the amount of workspace each GLIM data structure takes up. Vectors which have been defined, possibly through a $VARIATE, $FACTOR or $DATA directive, but which do not yet contain data values, will have a positive length but zero space. Note that the space taken up by a vector of a certain length will differ from computer to computer.

GLIM has upper limits on both the number of user identifiers which may be defined and on the amount of workspace which may be used. These limits will vary from installation to installation. The U keyword to the $ENVIRONMENT directive may be used to find out these limits.

```
$ENV U$
```

```
Usage:               used   left
  data space          184  15816
  identifiers           3     97
  model vecs.           1     30
  model terms           1    129
  PCS levels            1     15
```

We are interested in two of the lines of output. The first line is labelled "data space" and informs us of the amount of workspace used and the amount still available for use. Adding these two numbers will give the total amount of workspace. The second line of output gives the current number of user-defined identifiers and the number of user-defined identifiers which may still be defined. We see that in the above installation of GLIM, the workspace available is 16000, and the number of user-defined identifiers allowed is 100. These are adequate values for the statistical modelling used in this book. If your values are substantially below these limits, it may be because your computer can only support a smaller workspace. If you have either a large amount of data or very complex statistical modelling to carry out you would be well advised to extend the workspace and/or the number of identifiers if this is possible on your computer. In this case, the GLIM installer should be contacted as these modifications involve minor changes to the FORTRAN code.

We see that the amount of workspace used (184) is larger than the amount of space taken up by the three vectors (30). This is because the workspace is also used for the storage of scalars and for the storage of the current input line, as well as for temporary workspace when statistical models are being fitted.

The $DELETE directive was introduced in Section 1.5, when the system vector %RE was deleted from the workspace. It is also possible to delete user-defined variates.

```
$DELETE HAFTER$
$ENV D U$
```

We delete the vector HAFTER from the workspace, and re-examine the output from the $ENVIRONMENT directive, with two keywords D and U. The identifier has disappeared from the list of identifiers, and the amount of space used has also decreased. Space has been saved by deleting the structure, and the $DELETE directive is obviously useful for installations with small workspaces.

There are two further directives which it is important to introduce as soon as possible. The $END directive clears the workspace of all defined structures, and will also reset all internal settings back to the GLIM default values. It is used for starting a new piece of work or *job* while still remaining within the same GLIM session.

```
$END

-- new job begins
```

Some other effects of the $END directive are that the number of units will become unset, the help and warning switches will be reset to the default settings, all ordinary scalars will be reset to zero and all system scalars will be overwritten with their system values. The system scalar %JN (the Job Number) will be incremented by one. In later chapters, we will often assume that an $END directive has been issued to clear the workspace before a new dataset is read in and modelled.

The GLIM session is ended by using $STOP. The user is returned to the computer's operating system, and all structures defined are lost. It is possible to save a GLIM session before typing $STOP, and we discuss this in Section 1.9.

1.7 Introduction to calculation and data modification in GLIM

In this section, we examine the directive which is used for general calculation and modification of GLIM structures. We have already encountered the $CALCULATE directive in previous sections, when it was introduced without much explanation. We start by examining this directive in more detail.

The simplest use of the $CALCULATE directive (which we abbreviate to

$CALC) is as a calculator. Consider the following GLIM statement.

```
$CALC 5+6 $

11.00
```

The $CALC directive name is followed by an arithmetic expression "5+6", and the sum of the two numbers 5+6 is calculated and displayed. More complex expressions may be specified, and the displayed accuracy of the results may be controlled by the $ACCURACY directive. The following example converts the temperature of 29 degrees Celsius to degrees Fahrenheit. We multiply the Celsius temperature by 9, divide by 5, then add 32 to the result.

```
$ACCURACY 3$
$CALC  29*9/5 + 32 $

84.2
```

The symbols + * and / , when used in arithmetic expressions, are known as *arithmetic operators*, and represent addition, multiplication and divisions respectively. There are five arithmetic operators defined in GLIM, as follows:

Arithmetic operator	Function	Precedence	
+	Addition	3	low
-	Subtraction	3	
*	Multiplication	2	
/	Division	2	
**	Raise to the power of	1	high

Suppose now that we want to convert the temperature of 84.2 degrees Fahrenheit to degrees Celsius. We need to subtract 32 from the temperature and then multiply the result by 5, then divide by 9. We might therefore try the following expression

```
$ACCURACY 4 $CALC 84.2 - 32 *5/9 $

66.42
```

but, as we see, the result displayed is not the correct value. Each arithmetic operator, as we see from the above table, has a precedence associated with it. Multiplication and division have a higher precedence than addition and subtraction, and the expression 32*5/9 is calculated first. The result (16.78) is then subtracted from 84.2, and the incorrect result of 66.42 is displayed. We can force GLIM to evaluate the subtraction before the multiplication and division by using brackets in the arithmetic expression.

```
$CALC (84.2 - 32)*5/9 $

29.00
```

GLIM first evaluates any part of the arithmetic expression which is enclosed in brackets, then evaluates the rest of the expression according to the precedence rules. The subtraction 84.2 − 32 is therefore evaluated first, and as the remaining operations are of equal precedence, GLIM will evaluate the rest of the expression from left to right, carrying out first the multiplications, and then the division. If we were not sure in which order the multiplication and division were executed, we could force the multiplication to be carried out before the division by using a further pair of brackets:

```
$CALC ((84.2 - 32) * 5 )/ 9 $
```

Note that spaces may appear between items in a calculate expression. Any spaces used in this way are ignored by GLIM, and may be used to improve the visual appearance of a complex calculate expression.

The results of calculations, instead of being displayed on the screen, may be assigned to a scalar or vector. The equals sign, as we noted earlier, is used to assign the result of a calculation to a GLIM data structure. We first declare a standard length for vectors by using the $UNITS directive (Section 1.3)

```
$UNITS 8 $
$CALC %A = 5 $
$CALC A = 5 $
```

The first $CALC statement assigns the single number 5 to the scalar %A. What does the second $CALC statement do? GLIM first checks the list of identifiers to see if the identifier A has been declared. As it has not been declared, GLIM declares it to be a variate of length 8 (the standard length). GLIM then assigns the number 5 to all elements of the variate A. Note the difference between the above $CALCULATE statement and the $ASSIGN directive, which we introduced in Section 1.3.

```
$UNITS 8 $
$ASSIGN A = 5 $
```

will declare a variate called A of length 1, which contains the single number 5 as its only element. We return to the difference between these two directives later in this section.

It is possible to include both scalars and vectors in the calculate expression. We may create a new variate B, for example, to contain the current contents of the scalar %A multipied by 3.

```
$CALC B = %A*3 $
```

The variate B will have the default length of 8 units, and each element of B will contain the value 15. What happens if a vector appears in the calculate expression? We define a variate X containing the values 1,2,. . .,8 in ascending order.

```
$DATA X $READ 1 2 3 4 5 6 7 8$
```

We wish to create a new variate Y containing the values $1,2,2^2,\ldots\ldots,2^7$. For each value of X, the transformation $y=2^{(x-1)}$ will produce the required value of Y. Initially, we use the $CALC directive to display the result of the calculation.

```
$CALC  2 ** (X-1) $

1.000
2.000
4.000
8.000
16.00
32.00
64.00
128.0
```

The required result is achieved, with the calculation being performed separately for each element of X. We may now assign the result of this calculation to a new vector Y.

```
$CALC  Y = 2 ** (X-1) $
```

The variate Y is created with length 8, and each element of Y contains the transformation of the corresponding element of X. It is also possible to assign the result of the above calculation to a scalar instead of a vector. What is the effect of this?

```
$CALC %B = 2 ** (X-1) $PRINT *-1 %B$

128.
```

The scalar %B contains the value 128. It has taken the result of the calculation for the last element of the vector X. What has happened is that the calculate expression has been performed separately for each element of the vector X, beginning with the first element and ending with the last. The scalar %B therefore contains first the value 1, then 2, and so on, until the last element is reached and %B finally contains 128.

We may see this effect by examining a more complicated example.

```
$CALC %C=0 : %C = %C + X $
```

We might imagine that %C would contain the value 8, as the last element of X is 8, and %C has been initialized to the value 0. What is the result of the calculation?

```
$PRINT *-1 %C $

36.
```

The calculation has been performed *separately* for each element of the vector X in turn, starting with the first element. As each element of the vector X is processed in order, the scalar %C is updated. The scalar %C, having been initialized to zero, will therefore contain 1 after the first element of X has been dealt with, 1+2 after the second, and so on, until all eight elements of X have been treated and the scalar %C, having been updated eight times, contains 1+2+3+4+5+6+7+8, or 36. We have found the sum of the vector X. As an aside, the mean of the vector X may be found by dividing by the length of the vector, in this case the standard length or number of units, which is stored in the scalar %NU.

```
$CALC %M=%C/%NU $PRINT %M$

4.500
```

The contents of *defined* vectors may be overwritten at any time by using the $CALCULATE directive. Vectors may be transformed

```
$CALC Y = (Y+1)/2 $
```

or reassigned

```
$CALC Y = X**3 $
```

In this way, any existing vectors may be assigned new values. The deletion of the vector is not required, and the vector retains both its length and its type (that is, whether it is a variate or a factor).

So far, we have dealt with calculate expressions which contain vectors of standard length. What happens when we include vectors which are not of standard length? We increase the length of the vector X by using the $ASSIGN directive and adding two additional values to the end of the vector.

```
$ASSIGN X = X,9,10 $
```

The above $ASSIGN directive has the same effect as the alternative

```
$ASSIGN X=1,2,3,4,5,6,7,8,9,10 $
```

In both cases, the length of the vector X is increased from 8 to 10 units. The length of the vector Y has remained set at the standard length. We try combining the two vectors in a calculate expression.

```
$CALC Z = 2*X + Y$
```

A GLIM error message is produced.

```
** invalid or mixed lengths, at [= 2*X + Y$]

The vector Y has length 8.  This conflicts with the length 10 used elsewhere in
this directive.
```

All previously defined vectors appearing in the same calculate expression must have the same length. In a similar way, the directive

```
$CALC B = X/2 $
```

is illegal, as the vector X has length 10, whereas B has length 8. The length of B may not be redefined by the $CALC directive.

There is one exception to the above rule on the length of vectors. Let us declare a variate of length 1 called CONST:

```
$VAR 1 CONST$
```

Based on what we have seen above, we might expect the GLIM directive

```
$CALC CONST=X/2$
```

also to be illegal, as CONST and X are both vectors of different length. What GLIM does, however, is to treat vectors of length 1 as if they were scalars. The above expression is not illegal, and the vector CONST will contain the value of the last element of the vector X divided by 2.

```
$PRINT *-1 CONST$

5.
```

Apart from this exception, we have seen that all previously defined vectors must have the same length. It is, however, possible to define *new* variates which have a length other than the standard length. The GLIM directive

```
$CALC Z = X**2 $
```

will define a variate Z of length 10 (not the standard length). As the vector Z is undefined when the calculate expression is executed, the length of Z is defined to be the length of the vectors in the calculate expression, in this case 10 units. As we have seen earlier, if there are no vectors in the calculate expression, then the standard length is taken.

So far, we have covered only one assignment within the $CALCULATE directive. The syntax of the directive is quite general, and *multiple* assignments are allowed. A simple example is

```
$CALC %A=%B=10 $
```

where the scalar %B is set equal to 10, and the scalar %A is set equal to the scalar %B. The rightmost assignment is executed first, and execution of further assignments continues from right to left. Quite complex expressions may be built up in this way, although these are best avoided until some experience of the GLIM system has been obtained. We give a further example with no explanation. This stores the mean value of the elements of X in %M by combining two $CALC expressions given earlier in this section.

```
$CALC %C=0 $
$CALC %M = (%C=%C+X) / %NU $
$END
```

1.8 Function evaluation and suffixed vectors

In the previous section, we introduced the $CALCULATE directive and showed how it may be used to perform calculations to assign new values to existing scalars and vectors, and to define new vectors. We continue in this section to describe further features of the $CALC directive, in particular the use of *functions*, and introduce some new directives which may be used for data transformation and definition.

Feigl and Zelen (1965) presented data on 33 patients suffering from leukaemia. For each patient, the white blood cell count in thousands at the time of diagnosis and the time in weeks to death from initial diagnosis were recorded, together with the test result on the AG-factor at the time of diagnosis (positive or negative). The data are given below and are modelled in detail in Chapter 6.

AG-factor positive			AG-factor negative		
Patient number	White cell count (WBC)	Survival time (weeks)	Patient number	White cell count (WBC)	Survival time (weeks)
1	2.30	65	18	4.40	56
2	0.75	156	19	3.00	65
3	4.30	100	20	4.00	17
4	2.60	134	21	1.50	7
5	6.00	16	22	9.00	16
6	10.50	108	23	5.30	22
7	10.00	121	24	10.00	3
8	17.00	4	25	19.00	4
9	5.40	39	26	27.00	2
10	7.00	143	27	28.00	3
11	9.40	56	28	31.00	8
12	32.00	26	29	26.00	4
13	35.00	22	30	21.00	3
14	100.00	1	31	79.00	30
15	100.00	1	32	100.00	4
16	52.00	5	33	100.00	43
17	100.00	65			

Reproduced from: P. Feigl and M. Zelen, 'Estimation of exponential probabilities with concomitant information'. *Biometrics* **21**, 826–38 (1965). With permission from the Biometric Society.

There are two variates, the survival time in weeks (TIME) and the white blood cell count (WBC), and one factor (AG). We first clear the workspace by using the $END directive. In this example, we do not immediately read in the values of the factor.

We declare the number of units to be 33, define the two variates using the $DATA directive, and read the data into GLIM. Note the use of the $COMMENT directive after the data values. All text between the $COMMENT directive name and the terminating dollar symbol is treated as a comment and ignored. The comment may extend over more than one line, as in this example. The directive name may be abbreviated to $C.

```
$END
$UNITS 33 $
$DATA WBC TIME $READ
  2.30  65    .75 156    4.30 100   2.60 134    6.00 16 10.50 108
 10.00 121  17.00   4    5.40  39   7.00 143    9.40 56 32.00  26
 35.00  22 100.00   1  100.00   1  52.00   5  100.00 65  4.40  56
  3.00  65   4.00  17    1.50   7   9.00  16    5.30 22 10.00   3
 19.00   4  27.00   2   28.00   3  31.00   8   26.00  4 21.00   3
 79.00  30 100.00   4  100.00  43
$COMMENT
  VARIABLES
   WBC     ... WHITE BLOOD CELL COUNT IN THOUSANDS
   TIME    ... SURVIVAL TIME IN WEEKS FROM TIME OF DIAGNOSIS
$
```

We first plot the survival time against the white blood cell count.

```
$PLOT TIME WBC$
```

```
   190.0 |
   180.0 |
   170.0 |
   160.0 T
   150.0 |
   140.0 |    T
   130.0 |T
   120.0 |     T
   110.0 |     T
   100.0 |  T
    90.0 |
    80.0 |
    70.0 |TT                                              T
    60.0 | T  T
    50.0 |
    40.0 |  T                                             T
    30.0 |                  T                   T
    20.0 | T2 T               T
    10.0 |T                 T         T
     0.0 |    T   TTT T2                                  3
---------:---------:---------:---------:---------:---------:---------:
        0.0      20.0      40.0      60.0      80.0     100.0     120.0
```

The plot is difficult to interpret. A relationship may exist between the variates WBC and TIME, with the survival time of the patients decreasing as the white blood cell count increases, but the points are crowded close to the x- and y-axes. A plot of the log of the survival time against the log of the white blood cell count would be useful. We calculate a new variate in GLIM by using the %LOG function.

```
$CALC LWBC=%LOG(WBC)$
```

Each element of the new vector LWBC contains the LOG (to base *e*) of the corresponding element of the vector WBC. The function %LOG has one *argument*; this is usually a number, scalar or vector, but can be a calculate expression. The argument is enclosed in brackets after the function name.

```
%LOG(          )
          |
    argument to the
    %LOG function
```

We see that the name of the function %LOG begins with the % symbol. We have encountered this symbol before in the identifiers of ordinary scalars and system scalars and vectors, where we saw that it was used for identifiers which are not user-defined but known to GLIM at the start of a session. The names of GLIM functions also belong to this category, and thus start with the % symbol. The symbol is known formally as the function symbol.

Functions may be used flexibly in calculate expressions. The expression

```
$CALC %LOG(3) $
```

will calculate the log of 3 and display it. Calculate expressions may be used as arguments to any function.

```
$CALC LTD = %LOG(TIME*7) $
```

It is therefore possible for function calls to be nested, as follows:

```
$CALC LLT = %LOG(%LOG(TIME*7)) $
```

The log function is only defined for non-negative values and the log of zero is defined to be minus infinity. How does the GLIM %LOG function treat zero and negative values? Using the $ASSIGN directive, we create a small vector TEST containing positive, zero and negative values and see what happens.

```
$ASSIGN TEST = 3,2,1,0,-1,-2,-3 $
$CALC LTEST = %LOG(TEST) $

-- invalid function/operator argument(s)
```

A warning message appears, informing us that invalid function arguments have been used. However, when we examine the vector LTEST, we see that for the valid (positive) values of TEST, the correct log value has been stored.

```
$PRINT TEST : : LTEST$

3.000   2.000   1.000  0.       -1.000  -2.000  -3.000

1.099  0.6931  0.      0.       0.      0.      0.
```

The invalid values of TEST have produced values of zero for the vector LTEST. This is a feature of GLIM: if an argument to a function is invalid, then GLIM will assume that the result is zero, and continue with the calculation. We delete these two vectors and return to the white blood cell count data.

```
$DELETE TEST LTEST $
```

Our next task is to generate the values of the factor. We first declare the vector AG to be a factor with two levels. We may then use the function %GL to "Generate the Levels" of the factor.

```
$FACTOR AG 2 $
$CALC AG = %GL(2,17) $
$PRINT *-1 AG $

1.  1.  1.  1.  1.  1.  1.  1.  1.  1.  1.  1.  1.  1.  1.
1.  1.  2.  2.  2.  2.  2.  2.  2.  2.  2.  2.  2.  2.  2.
2.  2.  2.
```

The factor AG contains the correct values. The first 17 values are set to 1, and the last 16 values are set to 2. How has this been achieved?

The function %GL has two *arguments*, which are enclosed in brackets following the function name, and separated by a comma. We represent the two arguments by the letters k and n; these would be replaced by numbers or by scalar identifiers when the function is used. Both k and n need to be positive integers.

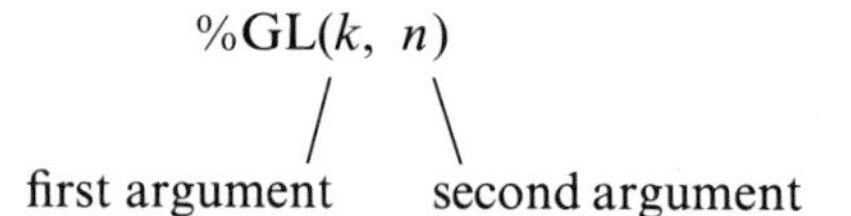

The %GL function is used to assign regular blocks of data values to a vector. The first n values of the vector are given the value 1, the next n values are given the value 2, and so on. This process continues until the value k is reached, then the process repeats itself, with the next n values being given the value 1, and so on, until all the values of the vector have been assigned.

There are many other useful arithmetical functions, including the exponential function %EXP, the sine function %SIN, the square root function

%SQRT, the normal deviate function %ND and the normal probability function %NP. A full list may be found in Appendix 3. A particularly useful function is the cumulative sum function %CU. The GLIM statement

```
$CALC %M = %CU(WBC) $
```

will calculate the sum of the values of the vector WBC and store the result in the scalar %M. If the result is assigned to a vector, rather than a scalar, the vector will contain the partial sums of WBC.

```
$CALC CWBC = %CU(WBC) $
$PRINT CWBC $

2.300   3.050   7.350   9.950   15.95   26.45   36.45   53.45   58.85
65.85   75.25   107.3   142.3   242.3   342.3   394.3   494.3   498.7
501.7   505.7   507.2   516.2   521.5   531.5   550.5   577.5   605.5
636.5   662.5   683.5   762.5   862.5   962.5
```

The function therefore provides a simpler method than that described in the last section for the calculation of sums and sums of squares of a vector. For example,

```
$CALC %M=%CU(WBC)/%NU : %V=%CU( (WBC-%M)**2 )/%NU $
```

will store the mean of the vector WBC in the scalar %M and its variance in the scalar %V.

The %CU function may also be used to construct an index (or observation number) vector—a vector which contains the values 1,2,. . .%NU. This is simply achieved by

```
$CALC IND=%CU(1) $PRINT *-1 IND$

1.   2.   3.   4.   5.   6.   7.   8.   9.   10.   11.   12.   13.   14.
15.   16.   17.   18.   19.   20.   21.   22.   23.   24.   25.   26.   27.
28.   29.   30.   31.   32.   33.
```

The vector IND is taken to be of standard length, and the cumulative sum of a vector of 1s of standard length is evaluated. Other numbers apart from 1 may be used as the argument to the %CU functions and the vector length may be defined explicitly, enabling a regularly spaced, ordered sequence of values to be produced easily. Such vectors are particularly useful in function evaluation and plotting. For example, we may evaluate the function $\sin(x)/x$ for $x = 0.25(0.25)10$ radians by the following GLIM statements:

```
$VAR 40 X $CALC X=%CU(0.25) : FX=%SIN(X)/X $PLOT FX X $
```

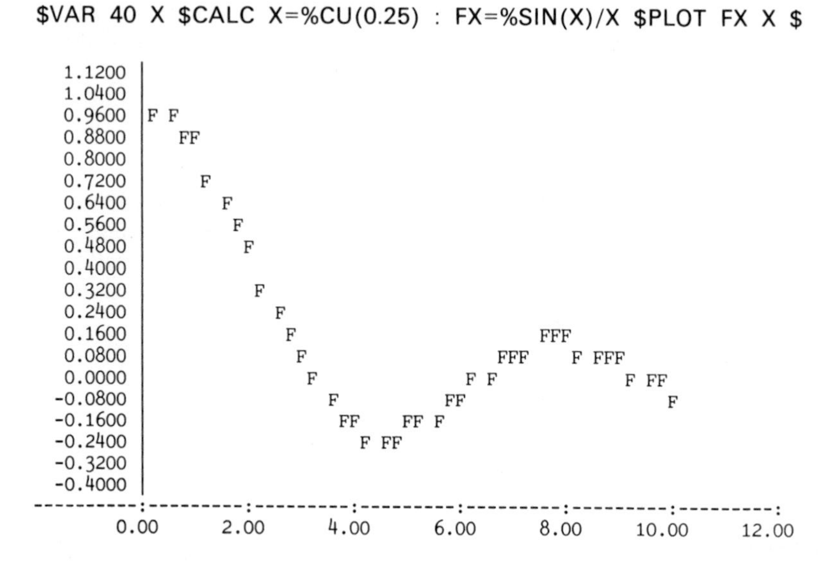

We return to the white blood cell data introduced above. The data are stored in vectors of length 33. For modelling purposes we may wish to identify a subset of the data, for example, all subjects with white blood cell count greater than 10.0 and with AG factor positive, and then to construct a new factor identifying these observations with a unique code. We may also wish to extract this subset of the data matrix into new vectors of the appropriate length. How do we do this in GLIM?

This problem involves both the identification of the relevant cases satisfying a certain criteria and the extraction of these cases from the data into new vectors. We achieve the first through the use of *logical functions.*

Logical functions are a particular type of function in GLIM which return only two values, zero (FALSE) and one (TRUE). We begin with a simple example.

```
$CALC A =%EQ(AG,1) $
$PRINT A : AG $

 1.000   1.000   1.000   1.000   1.000   1.000   1.000   1.000   1.000
 1.000   1.000   1.000   1.000   1.000   1.000   1.000   1.000  0.       0.
0.      0.      0.      0.      0.      0.      0.      0.      0.       0.
0.      0.      0.      0.
 1.000   1.000   1.000   1.000   1.000   1.000   1.000   1.000   1.000
 1.000   1.000   1.000   1.000   1.000   1.000   1.000   1.000   2.000
 2.000   2.000   2.000   2.000   2.000   2.000   2.000   2.000   2.000
 2.000   2.000   2.000   2.000   2.000   2.000
```

The function %EQ in the above calculate statement is a logical function, and is used to test for equality of two items. The items are specified as arguments to the function, and either item may be a vector, a scalar, a number or an expression which evaluates to a vector, scalar or number. In our example, if AG is equal to 1, then the %EQ function is TRUE and the value 1 will be stored in the relevant elements of the vector A. The remaining elements of A, corresponding to elements of AG which are not equal to 1, will contain zeros.

It is also possible to test for equality by using a *logical operator* rather than a logical function. The logical operator for testing for equality is a double equals sign (= =), and we may rewrite the above calculate expression.

```
$CALC A = (AG==1) $
```

The result is the same, with the vector A containing ones and zeros as before. The brackets in the expression are not necessary, but help to separate the assignment from the logical test. In this book, we will use logical functions rather than logical operators, since the latter are available only in GLIM 3.77.

There are six logical functions available. We give them below, with their equivalent logical operators. We assume here that A and B are expressions of equal length.

Logical function	Logical operator	Test
%EQ(A,B)	A==B	A equal to B
%NE(A,B)	A/=B	A not equal to B
%GT(A,B)	A>B	A greater than B
%LT(A,B)	A<B	A less than B
%GE(A,B)	A>=B	A greater than or equal to B
%LE(A,B)	A<=B	A less than or equal to B

Combinations of the logical functions may be made. *Multiplying* two logical functions will have the effect of a logical "and" operator. For instance, in the statement

```
$CALC AGWB = %EQ(AG,1)*%GT(WBC,10) $
```

the expression on the right hand side of the assignment will be true only if the AG factor is positive *and* the white blood cell count is greater than 10.0. We have identified the observations required for our example; their positions are the non-zero elements of vector AGWB.

```
$PRINT AGWB $

0.      0.      0.      0.      0.      1.000  0.      1.000  0.      0.
0.      1.000   1.000   1.000   1.000   1.000  1.000   0.     0.      0.
0.      0.      0.      0.      0.      0.     0.      0.     0.      0.
0.      0.      0.
```

The number of non-zero observations satisfying the above condition may be found by using the %CU function

```
$CALC %A=%CU(AGWB) $PRINT %A$
```

and the corresponding observation numbers may be found from the non-zero elements of the vector POS, calculated as follows:

```
$CALC POS=AGWB*%CU(1)$
$PRINT %A :   POS $
```

There are 8 of the 33 observations satisfying our condition, corresponding to observations 6, 8, 12, 13, 14, 15, 16 and 17. We may now construct a new factor, identifying those observations with white blood cell count greater than 10.0 and AG factor positive with code 1, and all other observations with code 2. This may be achieved by using an arithmetic calculate expression.

```
$CALC AW = 2 - AGWB $
$FACTOR AW 2 $
```

Alternatively, we may use the *conditional assignment* function %IF. The %IF function has three arguments, as follows:

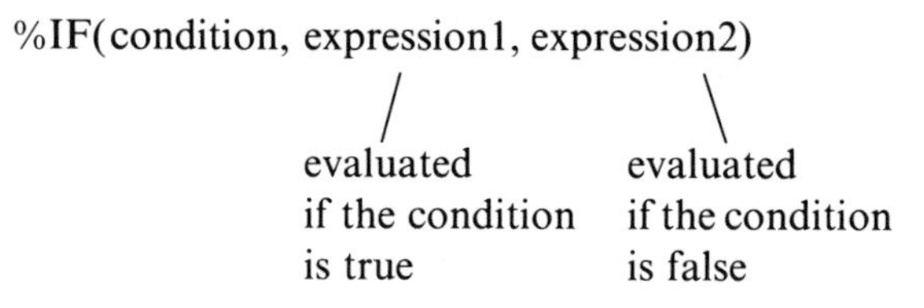

and we may rewrite the above calculate expression using this function.

```
$CALC AW = %IF( %GT(WBC,10.0)*%EQ(AG,1), 1, 2) $
$FACTOR AW 2$
```

For observations for which the condition (specified as the first argument) is true, the corresponding element of AW takes the value 1. All other elements of AW take the value 2.

The first argument of the %IF function which is usually a condition may be replaced by a vector or by a calculate expression, with the convention that zero values are false, and non-zero values are true. Thus, we could use the previously calculated vector AGWB to simplify the above expression.

```
$CALC AW = %IF( AGWB, 1, 2)$
$FACTOR AW 2$
```

How may we extract the eight observations satisfying our condition from the full data? We first declare the vectors which will contain the subset of observations:

```
$VAR %A SWBC STIM $
```

The vectors are declared as variates of length 8. We use the prefix S to the original vector names to indicate that the vectors are a subset of the original observations.

We now assign the appropriate elements of WBC and TIME to the elements of the vectors SWBC and STIM. A simple way to do this is to assign the values element by element. Individual elements of a vector may be assigned values or may be used in calculate expressions by using *suffixes*, that is by specifying the element or unit number in brackets following the vector name. Thus

```
$CALC SWBC(1) = 10.5 $
```

will set the first element of the new vector SWBC to 10.5. One method of extracting the subset is therefore to go through the elements required in turn, assigning individual elements of the full vectors to the appropriate elements of the new vectors as follows:

```
$CALC SWBC(1) = WBC(6)  : STIM(1) = TIME(6)
    : SWBC(2) = WBC(8)  : STIM(2) = TIME(8)
    : SWBC(3) = WBC(12) : STIM(3) = TIME(12)
    : SWBC(4) = WBC(13) : STIM(4) = TIME(13)
    : SWBC(5) = WBC(14) : STIM(5) = TIME(14)
    : SWBC(6) = WBC(15) : STIM(6) = TIME(15)
    : SWBC(7) = WBC(16) : STIM(7) = TIME(16)
    : SWBC(8) = WBC(17) : STIM(8) = TIME(17) $
```

This is not a satisfactory solution. A better method is to indicate the observations which are to be extracted from the full vectors by using *suffix vectors*. A suffix vector of length 33 is constructed which has zero elements for observations which are not to be extracted and with the remaining elements set to the new observation numbers of the reduced length vector.

In our example, we need to construct a suffix vector where the sixth element is 1, the eighth element is 2, the 12th element is 3, the 13th element is 4, the 14th is 5, the 15th is 6, the 16th is 7, the 17th is 8 and all other data values are zero. We also construct an index vector of length 33 containing the values 1,2,3,. . . .,33.

```
$CALC SUFF = AGWB*%CU(AGWB)  :  IND = %CU(1)
$PRINT SUFF $
```

```
0.      0.      0.      0.      0.      1.000  0.      2.000  0.      0.
0.      3.000   4.000   5.000   6.000   7.000  8.000   0.      0.      0.
0.      0.      0.      0.      0.      0.     0.      0.      0.      0.
0.      0.      0.
```

We may now extract the required subset of WBC and TIME into the new vectors SWBC and STIM by using the suffix vectors in two $CALCULATE statements.

```
$CALC SWBC(SUFF)=WBC(IND) : STIM(SUFF)=TIME(IND) $
$PRINT SWBC : : STIM$
```

```
10.50   17.00   32.00   35.00   100.0   100.0   52.00   100.0

108.0   4.000   26.00   22.00   1.000   1.000   5.000   65.00
```

The vectors SWBC and STIM contain the correct values. We note that the lengths of the vectors SWBC and WBC differ; it is the lengths of the *suffix vectors* SUFF and IND which must agree.

How has the calculate expression achieved its effect? The calculate expression SWBC(SUFF) = WBC(IND) is executed by taking the elements of the suffix vectors in turn. The expression is therefore internally expanded into 33 separate individual element assignments, as follows:

```
   Suffix vectors
  SUFF          IND
   |             |
SWBC(0)  =  WBC(1)
SWBC(0)  =  WBC(2)
SWBC(0)  =  WBC(3)
SWBC(0)  =  WBC(4)
SWBC(0)  =  WBC(5)
SWBC(1)  =  WBC(6)
SWBC(0)  =  WBC(7)
SWBC(2)  =  WBC(8)
SWBC(0)  =  WBC(9)
SWBC(3)  =  WBC(10)
```

and so on for all 33 elements. If a suffix of a vector is zero, then that element is not assigned.

As the length of SUFF agrees with the lengths of the vectors WBC and TIME, the above $CALC expressions may be simplified to

```
$CALC SWBC(SUFF) = WBC :  STIM(SUFF) = TIME $
```

with the index vector IND being removed from the expression. Further details of the use of suffix vectors may be found in the GLIM user guide.

We now consider the white blood cell count data in a slightly different way. For certain applications it may be useful to read in the data values so that each factor level is distinct and stored in separate vectors. The vectors defined will therefore be of different lengths. In our example, we define two vectors of length 17 to store the AG positive values of the white blood cell count and survival time, and two vectors of length 16 to store the AG negative variables. We do not define the number of units as there is no standard length.

```
$DATA 17 WBC1 TIM1 $READ
  2.30  65     .75 156    4.30 100    2.60 134    6.00 16 10.50 108
 10.00 121   17.00   4    5.40  39    7.00 143    9.40 56 32.00  26
 35.00  22  100.00   1  100.00   1   52.00   5  100.00 65
$DATA 16 WBC2 TIM2 $READ
  4.40  56    3.00  65    4.00  17    1.50   7    9.00 16  5.30  22
 10.00   3   19.00   4   27.00   2   28.00   3   31.00  8 26.00   4
 21.00   3   79.00  30  100.00   4  100.00  43
$COMMENT
VARIABLES
  WBC1    ... WHITE BLOOD CELL COUNT (000s) FOR AG +VE
                SUBJECTS
  WBC2    ... WHITE BLOOD CELL COUNT (000s) FOR AG -VE
                SUBJECTS
  TIM1    ... SURVIVAL TIME IN WEEKS FROM TIME OF
                DIAGNOSIS FOR AG +VE SUBJECTS
  TIM2    ... SURVIVAL TIME IN WEEKS FROM TIME OF
                DIAGNOSIS FOR AG -VE SUBJECTS
$
```

If the data have been read in this way, how do we combine the above vectors to produce the usual vectors of length 33? Concatenation of vectors is easily achieved through the $ASSIGN directive, which was introduced in Section 1.3.

```
$ASSIGN WBC  = WBC1,WBC2  : TIME = TIM1,TIM2 $
```

The length of the new vectors WBC and TIME is taken to be the sum of the lengths of the vectors on the right hand side of the assignment. The vectors WBC1 and WBC2 are assigned to the elements of WBC in order; the first seventeen elements will store the WBC1 values and the last sixteen elements will store the WBC2 values.

We may also combine numbers, scalars and vectors on the right hand side of the $ASSIGN directive. For example, if only the AG positive vectors WBC1 and TIM1 had been defined and assigned values, the vectors WBC and TIME could have been constructed by

```
$ASSIGN WBC  = WBC1, 4.40, 3.00, 4.00, 1.50, 9.00, 5.30, 10.00, 19.00,
               27.00, 28.00, 31.00, 26.00, 21.00, 79.00, 100.00, 100.00
   :    TIME = TIM1,56,65,17,7,16,22,3,4,2,3,8,4,3,30,4,43 $
```

1.9 Files and file handling

So far in this chapter, we have dealt with small datasets which can be easily typed into GLIM from the keyboard in a few minutes. The GLIM output produced and the commands typed in have appeared on the screen, and have eventually been replaced by further GLIM output, with no permanent record of the output being kept. In practice, this simple way of using GLIM is unsatisfactory and we would usually want to use *files* both to store sets of data and to store a permanent record of a GLIM session. The files may be physically stored on any one of a variety of storage media (hard disk, floppy disk, magnetic tape, etc.) depending on your computer system. Each file has a *file name* associated with it and the file naming conventions will depend on your computer's operating system. We have used simple example file names in this section; these will need to be replaced with file names suitable for your particular computer system.

Files which are to be used as *input* to GLIM need to exit before GLIM has been loaded. Each operating system has its own method of creating and storing information in a file, and you will need to consult your computer system's documentation to find out how to do this. Let us take as an example the white blood cell count data introduced in Section 1.8. We create a file called WBCDAT to store these data values. In this section, we represent a file as a rectangle; the contents of the file are shown within the rectangle.

data file WBCDAT

```
 2.30  65   0.75 156   4.30 100   2.60 134   6.00  16  10.50 108
10.00 121  17.00   4   5.40  39   7.00 143   9.40  56  32.00  26
35.00  22 100.00   1 100.00   1  52.00   5 100.00  65   4.40  56
 3.00  65   4.00  17   1.50   7   9.00  16   5.30  22  10.00   3
19.00   4  27.00   2  28.00   3  31.00   8  26.00   4  21.00   3
79.00  30 100.00   4 100.00  43
```

How do we inform GLIM that this data set is to be read in from a file rather than from the terminal? The $READ directive, which has been used until now to read data values from the terminal, is replaced by the $DINPUT directive (the directive name may be thought of as an abbreviation of DATA INPUT). An example of this directive is

```
$DINPUT 10$
```

The $DINPUT directive name is followed by a number—the *channel number* of the secondary input channel on which GLIM expects to find the data. In an interactive GLIM session, the primary input will be from the keyboard of the terminal, and the primary output will go to the screen. The primary input and output channel numbers are numbers which identify the particular primary input source and output destination. The primary input and output channel numbers used by a particular GLIM installation are stored in the system scalars %PIC and %POC. The value of these scalars may be discovered either by printing them

```
$PRINT %PIC %POC $
```

or by using the C option to the $ENVIRONMENT directive

```
$ENVIRONMENT C$
```

which also displays the channel numbers of many other GLIM input and output channels. Common values of the primary input and output channels are 5 and 6.

The $DINPUT directive above therefore tells GLIM to start reading data from channel 10. We have chosen the channel number 10 because it is a channel number which on our system is not being used by GLIM; this channel number should be substituted if necessary by a number appropriate for your installation of GLIM. Advice on which channel numbers to use for secondary input should be available in site-specific GLIM documentation. Most installations of GLIM have many spare channel numbers which may all be used for secondary input. A few installations of GLIM, however, specify explicitly which channels should be used for secondary input, and may also limit the number of secondary input channels which may be used to one or two.

How do we link the secondary input channel number with the file name? In most installations of GLIM3.77, when a new channel number is first encountered by the GLIM system, GLIM will prompt for a file name

```
$DINPUT 10 $
File name?
```

The file name will then be assigned permanently to that channel number for the duration of the GLIM session. If your installation of GLIM does not prompt for file names in this manner, or if you are using a previous release of GLIM, then the file needs to be assigned to the channel number *before* GLIM is loaded. The operating system commands needed to assign files to channel numbers are installation-specific, and again advice should be sought on how to do this for your version of GLIM. We will assume in the rest of this section that your installation supports the file name prompting facility described above.

We give the GLIM code needed to read in the white blood cell count data from the file WBCDAT, see (a) on page 49. The $UNITS directive defines a standard length of 33, and the data list is set up, defining two variates WBC and TIME. The $DINPUT directive instructs GLIM to read 66 numbers from channel 10. As channel 10 has been used for the first time in this GLIM session, GLIM prompts for a file name. After 66 data values have been read from the data file, control returns to the terminal.

Large datasets with many variables are usually stored in *fixed format*, with each variable occupying a fixed position within each record. In this case, the $DINPUT directive must be preceded by a $FORMAT directive which specifies the appropriate FORTRAN format. This format remains in force until changed by $FORMAT FREE$ or by another FORTRAN format.

Instead of storing data in a file, we may wish to store a set of GLIM directives. At an appropriate point in a GLIM session, we could then tell GLIM to read further GLIM statements from a file, rather than from a terminal. The GLIM statements stored in the file may contain embedded data, and this provides a useful method for storing datasets together with the necessary GLIM code to define the data.

Just as the $DINPUT directive is used for reading data values, the $INPUT directive is used to read GLIM statements from a secondary input channel. The directive name is followed by a suitably chosen channel number. To illustrate the use of this directive, we construct another file called WBCGLM, containing GLIM statements which define the white blood count data (see (b) on page 49). The directive $INPUT 11 $ is issued at the terminal, the file is assigned following the GLIM prompt, and directives are read from the file WBCGLM. Control is returned to the terminal by terminating the file with a $RETURN directive; if this directive were not present, then GLIM would attempt to read further directives from the file until the end of the file was reached. At this stage, a GLIM error would be produced, and control in most installations is returned to the terminal. For some installations and for earlier versions of GLIM, an operating system error would be generated, and the GLIM session would terminate prematurely.

If any of the GLIM statements in the file produce output, the output will be produced at the terminal as usual. The GLIM directives themselves, however, are not echoed to the screen, and this is sometimes a disadvantage. If a file of GLIM directives has been commented, none of the comments appear on the screen and if the file contains an illegal GLIM directive, the error message only is displayed. We may echo to the screen the contents of files by using the $ECHO directive. This directive, like the $HELP and $WARN directives encountered earlier, works as a switch, with the state of the echoing being changed from off to on, or from on to off, when the directive is used. GLIM is usually installed so that the echoing is off by default. It is therefore common to use the directive twice when reading a file, with the two instances of the

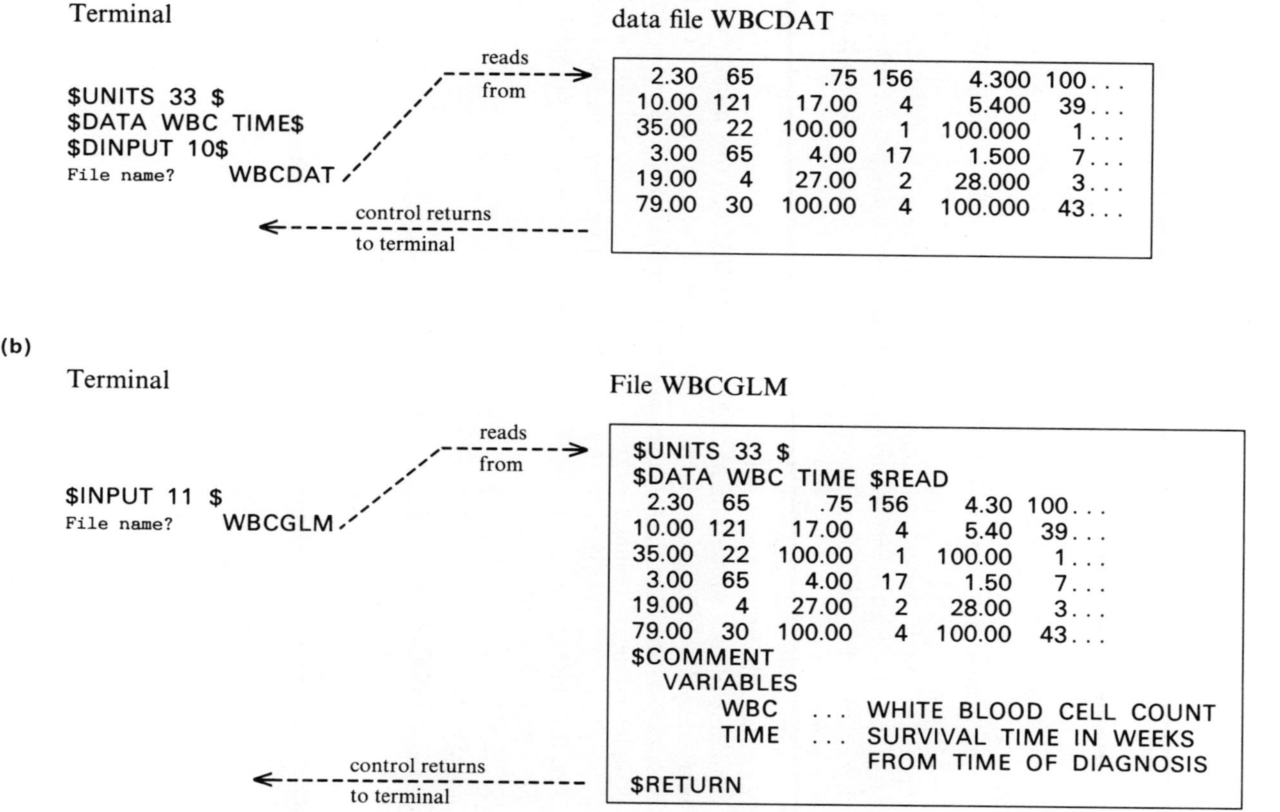
(a)
Terminal
$UNITS 33 $
$DATA WBC TIME$
$DINPUT 10$
File name? WBCDAT
reads from
control returns to terminal
data file WBCDAT
2.30 65 .75 156 4.300 100...
10.00 121 17.00 4 5.400 39...
35.00 22 100.00 1 100.000 1...
3.00 65 4.00 17 1.500 7...
19.00 4 27.00 2 28.000 3...
79.00 30 100.00 4 100.000 43...
(b)
Terminal
$INPUT 11 $
File name? WBCGLM
reads from
control returns to terminal
File WBCGLM
$UNITS 33 $
$DATA WBC TIME $READ
2.30 65 .75 156 4.30 100...
10.00 121 17.00 4 5.40 39...
35.00 22 100.00 1 100.00 1...
3.00 65 4.00 17 1.50 7...
19.00 4 27.00 2 28.00 3...
79.00 30 100.00 4 100.00 43...
$COMMENT
VARIABLES
WBC ... WHITE BLOOD CELL COUNT
TIME ... SURVIVAL TIME IN WEEKS
FROM TIME OF DIAGNOSIS
$RETURN

directive bracketing the $INPUT or $DINPUT directive as follows.

```
$ECHO
$INPUT 11 $
$ECHO
```

The echoing is switched on while the file is read, and switched off immediately afterwards. We switch the echoing off again as otherwise GLIM statements typed at the terminal will also be echoed and each GLIM statement will appear twice on the screen.

Two further directives are available to control input files. The $REWIND directive will rewind an input file assigned to a channel so that subsequent input from that channel will start again at the beginning of the file. The $REINPUT directive rewinds the file and rereads the GLIM statements on that channel. For both directives, the directive name is followed by a channel number.

An input file containing GLIM directives may be partitioned into a number of named subfiles, with each subfile containing a set of GLIM directives. The beginning of each subfile is marked by a $SUBFILE directive, and the end is marked by a $RETURN directive. In this way, many separate pieces of GLIM code may be stored in the same physical file. The $SUBFILE directive name is followed by a *subfile identifier* chosen to conform to GLIM naming conventions. If a physical file has a subfile structure, then the last GLIM directive in the file must be $FINISH. For example, the physical file DATASETS might contain

```
$SUBFILE DAT1
$C    FIRST DATASET
$UNITS 30 $
... further GLIM directives ...
$RETURN

$SUBFILE DAT2
$C       SECOND DATASET
$UNITS 160 $
... further GLIM directives ...
$RETURN

$SUBFILE MAC1
$C SUBFILE CONTAINING GLIM MACROS
... Macro definitions
$RETURN

$FINISH
```

The above file contains three GLIM subfiles called DAT1, DAT2 and MAC1. Two of the subfiles contain data (together with the GLIM code which defines the data) and the remaining subfile contains some GLIM macros (see Section 1.10).

To read a GLIM subfile into GLIM, we use the $INPUT or $REINPUT directives. The name of the subfile required is specified after the channel number. For example,

```
$INPUT 12 DAT2 $
```

will read in the subfile DAT2 from the file DATASETS. If the file has not been assigned to the input channel, GLIM will prompt for a file name in the usual manner. To find the subfile, GLIM searches through the file from the beginning until a $SUBFILE directive is encountered, then checks the name of the subfile found. If it is the desired subfile, then it is read in to GLIM, otherwise the process continues. We note that the file is terminated by a $FINISH directive. When a $FINISH directive is encountered, GLIM will continue the search for $SUBFILE directives from the beginning of the file.

Further subfiles may be read in to GLIM from the same input channel. The subfile DAT1 may be read by using the GLIM statement

```
$INPUT 12 DAT1 $
```

However, GLIM will now start searching for the subfile from the position at which the file was last accessed (in this example the $RETURN directive at the end of the subfile DAT2). As we know that subfile DAT1 comes at the beginning of the file, we may locate it more efficiently by using

```
$REINPUT 12 DAT1$
```

and avoiding the search to the end of the file.

We have spent some time in explaining GLIM subfiles as they are used extensively throughout this book. In the remaining chapters we shall be assuming that the examples file, which contains many GLIM subfiles containing datasets and macros, is available on your computer. Appendix 4 contains details of how to obtain this examples file in a computer-readable form, and also contains a listing of the contents of the file. We have also assumed in subsequent chapters that this file is assigned to input channel 1. If this channel is not available for secondary input on your system you will need to use a different channel.

We shall also be using many of the macros in the GLIM macro library (available only for GLIM3.77). The GLIM macro library also has a subfile structure, but it should be automatically assigned for you by the system at the start of every GLIM session. The channel number used by your installation is stored in the system scalar %PLC, or may be found by using the C option to the $ENVIRONMENT directive. For example, to read in the Box–Cox macros from the macro library, we should use

```
$INPUT %PLC BOXCOX $
```

The file which is most likely to be encountered in GLIM3.77 is the *transcript file*. This file will contain a record of the entire GLIM session, including all input to and output from the terminal, secondary input from files, and help, fault and warning messages. All output to the transcript file is written to the transcript file channel. When loading GLIM, your system may require you to specify a transcript file name but if not, the transcript file may need to be assigned before GLIM is entered.

Output to the transcript file is controlled by the $TRANSCRIPT directive. The directive name may additionally be followed by one or more options which specify the type of output which is to be written to the file. The options available include INPUT, OUTPUT, HELP, FAULT, and WARN and if an option is present, then that type of output is written to the transcript file. An option may be abbreviated to the initial letter of the word. Note that if no option is specified (i.e. $TRAN $) then all output to the transcript file is suppressed.

Facilities also exist for storing the current state of a GLIM session in a file. This is achieved by using the directive $DUMP. The names and contents of all GLIM structures, together with the values of all scalars and system settings, including model settings, are all written to a file, which needs to be assigned to the *dump channel*. This file is written in an internal code, and the file cannot therefore be read easily outside the GLIM system. At some future time or later on in the same GLIM session, the information in this file may be used to restart the GLIM session from the point at which the $DUMP directive was issued. The $RESTORE directive is used for this purpose.

When issuing a $DUMP directive, any further directives following it on the same line of input will both be executed and are also stored in the dump file and executed when the file is restored. We may take advantage of this when using dump files, as the following example demonstrates.

```
$DUMP $PRINT 'Dump from white blood cell count data ... run 1' $
File name?    WBCDMP

-- program dump completed
Dump from white blood cell count data ... run 1
```

The $DUMP directive is followed by a $PRINT directive, which is both executed and stored in the dump file. In a separate GLIM session the file is restored.

```
$RESTORE $
File name?   WBCDMP

-- program restarts from previous dump
Dump from white blood cell count data ... run 1
```

The $PRINT statement is again executed, and a message identifying the contents of the dump file is printed out.

Although this feature of GLIM is useful, care should be taken not to follow the $DUMP directive with inappropriate directives on the same line. For example, the sequence $DUMP $STOP should not be used, as if the dump file produced is restored, the $STOP directive would be executed and the GLIM session would terminate.

It is possible to restore a previously dumped GLIM session within the same session. In this case, the $REWIND directive is used to rewind the dump channel before restoring the GLIM session.

```
$REWIND $RESTORE $
```

Note that $REWIND directive assumes that the dump channel is to be rewound if no channel number is given.

```
$END
```

1.10 Macros and text handling

We now turn our attention to GLIM *macros*, which as we briefly mentioned in Section 1.6, may be used to store either text or GLIM code. At any point in a GLIM job, the name of the macro may be used to execute the defined piece of GLIM code, or to substitute a piece of text. The contents of macros are stored as a sequence of characters in the workspace. Macros will therefore compete with vectors for a share of the workspace, and this needs to be taken into account if many long macros are to be defined.

Our first example illustrates the use of macros for storing text strings. In Section 1.2 we saw that text may be printed by using the $PRINT directive. For instance, we may want to specify a heading which will appear in any output from the GLIM session. We may specify the text directly as an item to the $PRINT directive

```
$PRINT 'Analysis of White Blood Cell Count Data ....   Run 1 ' $
```

However, if the heading is required to be printed many times in a GLIM job, it is more convenient to use a macro to store the heading as a text string. The directives $MACRO and $ENDMACRO are used to define a macro, as follows:

```
$MACRO HEADING Analysis of White Blood Cell Count
Data ....   Run 1 $ENDMAC
```

We may now use the equivalent statement

```
$PRINT HEADING $
```

to produce the desired output.

We see that the $MACRO directive name is followed by the user-defined *macro identifier* HEADING, which in turn is followed by the contents of the macro. The macro definition is terminated by the $ENDMAC directive. Macro identifiers should conform to the same rules as all other identifiers in GLIM (Section 1.3). In particular, only the first four characters of a macro identifier are significant. The directive names $MACRO and $ENDMACRO may be abbreviated to $M and $E respectively.

In a similar way, we may define macros to contain text such as option lists or model formulae which it may be tedious to repeatedly type in at the terminal. For example, suppose that we wished to produce a series of plots with the x-limits, y-limits and style fixed to constant values. We could define a macro called OPT to contain the relevant option list

```
$MAC OPT (STYLE=1 XLIMITS=0,60 YLIMITS=0,60 ) $ENDMAC
```

We illustrate the use of this macro with the white blood cell count data, which may be read in from the subfile FEIGL which forms part of the examples dataset. We choose channel 1 to read in the subfile.

```
$ECHO $INPUT 1 FEIGL $ECHO
```

Three variables have been defined in the subfile: two variates (WBC and TIME) and one factor (AG). To produce a plot of WBC against TIME, with the options defined above, we need to substitute the contents of the macro OPT in the correct place in the $PLOT directive. This is achieved by using the *substitution symbol*, which is usually represented by #.

```
$PLOT #OPT WBC TIME $
      |
      |
      |
      substitute contents of macro OPT at this point.
```

The substitution symbol tells GLIM to start reading characters from the macro OPT rather than from the current input channel. Program control therefore resides in the macro OPT, and we say that the *level* of the *program control stack* has been increased by one level. When the end of the macro OPT is reached, program control returns to the current input line and the program control stack level is decreased by one. The macro OPT stays defined in the workspace and may be used as many times as required.

```
$PLOT #OPT WBC TIME '+*' AG $
```

While storage of text strings is a useful facility, the primary function of the macro structure is to store sequences of GLIM statements, which may then be executed at appropriate points in a GLIM session. We illustrate with an example to calculate the correlation coefficient of two variables **X** and **Y** with values x_i and y_i, for $i=1,2...n$. One common method given in textbooks for calculating the correlation coefficient r is given by

$$r=\frac{n\Sigma x_i y_i-\Sigma x_i\Sigma y_i}{\sqrt{[\{n\Sigma x_i^2-(\Sigma x_i)^2\}\{n\Sigma y_i^2-(\Sigma y_i)^2\}]}}$$

Assuming for the time being that the two variables are held in vectors X and Y of standard length, we can easily calculate the quantities Σx_i, Σy_i, $\Sigma x_i y_i$, Σx_i^2 and Σy_i^2 using the following calculate expressions, storing the results in scalars.

```
$CAL %A= %CU(X) $          ...Σx_i
  :   %B= %CU(Y) $          ...Σx_i
  :   %C= %CU(X*Y) $        ...Σx_i y_i
  :   %D= %CU(X*X) $        ...Σx_i^2
  :   %E = %CU(Y*Y) $       ...Σy_i^2
```

The standard length of the vectors is held in scalar %NU. We may incorporate these calculate expressions into a macro called CORR.

```
$MACRO CORR
$CAL %A = %CU(X)
  : %B = %CU(Y)
  : %C = %CU(X*Y)
  : %D = %CU(X*X)
  : %E = %CU(Y*Y)
  : %F = %NU*%C-%A*%B
  : %G = (%NU*%D-%A*%A)*(%NU*%E-%B*%B)
  : %R = %F/%SQRT(%G)
$PRINT %R $
$ENDMAC
```

The correlation coefficient of the two vectors X and Y is calculated and printed out by the macro. Note the terminating dollar after the last directive in the macro ($PRINT) which ensures that this directive is executed. We can now use the macro on the white blood cell count data to find the correlation of the two vectors TIME and WBC. We create two vectors X and Y, and copy the values of TIME into X, and WBC into Y. The macro is then executed by means of the $USE directive, with the name of the macro to be executed following the directive name. Note that it is also possible to initiate execution of the macro by using the substitution symbol (#CORR). We adopt the convention of using the substitution symbol only for text substitution where the text stored does not contain directives.

```
$CALC X=TIME : Y=WBC $
$USE CORR $
```

```
-0.3295
```

The two vectors TIME and WBC have a correlation coefficient of −0·3295. When the macro is executed, we note that none of the GLIM directives within the macro are echoed to the screen. This may sometimes be inconvenient, particularly if there is a mistake in the macro which needs to be corrected. The $VERIFY directive may be used to verify macro execution and to echo the directives to the screen line-by-line as they are executed. As with the $HELP, $WARN and $ECHO directives, the $VERIFY directive works as a switch.

As an aside, we note that it is not possible to edit macros in GLIM3. Macros may be redefined or overwritten, but if a mistake has been made in typing or in logic and a small correction needs to be made, there is nothing else to do apart from typing the macro in again from the beginning. For this reason, long macros are usually stored in files where the system editor may be used to change them. The macros may then be read in to GLIM by using the $INPUT directive.

How much workspace does the macro occupy? We use the $ENVIRONMENT directive with the D option (Section 1.6) as follows:

```
$ENV D$
```

```
Directory:   levels length  space
   HEAD          -1      0     32
   OPT           -1      0     24
   TIME           0     33     17
   WBC            0     33     17
   AG             2     33     17
   X              0     33     17
   Y              0     33     17
   CORR          -1      0    377
```

We see that the macro CORR occupies about 400 units of space (this number will differ depending on the particular installation of GLIM).

This macro works but can be improved in many ways. For general use, some of the following aspects may cause problems.

1. The macro occupies too much workspace. As macros are stored as a series of characters, the spaces between the end of the last directive on each line and the physical end of each line (defined by the primary input line length) are stored as part of the macro. This slows down execution of the macro and increases the amount of space necessary to store it.
2. The macro is clumsy to use, requiring the user to copy the values into vectors X and Y.
3. The algorithm used may cause numerical problems. The above formula is the usual method for hand calculation given in most introductory statistics text books; however, severe loss of accuracy may occur with certain sets of data.
4. The output from the macro is terse. An informative message could accompany the value of r.
5. The macro will produce a correct value only if vectors of standard length are used.

We deal with these points in turn.

The size of the macro may be reduced by placing multiple GLIM statements on each line. The *End-of-record* symbol (usually represented by an exclamation mark !) may also be used to ensure that no unnecessary white space is stored. The end-of-record symbol is placed after the last directive on each line of the macro definition. When GLIM encounters an end-of-record symbol, the remainder of the current line is ignored and a new line started. In a macro definition, no characters to the right of the end-of-record symbol on the same line are stored. This feature makes the end-of-record symbol extremely useful for the commenting of macros.

Macros may be written more generally by using the *argument* facility, which allows macros to be parameterized. In our example above, we would replace the vectors X and Y by the *formal arguments* %1 and %2 (the first and second formal arguments to the macro CORR). We construct a new macro COR1 using formal arguments and the end-of-record symbol.

```
$MACRO COR1 !
$CAL %A = %CU(%1) : %B = %CU(%2) : %C = %CU(%1*%2)!
  : %D = %CU(%1*%1) : %E = %CU(%2*%2) : %F=%NU*%C-%A*%B!
  : %G = (%NU*%D-%A*%A)*(%NU*%E-%B*%B) : %R = %F/%SQRT(%G)!
$PRINT %R $!
$ENDMAC
```

We may check using $ENV D$ that the macro COR1 occupies far less space than macro CORR.

Before the macro COR1 is executed, we need to specify *actual arguments* to replace the *formal arguments* in the macro contents. We do this either by using the $ARGUMENT directive

```
$ARGUMENT COR1 TIME WBC $USE COR1 $
                 |    |
     first actual    second actual
     argument to     argument to
            COR1     COR1
```

or (in GLIM3.77) by specifying the actual arguments after the macro name in the $USE directive.

```
$USE COR1 TIME WBC $

-0.3295
```

When the macro is executed the first formal argument (%1) will be replaced by the first actual argument (TIME) whenever it is encountered, while the second formal argument (%2) will be replaced by WBC in a similar manner. Once arguments to a macro are defined, they remain defined until a new $ARGUMENT directive or a new $USE directive resets the arguments to the macro. To illustrate this, we use COR1 to calculate the correlation of log(TIME) with WBC and log(WBC).

```
$CAL LTIM=%LOG(TIME) $USE COR1 LTIM$

-0.3863

$CAL LWBC=%LOG(WBC) $USE COR1 * LWBC$

-0.5232
```

In the first call to COR1, the second actual argument (WBC) remains set and does not need to be respecified. In the second call, the asterisk denotes that the first argument (LTIM) remains set.

We now turn to the algorithm used to calculate the correlation. We construct two variates X and Y of standard length which have a correlation of 1.

```
$CALC X=%CU(1)+100 : Y=%CU(1)+30000 $
```

We then execute the macro COR1, specifying X and Y as arguments.

```
$USE COR1 X Y $
```

We give the results of this calculation for various computers and their operating systems.

VAX11/785 / VMS	0.7147
IBM/PC / DOS	1.228
CDC7600 / SCOPE	1.000
WHITECHAPEL MG−1 / UNIX	1.227
PR1ME / PRIMOS	0.000†
ICL2900 / VME	0.000†

† Invalid function/operator arguments.

Large discrepancies between machines are found, with only one machine producing the correct value of 1.000. To avoid these discrepancies, we use the alternative formula

$$r = \frac{\Sigma(x_i - \bar{x})(y_i - \bar{y})}{\sqrt{\{\Sigma(x_i - \bar{x})^2 \ \Sigma(y_i - \bar{y})^2\}}}$$

A new macro COR2 is constructed based on the above algorithm. The macro also improves the terminating $PRINT statement by including an informative message.

```
$MACRO COR2 !
$CAL %A = %CU(%1)/%NU : %B = %CU(%2)/%NU!
  : %C = %CU((%1-%A)*(%2-%B))!
  : %D = %CU((%1-%A)**2) : %E = %CU((%2-%B)**2)!
  : %R = %C/%SQRT(%D*%E)!
$PRI : 'The correlation coefficient of ' *N %1 ' and ' *N %2 ' is ' %R $!
$ENDMAC

$USE COR2 WBC TIME$

The correlation coefficient of WBC and TIME is  -0.3295

$USE COR2 X Y$

The correlation coefficient of X and Y is    1.000
```

The macro COR2 produces the same correlation coefficient for the variates WBC and TIME, and the correct value of 1.0 for the variates X and Y.

While the macro is algorithmically correct, it may be developed further. The macro uses the scalar %NU, and so it may only be used for vectors of standard length. Checks may also be incorporated for the following events, all of which would cause the macro to fail:

(1) failure to set either or both of the two arguments;

(2) zero values of the sums of squares about the mean for either vector (scalars %D and %E);

(3) vectors of different length specified as arguments to the macro.

While the checks are probably not essential, they enable common errors to be trapped by the macro rather than by GLIM, and allow appropriate action (usually the display of an informative error message) to be taken. We incorporate a check for the third of these possibilities into the macro. At the same time, we make the macro able to deal with vectors of arbitary length.

A new macro COR3 is introduced, as are two previously unencountered GLIM directives $SWITCH and $EXIT.

```
$MACRO COR3 !
$CAL %M=%CU(%EQ(%1,%1)) : %N=%CU(%EQ(%2,%2))!
 : %Q =%NE(%M,%N) !
$ARG CORX %1 %2 $SWITCH %Q CORX $ !
$CAL %A = %CU(%1)/%N : %B = %CU(%2)/%N!
 : %C = %CU((%1-%A)*(%2-%B))!
 : %D = %CU((%1-%A)**2) : %E = %CU((%2-%B)**2)!
 : %R = %C/%SQRT(%D*%E)!
$PRI : 'The correlation coefficient of ' *N %1 ' and ' *N %2 ' is ' %R $!
$ENDMAC

$MACRO CORX !
$PRI 'The lengths of the vectors ' *N %1 ' [' *I %M '] and '!
*N %2 ' [' *I %N '] are unequal. ' : 'Macro COR3 abandoned' !
$EXIT 2$ !
$ENDMAC

$USE COR3 WBC TIME$
```

```
The correlation coefficient of WBC and TIME is  -0.3295
```

We illustrate the error handling of the latest version of our macro by constructing two vectors A and B of different length.

```
$ASSIGN A=1,3,5,7,9 : B = 2,8,11,16
$USE COR3 A B$
```

```
The lengths of the vectors A [5] and B [4] are unequal.
Macro COR3 abandoned
```

The first two calculate expressions calculate the length of the vectors corresponding to the formal arguments %1 and %2. The scalar %Q is then set to 1 if the lengths of the vectors are unequal, and zero if they are not. The $SWITCH directive has the syntax

$SWITCH scalar list of macro names $

and has the function of *conditionally executing* one of a set of macros depending on the value of the scalar given as the first item. If the scalar is zero or negative, no macro is executed. If the scalar has the value 1, the first macro in the list is executed, if 2, then the second macro in the list is executed, and so on. Here the $SWITCH directive has the effect of executing the macro CORX if the lengths of the vector arguments are unequal, and continuing with the next directive in macro COR3 if they are equal.

When the macro CORX is entered from within the macro COR3, the program control stack level again increases by one. Macro CORX contains a $PRINT directive which displays the error message and an $EXIT directive. The $EXIT directive name is followed by the number 2, which indicates that the program control stack level is to be decreased by two. Practically, this will have the effect of abandoning both CORX and COR3, and control will return to the terminal.

The macro may be developed further along the lines suggested but this is left as an exercise for the interested reader.

Repeated use of macros can be achieved using the $WHILE directive. This is a powerful facility for general statistical programming and many of the macros discussed in subsequent chapters of this book make use of this directive.

1.11 Sorting and tabulation

A sorting algorithm is provided in GLIM through the $SORT directive. As well as providing a flexible means of sorting vectors, the directive may also be used for obtaining the ranks of a vector or for lagging the values of a vector. We do not elaborate here on these more advanced features.

We illustrate the simple use of the $SORT directive with the white blood cell count data, which are stored in the subfile FEIGL. We will need to read this data set into GLIM if this has not been previously done.

```
$INPUT 1 FEIGL $
```

The simplest use of the $SORT directive is

```
$SORT TIME $
$PRINT *-1 TIME $
```

```
1.   1.   2.   3.   3.   3.   4.   4.   4.   4.   5.   7.   8.   16.   16.
17.   22.   22.   26.   30.   39.   43.   56.   56.   65.   65.   65.   100.
108.   121.   134.   143.   156.
```

The vector TIME has been sorted into ascending order (that is, the first element of the sorted vector contains the lowest value). The vector is sorted into itself, and the original order of the values is lost. Usually, both the original and the sorted versions of a vector are required, and sorting from one vector into another is accomplished by specifying two items to the $SORT directive. We recover the unsorted values of the variate TIME by using the $REINPUT directive.

```
$REINPUT 1 FEIGL $
$SORT OTIM TIME $
        |     |
   Sorted     Original
   vector     vector
```

The newly-created vector OTIM will now contain the sorted (ordered) values of TIME. It is also possible to sort one vector into another according to the values stored in the third. We may use this to create new versions of the factor AG and the variate WBC which have been ranked according to the values of the variate TIME. This is achieved by:

```
$SORT OWBC    WBC      TIME $   :  OAG AG TIME $
         |      |        |
     store    the        sorted
        in    sorted     according
       1st    values     to the
 vector....   of 2nd     values of
              vector.... 3rd vector
```

The three vectors WBC, TIME and AG have been sorted into ascending TIME order, and stored in the new variates OWBC, OTIM and OAG.

The $SORT directive may be used to find the maximum, minimum and median values of a vector by extracting the relevant elements of the *sorted* vector. For example,

```
$CAL %A = OTIM(1) : %B=OTIM(33) : %C=OTIM(17)
```

would store the minimum value of the vector TIME in %A, the maximum in %B and the median in %C.

However, a more convenient way of extracting this information which avoids the use of calculate expressions is to use the facilities of the $TABULATE directive (GLIM3.77 only), which we now introduce.

```
$TABULATE THE TIME SMALLEST : THE TIME LARGEST
  : THE TIME FIFTY $
```

```
1.000
156.0
22.00
```

The smallest, largest and median values are displayed on the screen. In this use of the $TABULATE directive, the directive name is followed by an *input phrase* which consists of the *phraseword* THE, the name of the variate and the name of the statistic which is required. Only the first letter is needed for both the statistic name and the phraseword. As the initial letter of each statistic name has to be unique and the statistic name MEAN exists, this explains the strange statistic name of FIFTY for the median of a vector. We list some of the statistics available below.

DEVIATION	Standard deviation
FIFTY	Median
LARGEST	Maximum
MEAN	Mean
PERCENTILE <number>	<number> th Percentile
SMALLEST	Minimum
TOTAL	Total
VARIANCE	Variance

The result of any input phrase can be stored in a vector rather than displayed on the screen by using the appropriate *output phrase*. The *output phraseword* INTO is matched to the input phraseword THE.

```
$TAB THE TIME LARGEST INTO TMAX $
   :  THE TIME DEVIATION INTO TSDV $
```

```
$PRINT TMAX : TSDV $
```

$TABULATE creates TMAX and TSDV as variates of length 1 but does not now display the output.

We may also use the $TABULATE directive to produce a table of counts.

```
$TABULATE FOR AG$
```

```
      1      2
[]  17.00  16.00
```

The FOR input phrase is used to produce a table containing two cells. The output shows that seventeen subjects have AG-factor positive (coded 1) and sixteen subjects have AG-factor negative (coded 2). How many of each group have their white blood cell count above 10 000? Reminding ourselves that WBC is measured in thousands, we may find this out by constructing a new variate W which has the value 1 for those subjects with WBC greater than 10, and 0 otherwise.

```
$CALC W = %GT(WBC,10) $
```

We may now use the $TABULATE directive to obtain the required information. As above, we use the FOR phraseword to produce the table, but this time we specify a STYLE of 1 as an option to the $TABULATE directive.

```
$TAB(STYLE=1)  FOR  AG;W $

        +-----------------+
        |  0.000   1.000  |
+-------+-----------------+
|     1 |  9.000   8.000  |
|     2 |  7.000   9.000  |
+-------+-----------------+
```

A table with four cells is produced, and we see that eight of the AG-positive group and nine of the AG-negative group have white blood cell count above 10 000. Note that it is not necessary for W to be declared as a factor, or even to be coded as a factor.

As in the earlier example, it is possible to specify an output variate: the USING output-phrase is matched to the FOR input-phrase.

```
$TAB FOR AG;W USING COUNT$
```

COUNT is constructed as a variate of length 4. We introduce the $TPRINT directive to print this variate as a two by two table.

```
$TPRINT COUNT 2;2 $

      1      2
1  9.000  8.000
2  7.000  9.000
```

The table is not labelled informatively. We may achieve this by specifying output vectors using the BY output phrase to contain the classification of the table. These vectors, which we name AGZ and WZ, will also be of length 4.

```
$TAB FOR AG;W USING COUNT BY AGZ;WZ $
$LOOK COUNT AGZ WZ $
```

```
     COUN      AGZ       WZ
1   9.000    1.000    0.000
2   8.000    1.000    1.000
3   7.000    2.000    0.000
4   9.000    2.000    1.000
```

We see that the vector AGZ contains the values 1 and 2, whereas WZ contains the values 0 and 1. AGZ and WZ *cross-classify* the vector of counts, stored in COUNT. As AG is a factor, AGZ is automatically declared as a factor. WZ, however, is declared as a variate. The types of the vectors specified in the FOR phrase are transferred to the output vectors specified in the BY phrase.

We may now use the $TPRINT directive to print a more informative version of the above table. We specify a STYLE of 1 in the option list.

```
$TPRINT (STYLE=1) COUNT AGZ;WZ $
```

```
       +-----------------+
    WZ |  0.000   1.000  |
   AGZ |                 |
+------+-----------------+
|    1 |  9.000   8.000  |
|    2 |  7.000   9.000  |
+------+-----------------+
```

Combinations of input phrases may be used to produce tables of summary statistics. For example, to produce a table of TIME means rather than counts, we use

```
$TAB THE TIME MEAN FOR AG;W $
```

```
   0.000  1.000
1  92.22  29.00
2  26.57  11.22
```

As before, output phrases may be used to store the vector of means, the associated vector of counts and the cross-classifying vectors. All the output vectors will again be of length 4.

```
$TAB THE TIME MEAN FOR AG;W INTO
 TMEAN USING COUNT BY AGZ;WZ$
```

We may then use $TPRINT to print both the cell means of TIME and the associated cell counts in one table.

```
$TPRINT (S=1) TMEAN;COUNT AGZ;WZ $
```

```
            +-----------------+
         WZ |  0.000   1.000  |
  AGZ       |                 |
+-----------+-----------------+
|    1 TMEA |  92.22   29.00  |
|      COUN |  9.000   8.000  |
-------------------------------
|    2 TMEA |  26.57   11.22  |
|      COUN |  7.000   9.000  |
+-----------+-----------------+
```

Many other more complex uses of the $TABULATE directive exist, and these are covered in detail in the GLIM3.77 reference guide, Baker (1985). The $TABULATE directive is used extensively in Chapters 4 and 5 to aggregate individual level data into contingency tables and to collapse existing contingency tables.

2
Statistical modelling and statistical inference

2.1 Statistical models

In this book we will be concerned with the statistical analysis of data from experimental and observational studies. In experimental studies *randomization* or random assignment of individual experimental units (for example, human or animal subjects, agricultural plots) to the experimental treatments plays a fundamentally important role.

The experimental units may themselves be sampled randomly according to some sample design from a larger population, in which case conclusions from the experiment can be drawn about the larger population, or they may be all that is available for the experiment (for example, all patients with a particular disease being assigned to treatments in a hospital clinical trial), in which case conclusions may not be generalizable to a broader population.

In observational studies, observational units are drawn from a population according to some random sample design (which may be the complete examination of the entire population), and conclusions are to be drawn about the complete population.

In both kinds of study, the random selection or allocation of observations is critical to the analysis and interpretation of the data. Studies which are not based on sample or experimental designs, that is, in which the data collected are accidental, or are subjectively chosen, are particularly hazardous to interpret because of the possibility that inclusion in the study is systematically related to important variables, so that the data are not representative of the population about which conclusions are to be drawn.

The object of statistical modelling is to present a simplified or *smoothed* representation of the underlying population. This is done by separating *systematic* features of the data (for example, sex differences in height or weight) from *random variation* (natural variability in height or weight within sex). The systematic features are represented by a *regression function* involving parameters which can be simply related to the structure of the sample and to important variables measured on each experimental or observational unit. The random variation unrelated to important variables is represented by a *probability distribution* depending on a small number of parameters, typically one or two. Interpretation of the data, and conclusions about the population, can then be based on the regression function, with no attempt to interpret random variation.

How do we distinguish between random and systematic variation? This is a familiar problem in *hypothesis testing*: we suppose first that variation is random, and then test whether the introduction of a systematic effect into the model improves the model, in a well-defined way. If the improvement is substantial, the systematic effect is retained and can be interpreted: if it is only minor, the systematic effect is not retained and is not interpreted. How large an improvement is substantial is partly a subjective matter, but well-established rules in hypothesis testing provide a good guide. The importance of hypothesis testing results from a general philosophical approach to scientific inference by statisticians and other scientists, based on the *principle of parsimony*, known historically as Occam's Razor (from William of Occam or William Ockham, 1280–1349): "entia non sunt multiplicanda praeter necessitatem"— entities should not be needlessly multiplied. Systematic effects should be included in a model only if there is convincing evidence of the need for them: we should not spend time and effort interpreting effects which could just as well be random variation. The principle of parsimony provides an important guide to *model simplification*: though we may use very complex models for complex sample survey designs, our aim is always to simplify the model as much as possible, while remaining consistent with the observed data.

An important distinction needs to be made between the use of models for *representation* of an existing population, as we have just described, and their use for *prediction* of values of variables for future observations. These uses of models are often confused. The use of models for prediction will be considered in Chapter 3.

It is assumed in subsequent discussions that the data being modelled consist of a random sample of some kind from a population. In some cases, as noted earlier, the data available are a complete enumeration of a population. In such cases the simplification of models using hypothesis testing seems quite artificial, because no sampling is involved, and we already know whether or not the hypothesis is true.

There are two possible approaches to the statistical modelling of complete populations. One is to regard the population as a sample from a random process generating this population, and to model the population as though it were a sample. Such an approach is called *superpopulation modelling* (Smith, 1976) although no concept of sampling the observed population from a hypothetical superpopulation of populations is actually necessary.

The other approach, which we shall follow, is to regard models simply as smooth approximations to the rough, irregular complete population. The irregular variations about the smooth structure are treated as non-systematic variation to be ignored. The degree to which the population can be smoothed is now entirely subjective, since the formal rules of hypothesis testing are not appropriate, but smoothing to the extent appropriate if we had a sample, and not a population, will often be reasonable unless the population size is small.

2.2 Types of variables

The data for which models can be constructed in GLIM consist of values of a *single response variable*, which we will denote by y, and a set of *explanatory* or *predictor* variables $x_1, x_2, \ldots, x_p$, which are assumed to be measured without error and recorded without missing values. These are important restrictions which are discussed in Chapter 3. GLIM cannot fit models to multiple response variables, except for multiple-category variables (Chapter 5). The terms "dependent" and "independent" variables which are often used are confusing (since the "independent" variables are not statistically independent) and will not be used in this book. The response and explanatory variables can take a wide range of forms, set out below.

Variable type	Examples
Categorical, two categories (binary)	Male/female, alive/dead, presence/absence of a disease, inoculated/not inoculated
Categorical, more than two unordered categories	Blood group, cause of death, type of cancer, political party vote, religious affiliation
Categorical with ordered categories	Severity of symptoms or illness, strength of agreement or disagreement, class of University degree
Discrete count	Number of children in a family, number of accidents at an intersection, number of ship collisions in a year
"Continuous"	Height, weight, response time, survival time, Stanford-Binet IQ.

Quotation marks are used around "continuous" because all such variables are in practice measured to a finite precision, so they are actually discrete variables with a large number of numerical values. For example, height may be measured to the nearest half-inch or centimetre, survival time to the nearest day, week or month, and Stanford–Binet IQ may be given as an integer though it is defined as the ratio of mental age to chronological age multiplied by 100. Models for "continuous" data usually ignore this measurement precision, but if grouping is very coarse it may need to be incorporated in the model (see Section 2.5 and Appendix 1).

Discussions of measurement in the social sciences frequently use Stevens's (1946) classification of nominal, ordinal, interval and ratio scales. Nominal scales correspond to the first two variable types above and ordinal to the third. The distinction between interval and ratio scales (the latter has a fixed zero, the former does not) is not very useful, since it is often not relevant to the kind of analysis which is appropriate.

2.3 An example

We consider a practical example from a psychological study. A sample of twenty-four children was randomly drawn from the population of fifth-grade children attending a state primary school in a Sydney suburb. Each child was assigned to one of two experimental groups, and given instructions by the experimenter on how to construct, from nine differently coloured blocks, one of the 3×3 square designs in the Block Design subtest of the Wechsler Intelligence Scale for Children (WISC), see Wechsler (1949). Children in the first group were told to construct the design by starting with a row of three blocks (row group), and those in the second group were told to start with a corner of three blocks (corner group). The total time in seconds to construct four different designs was then measured for each child.

Before the experiment began, the extent of each child's "field dependence" was tested by the Embedded Figures Test, which measures the extent to which subjects can abstract the essential logical structure of a problem from its context (high scores corresponding to high field dependence and low ability).

The data are given below. The experimenter was interested in whether the different instructions produced any change in the average time required to construct the designs, and whether this time was affected by field dependence.

In practical data analysis, we usually begin with a plot of the data. Throughout this book it is assumed that the data sets in Appendix 4 are on a computer file, and can be read in to GLIM on channel 1 using the $INPUT directive. Your version of GLIM may use a different channel for secondary

Row group

TIME:	317	464	525	298	491	196	268	372	370	739	430	410
EFT:	59	33	49	69	65	26	29	62	31	139	74	31

Corner group

TIME:	342	222	219	513	295	285	408	543	298	494	317	407
EFT:	48	23	9	128	44	49	87	43	55	58	113	7

input; you should consult local documentation for information on which channels are available and the method of attaching files to channels—see also Section 1.9 on file handling in GLIM. The data file listed in the Appendix contains a subfile SOLV which contains the data above and descriptions of the variables, including the factor GROUP taking values 1 and 2 for the row and corner groups respectively.

```
$INPUT 1 SOLV $
$PLOT TIME EFT $
```

```
 750.0 |                                                       T
 720.0 |
 690.0 |
 660.0 |
 630.0 |
 600.0 |
 570.0 |
 540.0 |           T T
 510.0 |                                                    T
 480.0 |                T  T
 450.0 |         T
 420.0 | T       T               T   T
 390.0 |
 360.0 |         T         T
 330.0 |            T    T                   T
 300.0 |           TT T    T
 270.0 |         T
 240.0 |
 210.0 |  T    TT
 180.0 |
----------:---------:---------:---------:---------:---------:---------:
         0.0      30.0      60.0      90.0     120.0     150.0     180.0
```

There appears to be a general increase in time required as EFT increases, with much individual variation. The plot does not distinguish between the two experimental groups: this is achieved with the option in $PLOT for user-defined plotting characters. Using R for row group and C for corner group observations, we have

```
$PLOT TIME EFT 'R C' GROUP $
```

Separate plots for the two groups can be obtained with the system vector %RE:

```
$CALC %RE = %EQ(GROUP,1) $
$PLOT TIME EFT '+' $
```

This plots the row group data. For the corner group data,

```
$CALC %RE = %EQ(GROUP,2) $
$PLOT TIME EFT '+' $
```

The + character gives a more attractive appearance to the plot. Following plots of subsets of the data in this way, it is good practice to delete %RE, so that subsequent plots are not restricted unintentionally:

```
$DEL %RE $
```

It looks as though the corner group take less time to construct the designs than the row group, but the difference is not marked and may be due simply to sample fluctuations. The assessment of this question is based on a statistical model.

The usual approach to the analysis of the data above is by multiple regression, or analysis of covariance. We formulate the model, expressed in a traditional form,

$$y_i = \beta_0 + \beta_1 x_{1i} + \beta_2 x_{2i} + \varepsilon_i \qquad i = 1, \ldots, 24$$

where y_i is the time taken by the i-th child, x_{1i} is the EFT score of the i-th child, and x_{2i} is a dummy variable, with $x_{2i}=0$ for the row group children and $x_{2i}=1$ for the corner group children. The "error" variables ε_i are assumed to be independent and normally distributed in repeated sampling with mean zero and common variance σ^2, written $\varepsilon_i \sim N(0,\sigma^2)$. Dummy variables for categorical (here binary) explanatory variables are automatically defined by GLIM if the variables are declared to be factors. In the listing of the SOLV file in Appendix 4, this is achieved by the directives

```
$CALC GROUP = %GL(2,12) $
$FACTOR GROUP 2 $
```

Thus labels 1 and 2 for the group values do not have to be read in as data: they can be calculated instead. The number of levels of the group factor has to be specified. The dummy variable used by GLIM for x_2 in the regression is denoted by GROU(2). For a factor **A** with k levels ($k \geq 2$), GLIM defines $(k-1)$ dummy variables **A**(2), . . ., **A**(k) with **A**(j) = 1 for the j-th level of **A**, and zero otherwise. These variables cannot be accessed by the user, but it is easy to define such variables explicitly, using the calculate directive. Thus to define a dummy variable for the second level of the factor **A**, we use

```
$CALC A2 = %EQ(A,2) $
```

and similarly for the other levels of **A**.

The systematic part of the model is the regression function $\beta_0 + \beta_1 x_1 + \beta_2 x_2$ for the mean time taken, and the random part is the normal error variation about the regression with variance σ^2. If we write μ_i for the mean time taken for the i-th child, then

$$\mu_i = \beta_0 + \beta_1 x_{1i} + \beta_2 x_{2i}, \qquad y_i = \mu_i + \varepsilon_i, \qquad \varepsilon_i \sim \mathrm{N}(0, \sigma^2),$$

which we will call *Model 2*: in all we will consider five models for these data. In the remainder of this chapter we will suppress the specification of the *random* part of the model as it is the same for all the regression models for μ_i.

Since x_2 is a dummy variable, we may write this regression function equivalently as

$$\mu_i = \beta_0 + \beta_1 x_{1i} \qquad x_{2i} = 0\text{: Row group}$$

$$\mu_i = \beta_0 + \beta_2 + \beta_1 x_{1i} \qquad x_{2i} = 1\text{: Corner group}$$

The two regression lines for the row group and the corner group for the five models discussed are shown in Fig. 2.1. The regression lines for model 2 are parallel, with slope β_1 representing the increase in mean time taken for a one-point increase in EFT score, β_2 representing the difference in mean time taken between the corner and row groups with the same value of EFT, and β_0 the (hypothetical) mean time taken for a child with EFT score zero in the row group.

If the instructions have no differential effect on completion time, then $\beta_2 = 0$, and model 2 reduces to *Model 3*:

$$\mu_i = \beta_0 + \beta_1 x_{1i},$$

identical regressions for the two instruction groups. If the instructions have a different effect, but EFT is unrelated to completion time, then $\beta_1 = 0$, and model 2 reduces to *Model 4*:

$$\mu_i = \beta_0 + \beta_2 x_{2i}.$$

It is also possible that neither EFT nor instructions has any effect, in which case the null *Model 5* results:

$$\mu_i = \beta_0.$$

A further possibility is that the regressions of mean time on EFT score for the two instruction groups are linear, but not parallel. This can be represented by the *interaction Model 1*:

$$\mu_i = \beta_0 + \beta_1 x_{1i} + \beta_2 x_{2i} + \beta_3 x_{1i} x_{2i},$$

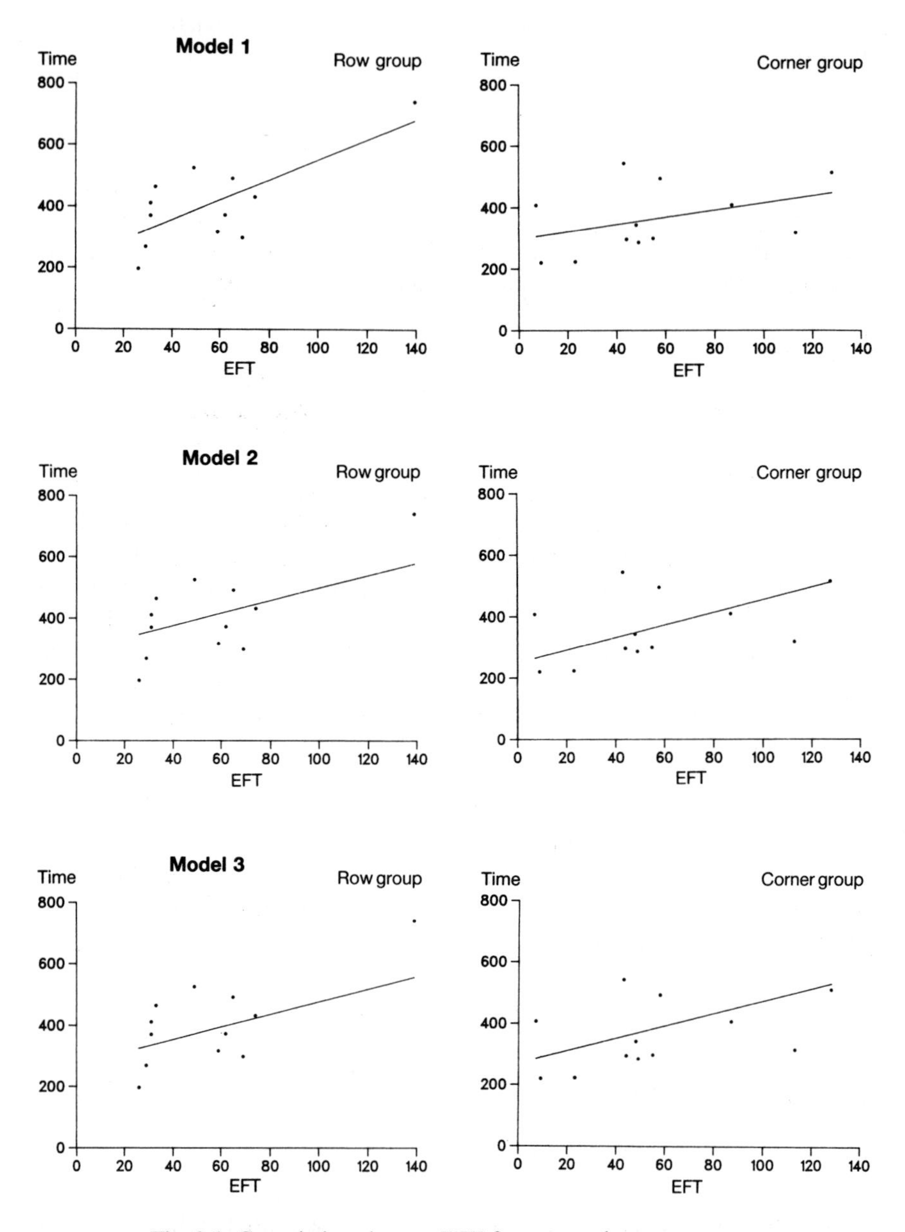

Fig. 2.1. Completion time vs EFT for row and corner groups

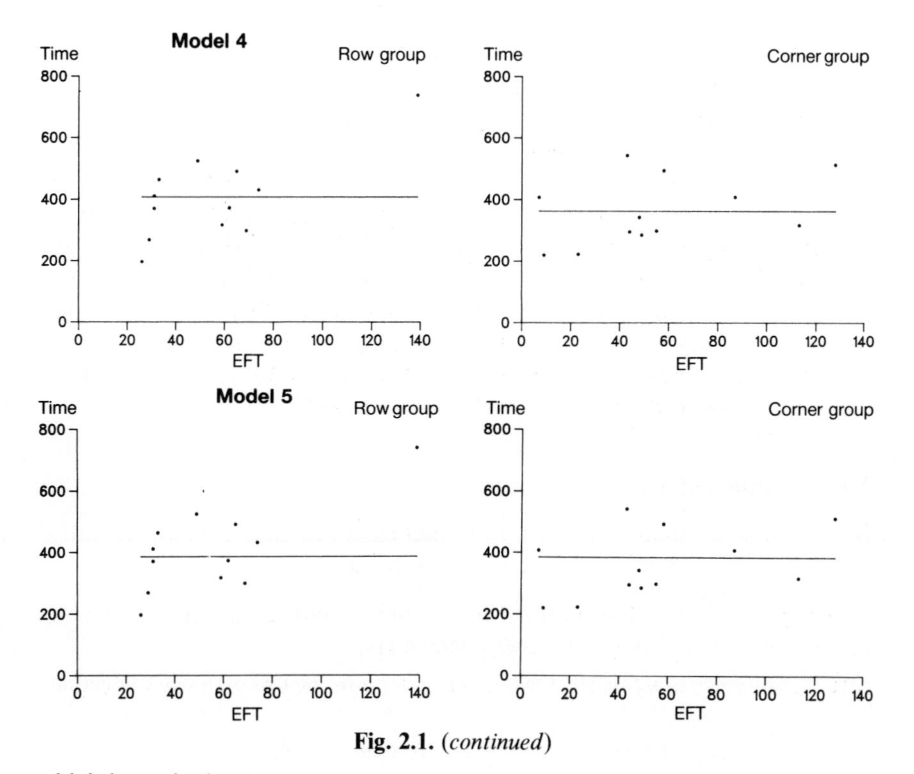

Fig. 2.1. *(continued)*

which is equivalent to

$$\mu_i = \beta_0 + \beta_1 x_{1i} \qquad x_{2i} = 0\text{: Row group}$$
$$\mu_i = (\beta_0 + \beta_2) + (\beta_1 + \beta_3) x_{1i} \qquad x_{2i} = 1\text{: Corner group.}$$

For model 1, the difference between mean completion times in the two experimental groups, for the same EFT score x, is

$$(\beta_0 + \beta_2) + (\beta_1 + \beta_3)x - (\beta_0 + \beta_1 x) = \beta_2 + \beta_3 x$$

a linear function of x. Thus the effect of different instructions on mean completion time depends on EFT score, a much more complicated interpretation than that of model 2, where the difference is a constant β_2 because β_3 is zero.

The interpretation of the experiment depends, therefore, on which of the five (or possibly other) models is the "best" representation of the data, in the sense of "simplest consistent with the data". In the subsequent sections of this chapter we develop the general theory for this question.

It is worth stressing that none of the models is a true representation of the population. If we could take a complete census of the population of fifth-grade children in the school, and administer the EFT and WISC tests to all of them, we would find that the mean completion time for children with each EFT score in each experimental group did not lie on a straight line: the means would be quite irregular about a general trend. Our model ignores the variation, and uses only a linear trend: in a sample we would expect sample variations about a straight line even if the population corresponded exactly to the model. The model thus represents an idealized or smooth version of the population. This is most obvious when we are in fact modelling complete populations: we still smooth the complete population using the model so that the gross systematic features of the population are retained, but the minor irregularities or fine structure are lost, being represented as random variation.

We turn now to the formalities of statistical models.

2.4 Definition of a statistical model

In this book we shall consider a restricted class of statistical models, called *generalized linear models*, defined by three components:

(1) a probability distribution $f(y)$ for the single response variable y depending on the mean μ, and possibly other parameters;

(2) a *linear regression function* or *linear predictor* in the p explanatory variables

$$\eta = \boldsymbol{\beta}'\mathbf{x} = \beta_0 x_0 + \beta_1 x_1 + \beta_2 x_2 + \ldots + \beta_p x_p,$$

where x_0 is identically 1;

(3) a *parameter transformation* or *link function* $g(\mu)$ which relates the linear predictor η to the mean μ:

$$\eta = g(\mu).$$

For certain distributions f and link functions g, these models can easily be fitted in GLIM. We have to specify the response variable y to be modelled, the probability (error) distribution for y, the link function and the regresssion model to be fitted. The response variable is specified by the directive $YVAR whose argument is the name of the response variable. The error distribution is specified by the directive $ERROR, which has possible arguments N (normal), B (binomial), P (Poisson) or G (gamma). The link function is specified by the $LINK directive which has to follow the $ERROR directive: its possible arguments depend on the error distribution. The regression model is specified as the argument of the $FIT directive: examples of its use are given in Section 2.7.

Much more general models (for example, non-linear in the parameters) can be fitted by generalizations of the computational algorithm used in GLIM;

details are given in McCullagh and Nelder (1983). We shall concentrate on the standard models in GLIM, with extensions to other distributions (like the exponential, Weibull and extreme value) which can also be fitted with only minor extensions of the standard models.

The standard analysis of the data of Section 2.3 by regression analysis or analysis of covariance corresponds to the fitting of a series of models with the following specifications:

(1) the probability distribution for y is normal $N(\mu,\sigma^2)$ with mean μ and variance σ^2;

(2) the linear predictor is $\eta=\beta_0+\beta_1x_1+\beta_2x_2$, where x_1 is EFT and x_2 group, or one of the other models 1–5;

(3) the link function is the *identity* $\eta=g(\mu)=\mu$.

This description of the model may seem unfamiliar and cumbersome. Why not use

$$y=\beta_0+\beta_1x_1+\beta_2x_2+\varepsilon, \qquad \varepsilon\sim N(0,\sigma^2),$$

the familiar form used in Section 2.3? There are good reasons for this. For probability distributions other than the normal and extreme value (Chapter 6) the representation of the model with an "additive error" term ε is not satisfactory, because ε does not have any simple distribution. In many models the scale of μ on which the linear predictor operates can have several different forms, so the explicit definition of a link function is needed to separate the choice of scale from the linear predictor. Examples are given in Chapter 3 of a normal distribution for y with a log link function.

Given the specification of the model, how is it to be fitted to the data? Since several different regression functions might be used, how do we decide which is the most appropriate? These questions are answered by the use of the *likelihood function* which we now consider.

2.5 The likelihood function and maximum likelihood estimation

The probability distribution for y is an essential component of the statistical model because the optimal fitting of the model depends on the form of the probability distribution. Denote by $f(y|\mu,\boldsymbol{\phi})$ the probability density or mass function for y, depending on the mean μ, to which the linear predictor is related, and a vector of other parameters $\boldsymbol{\phi}$ which are constant and not related to the linear predictor. (For some probability distributions there are no $\boldsymbol{\phi}$ parameters.)

Let the i-th observation on the response and explanatory variables $(i=1,2,\ldots,n)$ be represented by $(y_{1i},\mathbf{x}_i)$ where $\mathbf{x}_i=(x_{0i}, x_{1i},\ldots, x_{pi})$ is a vector of length $(p+1)$ whose first component x_{0i} is identically 1. Let μ_i be the value of the mean μ for the i-th observation. If the data collection process leading to the sample is simple random sampling (other forms of sampling are considered in Section 2.9) then the probability of obtaining the observations $y_1, \ldots, y_n$ is

$$f(y_1|\mu_1,\phi)f(y_2|\mu_2,\phi)\ldots f(y_n|\mu_n,\phi)$$

if y is discrete, and

$$f(y_1|\mu_1,\phi)f(y_2|\mu_2,\phi)\ldots f(y_n|\mu_n,\phi)\mathrm{d}y_1\mathrm{d}y_2\ldots\mathrm{d}y_n$$

if y is continuous, provided the sampling fraction is small. Once the observations are obtained, the *likelihood function* $L=L(\mu_1, \ldots, \mu_n,\phi)$ is defined by

$$L(\mu_1, \ldots, \mu_n,\phi) = \prod_{i=1}^{n} f(y_i|\mu_i,\phi)$$

that is, the likelihood function is (proportional to) the probability of obtaining the given sample observations, *regarded as a function of the unknown parameters* $\mu_1, \ldots, \mu_n, \phi$.

The definition of the likelihood function is not consistent in statistical theory: in some definitions multiplicative constants which are not functions of the parameters are omitted, since this makes no difference to conclusions about regression model choice based on the likelihood ratio test—see Section 2.7. However, when different probability distributions are being compared as models, it is essential to keep all the multiplicative constants in the likelihood for each distribution—see Section 3.1 and Chapter 6. The differential element $\mathrm{d}y_1, \ldots \mathrm{d}y_n$ is always omitted from the likelihood for continuous distributions; see Appendix 1 for further discussion.

Before proceeding further we consider the simple example of the completion times discussed in Section 2.3. The probability distribution for time $f(y)$ is normal with mean μ and variance σ^2, and the linear predictor for the 'null' model 5 is just a *constant*:

$$\eta_i = \beta_0 = \mu,$$

so that $\mu_1 = \mu_2 = \ldots = \mu_n = \mu$, corresponding to $\mathbf{x}_i = \mathbf{1}$.

This model implies that EFT and instructions are *unrelated* to completion time: the observed variations in completion time with EFT and instructions are just random normal variations about a constant mean μ with variance σ^2.

The probability density function for y is

$$f(y|\mu,\sigma) = \frac{1}{\sigma\sqrt{(2\pi)}} \exp\left\{-\frac{(y-\mu)^2}{2\sigma^2}\right\}$$

and the likelihood function is therefore

$$\begin{aligned} L(\mu,\sigma) &= \prod_{i=1}^{n} \left[\frac{1}{\sigma\sqrt{(2\pi)}} \exp\left\{-\frac{(y_i-\mu)^2}{2\sigma^2}\right\}\right] \\ &= \frac{1}{\{\sigma\sqrt{2\,\pi)}\}^n} \exp\left\{-\frac{1}{2\sigma^2}\sum_{i=1}^{n}(y_i-\mu)^2\right\}. \end{aligned}$$

(This assumes that y is measured with high precision relative to σ: see Appendix 1 for discussion of the likelihood function when this is not the case.)

The parameters μ and σ are unknown, and the process of model fitting is that of choosing the best-fitting values of the unknown parameters. The criterion of best fit used generally in statistical theory and throughout this book is that of maximum likelihood.

Definition

For a likelihood function $L(\boldsymbol{\theta})$ of a vector of k parameters $\boldsymbol{\theta}$, the *maximum likelihood estimate* $\hat{\boldsymbol{\theta}}$ of $\boldsymbol{\theta}$ is defined as the value of $\boldsymbol{\theta}$ for which

$$L(\hat{\boldsymbol{\theta}}) \geq L(\boldsymbol{\theta})$$

for all $\boldsymbol{\theta}$. For the statistical models considered in this book, the likelihood function has a unique maximum and no minima (though this is not true generally), and the maximum can usually be found by partial differentiation of the log likelihood function as the solution of

$$\frac{\partial \ell}{\partial \boldsymbol{\theta}} = \frac{\partial \log L}{\partial \boldsymbol{\theta}} = \frac{1}{L}\frac{\partial L}{\partial \boldsymbol{\theta}} = \mathbf{0}.$$

We illustrate using the normal distribution example discussed above.

$$\begin{aligned} \ell(\mu,\sigma) &= \log L(\mu,\sigma) \\ &= -\frac{n}{2}\log(2\pi) - n\log\sigma - \sum_{i=1}^{n}(y_i-\mu)^2/2\sigma^2 \end{aligned}$$

In the rest of this book, summation will be assumed to be over the index i from 1 to n unless otherwise stated.

$$\frac{\partial \ell}{\partial \mu} = \Sigma(y_i-\mu)/\sigma^2$$

$$\frac{\partial \ell}{\partial \sigma} = -n/\sigma + \Sigma(y_i-\mu)^2/\sigma^3.$$

Equating the derivatives to zero gives

$$\hat{\mu} = \bar{y}$$
$$\hat{\sigma}^2 = \Sigma(y_i - \hat{\mu})^2/n$$
$$= \Sigma(y_i - \bar{y})^2/n.$$

The value of the log likelihood function at its maximum is

$$\ell(\hat{\mu},\hat{\sigma}) = -(n/2)\log(2\pi) - n\log\hat{\sigma} - n/2$$
$$= \mathrm{c} - (n/2)\log(\mathrm{RSS})$$

where

$$\mathrm{c} = n\{\log n - \log(2\pi) - 1\}/2$$

and

$$\mathrm{RSS} = n\,\hat{\sigma}^2 = \Sigma(y_i - \bar{y})^2$$

is the *residual sum of squares* obtained by minimizing the "sum of squared deviations from the mean" $\Sigma(y_i - \mu)^2$ with respect to μ.

The above statistical model is fitted very simply in GLIM. We declare TIME to be the response variable, the probability distribution to be normal, the link function to be the identity, and the regression model to be a constant:

```
$YVAR TIME $ERROR N $LINK I $FIT $
```

The directive $FIT without an argument always means that the null model is fitted. The identity link is the default link for the normal distribution: if the link directive is omitted, the identity is assumed. The normal is the default error distribution: if the error directive is omitted, the normal is assumed.

The standard output from $FIT is restricted to the *deviance*, which for the normal distribution is the residual sum of squares, and its degrees of freedom. Additional output can be displayed using the $DISPLAY directive, which we will abbreviate to $D. Printing of the output is controlled by a sequence of option letters. For example option E displays the estimated regression coefficients and their standard errors, and option R displays the residuals and fitted values. Many other options are available: Appendix 2 gives a list.

```
$D E $

     estimate          s.e.      parameter
1       384.3         25.80      1
scale parameter taken as  15980.
```

For the null model fitted above, the estimated regression coefficient is just the sample mean, which is 384.3, and the residuals are just the deviations

$y_i - \bar{y}$. The residual sum of squares is 367537. (In GLIM 3.12 the deviance is printed only to four significant figures — 0.3675×10^6. It can be printed more accurately by $PRINT %DV $.) The maximum likelihood estimate of σ^2 is 15314 (i.e. 367537/24), and of σ is 123.8. GLIM calculates the unbiased estimate $s^2 = \text{RSS}/(n-1) = 15980$ and prints it as the SCALE PARAMETER beneath the table of parameter estimates.

We would usually want to attach a standard error to the sample mean: this is given under the "S.E" heading in the print out – 25.80. We now consider standard errors in general.

2.6 The accuracy of parameter estimation

The second derivative matrix of the likelihood function plays a central role in assessing the accuracy of parameter estimation. We write

$$D(\boldsymbol{\theta}) = -\frac{\partial^2 \ell}{\partial \boldsymbol{\theta} \partial \boldsymbol{\theta}'}.$$

Then the *observed information matrix* $I(\hat{\boldsymbol{\theta}})$ is $D(\boldsymbol{\theta})$ evaluated at the maximum likelihood estimate:

$$I(\hat{\boldsymbol{\theta}}) = D(\hat{\boldsymbol{\theta}})$$

and the *expected information matrix* is

$$\mathscr{I}(\boldsymbol{\theta}) = E[D(\boldsymbol{\theta})]$$

where the expectation is over the distribution of $\mathbf{y}$. The *estimated expected information matrix* is $\mathscr{I}(\hat{\boldsymbol{\theta}})$.

Standard large-sample theory gives the result that in repeated sampling, under suitable regularity conditions,

$$(\hat{\boldsymbol{\theta}} - \boldsymbol{\theta})' \mathscr{I}(\boldsymbol{\theta})(\hat{\boldsymbol{\theta}} - \boldsymbol{\theta}),$$

$$(\hat{\boldsymbol{\theta}} - \boldsymbol{\theta})' \mathscr{I}(\hat{\boldsymbol{\theta}})(\hat{\boldsymbol{\theta}} - \boldsymbol{\theta})$$

and

$$(\hat{\boldsymbol{\theta}} - \boldsymbol{\theta})' I(\hat{\boldsymbol{\theta}})(\hat{\boldsymbol{\theta}} - \boldsymbol{\theta})$$

are all asymptotically distributed as χ^2_k. In large samples, we may take $\hat{\boldsymbol{\theta}}$ as approximately multivariate normally distributed with mean $\boldsymbol{\theta}$ and covariance matrix $V(\hat{\boldsymbol{\theta}}) = A^{-1}$, where A may be taken as $\mathscr{I}(\boldsymbol{\theta})$, $\mathscr{I}(\hat{\boldsymbol{\theta}})$ or $I(\hat{\boldsymbol{\theta}})$. The diagonal element v_{jj} of V is then the approximate variance of the maximum likelihood estimator $\hat{\theta}_j$, regarded as a random variable in repeated sampling. Approximate confidence intervals for θ_j can be based on $\hat{\theta}_j \pm \lambda\, v_{jj}^{\frac{1}{2}}$, where λ is the appropriate percentage point of the normal distribution.

GLIM provides standard errors for the parameters in a fitted regression model based on the estimated expected information matrix. It is also possible to calculate in GLIM standard errors based on the observed information matrix; there is increasing evidence that the observed information matrix is more appropriate (see e.g. Efron and Hinkley, 1978), and it represents the curvature of the actual log likelihood function. In some cases the two information matrices are identical.

The value of all these standard error estimates of precision of the maximum likelihood estimate of $\boldsymbol{\theta}$ depends on the goodness of the approximation of the log likelihood function by a quadratic function of $\boldsymbol{\theta}$. If the log likelihood is highly skewed, all the standard error estimates may be quite misleading, and it may be preferable to tabulate the likelihood function directly. An example of the need for this procedure is discussed in Section 2.15.

In the normal example above, with

$$\ell(\mu,\sigma) = -n/2 \log(2\pi) - n \log \sigma - \Sigma(y_i-\mu)^2/2\sigma^2,$$

we have

$$\frac{\partial^2 \ell}{\partial \mu^2} = -n/\sigma^2$$

$$\frac{\partial^2 \ell}{\partial \sigma^2} = n/\sigma^2 - 3\Sigma(y_i-\mu)^2/\sigma^4$$

$$\frac{\partial^2 \ell}{\partial \mu \partial \sigma} = -2\Sigma(y_i-\mu)/\sigma^3.$$

Taking expectations, we have

$$\mathscr{I}(\mu,\sigma) = \begin{bmatrix} n/\hat{\sigma}^2 & 0 \\ 0 & 2n/\hat{\sigma}^2 \end{bmatrix}.$$

Evaluating the derivatives at $\hat{\mu}$, $\hat{\sigma}$, we have

$$I(\hat{\mu},\hat{\sigma}) = \begin{bmatrix} n/\hat{\sigma}^2 & 0 \\ 0 & 2n/\hat{\sigma}^2 \end{bmatrix}.$$

Thus in this case $\mathscr{I}(\hat{\mu},\hat{\sigma}) = \mathrm{I}(\hat{\mu},\hat{\sigma})$, that is the observed and estimated expected information matrices are the same.

The concept of information in a sample was introduced by Fisher (1922). For both μ and σ, the information about these parameters provided by the data is directly proportional to n—larger samples give more information and higher precision—and inversely proportional to $\hat{\sigma}^2$—larger sample variances give less information and lower precision. The covariance matrix of the parameter estimates is

$$V(\hat{\mu},\hat{\sigma}) = \begin{bmatrix} \hat{\sigma}^2/n & 0 \\ 0 & \hat{\sigma}^2/2n \end{bmatrix}$$

The approximate variance of $\hat{\mu}=\bar{y}$ would then be $\hat{\sigma}^2/n$, and an approximate 95% confidence interval for μ would be $\bar{y} \pm 1.96\hat{\sigma}/\sqrt{n}$, or 384.3 ± 1.96 (25.26), that is 384.3 ± 49.5.

These approximate results can be improved for the normal model by using the t-distribution. An exact 95% confidence interval for μ (assuming the correctness of the assumption of a normal distribution for y) is given by

$$\bar{y} \pm t_{0.025,\, n-1}\, s/\sqrt{n},$$

where

$$s^2 = \mathrm{RSS}/(n-1) = n\hat{\sigma}^2/(n-1).$$

Thus the t-interval is $\bar{y} \pm t_{0.025,\,23}\, \hat{\sigma}/\sqrt{(n-1)}$, which is 384.3 ± 2.07 (25.80), that is 384.3 ± 53.4. The interval based on the asymptotic normal distribution is too short, and the discrepancy increases in smaller samples. GLIM gives the standard error of 25.80 based on the unbiased variance estimate s^2 instead of $\hat{\sigma}^2$.

For probability distributions other than the normal, exact results of this kind are rarely obtainable, and so we have to rely on the asymptotic results even in small samples. Direct plotting of the likelihood function can often be used to improve the asymptotic results: examples are given in Section 2.15 and in other chapters.

We are not usually interested in confidence intervals for the standard deviation σ: since this is assumed constant in the model it is of interest only for describing the magnitude of random variation. Inferences about the regression parameters depend on σ, which is therefore called a *nuisance parameter*. Confidence intervals based on $\hat{\sigma} \pm \lambda\hat{\sigma}/\sqrt{(2n)}$ present difficulties: if $n < \lambda^2/2$ the interval will include negative values. In practice intervals are based on the $\sigma^2\chi^2_{n-1}$ distribution of RSS (see Section 2.15 for a discussion of confidence intervals based directly on the likelihood function).

We now turn to the choice between possible regression functions.

2.7 The likelihood ratio test and the choice among regressions

For the example of Section 2.3, we have already discussed five different regression functions as candidates for the final model. We present them below with their GLIM model formulae as used in the $FIT directive:

Model	Model formula	Regression function
1	EFT+GROUP+EFT.GROUP (or EFT*GROUP)	$\beta_0+\beta_1x_1+\beta_2x_2+\beta_3x_1x_2$
2	EFT+GROUP	$\beta_0+\beta_1x_1+\beta_2x_2$
3	EFT	$\beta_0+\beta_1x_1$
4	GROUP	$\beta_0 \qquad +\beta_2x_2$
5	1	β_0

The GLIM model formulae are easily understood. Main effect terms in the model are represented by variate or factor names linked by a plus sign. Interactions are represented by dot products like EFT.GROUP. Any number of factors may be included in an interaction, but at most one variate. The star product (as in EFT*GROUP) is a shorthand for the full model with main effects of each variable and their interactions of all orders. The null model is conveniently represented by the value 1 of the implicit variable x_0, the only variable in the model.

The choice among the five models (and any others) is based on the *likelihood ratio test*. Each of the models 2–5 is a special case of model 1 in which a subset of regression coefficients in this model is zero. Consider two candidate models for the data: the full model $\eta_f=\boldsymbol{\beta}'\mathbf{x}$ and a reduced model $\eta_r=\boldsymbol{\beta}_r'\mathbf{u}$, where $\mathbf{u}$ is a vector of length p_1+1, made up of the constant 1 plus p_1 of the explanatory variables in $\mathbf{x}$, and $\boldsymbol{\beta}_r$ is the corresponding vector of regression coefficients. Write $\mathbf{v}$ for the vector of explanatory variables omitted from $\mathbf{x}$ in defining $\mathbf{u}$, and $\boldsymbol{\beta}_s$ for the corresponding vector of regression coefficients, so that

$$\boldsymbol{\beta}'\mathbf{x} = \boldsymbol{\beta}_r'\mathbf{u} + \boldsymbol{\beta}_s'\mathbf{v}$$

Then if $\boldsymbol{\beta}_s=\mathbf{0}$, the full model η_f can be simplified to the reduced model η_r, and so the use of the reduced model is justified if the evidence against $\boldsymbol{\beta}_s=\mathbf{0}$ is not strong. This evidence is provided by the likelihood ratio test of the hypothesis $\boldsymbol{\beta}_s=\mathbf{0}$, which compares the maximized likelihood functions under the two models.

The likelihood function for the full model is

$$\begin{aligned} L(\boldsymbol{\beta},\sigma) &= \prod_{i=1}^{n}\left[\frac{1}{\sigma\sqrt{(2\pi)}}\exp\left\{-\frac{1}{2}\frac{(y_i-\mu_i)^2}{\sigma^2}\right\}\right] \\ &= \frac{1}{(\sigma\sqrt{(2\pi)})^n}\exp\left\{-\frac{1}{2\sigma^2}\Sigma(y_i-\mu_i)^2\right\} \end{aligned}$$

where

$$\mu_i = \boldsymbol{\beta}'\mathbf{x}_i = \mathbf{x}_i'\boldsymbol{\beta}.$$

The log likelihood and its derivatives are

$$\ell(\boldsymbol{\beta},\sigma) = -\frac{n}{2}\log(2\pi) - n\log\sigma - \frac{1}{2\sigma^2}\Sigma(y_i - x_i'\boldsymbol{\beta})^2$$

$$\frac{\partial\ell}{\partial\boldsymbol{\beta}} = \Sigma\mathbf{x}_i(y_i - \mathbf{x}_i'\boldsymbol{\beta})/\sigma^2$$

$$\frac{\partial\ell}{\partial\sigma} = -\frac{n}{\sigma} + \Sigma(y_i - \mathbf{x}_i'\boldsymbol{\beta})^2/\sigma^3$$

$$\frac{\partial^2\ell}{\partial\boldsymbol{\beta}\partial\boldsymbol{\beta}'} = -\Sigma\mathbf{x}_i\mathbf{x}_i'/\sigma^2$$

$$\frac{\partial^2\ell}{\partial\boldsymbol{\beta}\partial\sigma} = -2\Sigma\mathbf{x}_i(y_i - \mathbf{x}_i'\boldsymbol{\beta})/\sigma^3$$

$$\frac{\partial^2\ell}{\partial\sigma^2} = \frac{n}{\sigma^2} - 3\,\Sigma(y_i - \mathbf{x}_i'\boldsymbol{\beta})^2/\sigma^4.$$

For a maximum of the likelihood,

$$\frac{\partial\ell}{\partial\boldsymbol{\beta}} = \mathbf{0}, \qquad \frac{\partial\ell}{\partial\sigma} = 0$$

giving

$$\hat{\sigma}^2{}_f = \Sigma(y_i - \mathbf{x}_i'\hat{\boldsymbol{\beta}}_f)^2/n = \Sigma(y_i - \hat{\mu}_i)^2/n = \text{RSS}_f/n,$$

where RSS_f is the residual sum of squares from the full model, and

$$\hat{\boldsymbol{\beta}}_f = (\Sigma\mathbf{x}_i\,\mathbf{x}_i')^{-1}\,\Sigma\mathbf{x}_i y_i,$$

which is usually written in the form

$$\hat{\boldsymbol{\beta}}_f = (X'X)^{-1}X'\mathbf{y}$$

where

$$X' = [\mathbf{x}_1 \ldots \mathbf{x}_n].$$

The observed and estimated expected information matrices are

$$I(\hat{\boldsymbol{\beta}}_f,\hat{\sigma}_f) = \mathscr{I}(\hat{\boldsymbol{\beta}}_f,\hat{\sigma}_f) = \begin{bmatrix}(X'X)/\hat{\sigma}_f^2 & 0\\ 0 & 2n/\hat{\sigma}_f^2\end{bmatrix}$$

so that

$$V(\hat{\boldsymbol{\beta}}_f,\hat{\sigma}_f) = \begin{bmatrix}\hat{\sigma}_f^2(X'X)^{-1} & 0\\ 0 & \hat{\sigma}_f^2/2n\end{bmatrix}$$

The parameter standard errors printed out by GLIM are the square roots of the diagonal elements of the matrix $s^2(X'X)^{-1}$, where

$$s^2 = \Sigma(y_i - \mathbf{x}_i'\hat{\boldsymbol{\beta}}_f)^2/(n-p-1) = \text{RSS}_f/(n-p-1)$$

is the unbiased estimate of σ^2.

The maximized log likelihood is, as in Section 2.5,

$$\ell(\hat{\boldsymbol{\beta}}_f, \hat{\sigma}_f) = \text{c} - n/2 \log \text{RSS}_f$$

where

$$c = n\{\log n - \log(2\pi) - 1\}/2.$$

Now consider the same estimation procedure for the reduced model. We obtain

$$\hat{\boldsymbol{\beta}}_r = (U'U)^{-1}U'\mathbf{y}$$

where

$$U' = [\mathbf{u}_1 \ldots \mathbf{u}_n]$$

and

$$\hat{\sigma}^2{}_r = \Sigma(y_i - u_i'\hat{\boldsymbol{\beta}}_r)^2/n = \text{RSS}_r/n$$

where RSS_r is the residual sum of squares from the reduced model. The maximized log likelihood is

$$\ell(\hat{\boldsymbol{\beta}}_r, \hat{\sigma}_r) = \text{c} - n/2 \log \text{RSS}_r.$$

A formal comparison of the two models is expressed in terms of the likelihood ratio test of the hypothesis that the variables omitted from $\mathbf{x}$ have zero regression coefficients and do not contribute to the regression function, so that $\boldsymbol{\beta}_s = \mathbf{0}$. The *likelihood ratio test statistic* (LRTS) for this hypothesis is

$$\begin{aligned}\lambda &= -2\{\ell(\hat{\boldsymbol{\beta}}_r, \hat{\sigma}_r) - \ell(\boldsymbol{\beta}_f, \hat{\sigma}_f)\}\\ &= n\{\log \text{RSS}_r - \log \text{RSS}_f\}\\ &= n \log(\text{RSS}_r/\text{RSS}_f).\end{aligned}$$

(Note that the constant c disappears from the LRTS. For this reason numerical constants which are not functions of the parameters can be omitted in the definition of the likelihood function, as noted in Section 2.5.)

If the hypothesis is true, then the residual sum of squares for the reduced model will be not much larger than that for the full model, and so λ will be small. If the hypothesis is false, then λ will be large.

For the hypothesis that a single component β_j of $\boldsymbol{\beta}$ is zero, an alternative (the

"Wald") test compares $\hat{\beta}_j$ with its asymptotic standard error $v_{jj}^{\frac{1}{2}}$, and treats $t_j = \hat{\beta}_j / \hat{v}_{jj}^{\frac{1}{2}}$ as approximately $N(0,1)$ under the hypothesis. For large samples in which the likelihood function is normal in the parameters, this test will be closely equivalent to the likelihood ratio test, with $t_j^2 \approx \lambda$. For small samples and for regression models in which the response variable is not normally distributed, the likelihood may be far from normal and the value of t_j^2 may be substantially different from λ. An example of this occurs in Section 4.3 for a logit model with a small sample.

Critical values for λ are based on its asymptotic distribution when the null hypothesis is true, which is χ_q^2, where $q = p - p_1$ if both X and U are of full rank (if not, $q = \text{rank}(X) - \text{rank}(U)$).

For the normal distribution this asymptotic result is not used, since exact results are available. If we define

$$\text{HSS} = \text{RSS}_r - \text{RSS}_f$$

then

$$\begin{aligned}\lambda &= n \log(1 + \text{HSS}/\text{RSS}_f)\\ &= n \log\left[1 + \frac{q}{n-p-1}\left\{\frac{\text{HSS}/q}{\text{RSS}_f/(n-p-1)}\right\}\right]\\ &= n \log(1 + qF/(n-p-1))\end{aligned}$$

where

$$F = \left[\frac{\text{HSS}/q}{\text{RSS}_f/(n-p-1)}\right]$$

is the usual F-statistic for the test of the hypothesis $\boldsymbol{\beta}_s = 0$ in multiple regression or the analysis of variance. The exact distribution of λ could therefore be obtained from that of F (or from the beta distribution of $\text{RSS}_f/\text{RSS}_r$), but it is simpler to use the F-distribution itself which is well tabulated.

Under the hypothesis that a single component β_j of $\boldsymbol{\beta}$ is zero, the distribution of $t_j = \hat{\beta}_j / \hat{v}_{jj}^{\frac{1}{2}}$ is exactly t_{n-p-1}, which is directly related to F since the distribution of t_{n-p-1}^2 is $F_{1,n-p-1}$.

GLIM provides, under the "deviance" heading, the residual sum of squares for the fitted model. It does not provide F-tests (or t-tests for regression coefficients) or analysis of variance tables. Such tables can be constructed by successive differencing of the residual sum of squares of a *hierarchical* sequence of models of increasing complexity.

The residual sums of squares for all the models are quickly obtained by a sequence of $FIT statements:

```
$FIT : +EFT : GROUP : +EFT : +EFT.GROUP $
```

The continuation sign : replaces the repeated $FIT directive. The use of

+EFT avoids the complete rewriting of the previous model formula. Terms can be removed from the model in the same way using −. The residual sums of squares are 367537, 257274, 355522, 245531 and 218076. Listing models 5, 4, 2, 1 in order and differencing their residual sum of squares gives sums of squares attributable to the various terms being considered.

Model		RSS	Residual df	Source	SS	df
5	1	367537	23			
4	GROUP	355522	22	GROUP	12015	1
2	+EFT	245531	21	EFT	109991	1
1	+EFT.GROUP	218076	20	Interaction	27455	1
				Residual	218076	20
				s^2=10903.8		

The residual in the above table is simply the residual sum of squares from the final model in the sequence and corresponds to the residual sum of squares from the full model. The usual approach in analysis of variance tables is to obtain mean squares by dividing each sum of squares by its degrees of freedom and then to assess the importance of each effect by testing its mean square against the residual mean square. This procedure is of little use in the analysis of survey data because the sum of squares for each effect generally depends on its order of fitting in the model. Only for orthogonal experimental designs is this not the case. Thus, if we list the models in the order 5, 3, 2 and 1, and difference the residual sums of squares, we obtain a different table:

Model		RSS	Residual df	Source	SS	df
5	1	367537	23			
3	EFT	257274	22	EFT	110263	1
2	+GROUP	245531	21	GROUP	11743	1
1	+EFT.GROUP	218076	20	Interaction	27455	1
				Residual	218076	20
				s^2=10903.8		

The analysis of variance tables differ because the variables EFT and GROUP are slightly correlated, this correlation being due to the small difference in EFT means between the row and corner groups. The difference is only small because of the random allocation of children to groups. In observational studies such differences can be large, as shown in Section 2.15 and discussed further in Chapter 3. Qualitatively, we can see that the EFT mean square is large (more than 10 times the residual mean square in either model) while the GROUP mean square is small (about the same size as the residual mean square). The interaction mean square is the same in both tables, since it is always the last term fitted, and is 2.52 times the residual mean square.

We do not formalize these comparisons as F-tests because a more general approach is necessary for more complex models, based on a *simultaneous test procedure* (Aitkin, 1974, 1978). Consider any submodel $\eta_r = \boldsymbol{\beta}_r'\mathbf{u}$ where the subset $\mathbf{u}$ of included variables is chosen in any way, for example by inspection of the data. Then a level α-test of the hypothesis $\boldsymbol{\beta}_s = \mathbf{0}$, valid *simultaneously* for all possible selections of $\mathbf{u}$ from $\mathbf{x}$, rejects the hypothesis if

$$\text{RSS}_r/\text{RSS}_f > 1 + pF_{\alpha,p,n-p-1}/(n-p-1).$$

The right hand side of the inequality is constant for all submodels. If the subset $\mathbf{u}$ were specified *in advance*, and only the one hypothesis were being tested, the ratio on the left hand side would be compared with

$$1 + qF_{\alpha,q,n-p-1}/(n-p-1)$$

with $q = p - p_1$. The larger critical value above allows for the simultaneous testing of all possible models. Such simultaneous test procedures are very conservative if p is large and α is fixed at conventional levels like 0.05 or 0.01. We will generally allow α to increase with p; a value of α between 0.25 and 0.5 will often be reasonable if p is large: if it is not, a value near $1-(0.95)^p$ may be used. This procedure often requires unusual percentage points of the F-distribution: these can be obtained from the Tables of the Incomplete Beta Function (Pearson, 1968) or from calculator or computer subroutines.

In our example $p=3$ (assuming that no more complicated model is necessary), and $n=24$. The value of $1-0.95^3$ is 0.143. Rounding this to $\alpha = 0.15$, the critical value for the ratio $\text{RSS}_r/\text{RSS}_f$ is $1 + 3F_{0.15,3,20}/20 = 1.30$, so the critical value for RSS_r is 1.30×218076 or 283499. Any model with a residual sum of squares less than this is an adequate (α) model in the sense that we could not reject the hypothesis at (simultaneous) level α that the variables omitted from the model had zero regression coefficients. Thus the models including EFT are adequate (0.15), but those models excluding it are not. The simplest such model is that using EFT alone: such a model which is itself *adequate* (α), but for which no submodel is adequate, is called *minimal adequate* (α).

The principle of parsimony then implies that the best model to represent the data is a minimal adequate model. It is important to recognize that there may be more than one minimal adequate model, sometimes involving quite different variables, with correspondingly different interpretations. In such cases considerable caution is needed in drawing firm conclusions from the data.

In the present case the only minimal adequate model in the set 1–5 is EFT, model 3. The fitted values of the response variable are stored in the system vector %FV, and can be plotted with the observed times as follows:

```
$FIT EFT  $D E  $PLOT TIME %FV EFT '*+' $
```

```
    750.0 |                                             *
    720.0 |
    690.0 |
    660.0 |
    630.0 |
    600.0 |
    570.0 |
    540.0 |             * *                          +  +
    510.0 |                                     +    *
    480.0 |                  *  *
    450.0 |          *                 +
    420.0 | *       *            + 2   *
    390.0 |                 +++++
    360.0 |         *   ++3    *
    330.0 |       ++3+    *   *                 *
    300.0 | ++           ** *    *
    270.0 |         *
    240.0 |
    210.0 |  *    **
    180.0 |
----------:---------:---------:---------:---------:---------:---------:
         0.0      30.0      60.0      90.0     120.0     150.0     180.0
```

The fitted regression is

$$\hat{\mu} = 271.1 + 2.04 \text{ EFT},$$

a one-point increase in EFT score being associated with a two-second increase in mean time. Thus field dependence, as measured by the Embedded Figures Test, is associated with increased completion time, but the type of instruction given is not. Completion times vary randomly about the mean with variance estimated by $257274/22 = 11694$, or standard deviation $s = 108$ seconds.

To readers used to classical analysis of variance and covariance, this conclusion may seem unsatisfying. Where is the estimate of the treatment effect and its standard error? In simplifying the model, we have set the treatment effect (the coefficient β_2 of GROUP) equal to zero because there is no strong evidence from the data against this value. If we are particularly interested in certain parameters of the model because these represent experimentally controlled variables, then adequate models containing these variables may be presented as the final summary of the data. The parameter estimates, with standard errors, for the fitted model GROUP+EFT, are shown below.

$$\hat{\mu} = \underset{(48.4)}{293.4} + \underset{(0.665)}{2.038}\ \text{EFT} - \underset{(44.14)}{44.24}\ \text{GROU(2)}$$

The residual mean square estimate of σ^2 is 11692 with 21 df. A 95% confidence interval for β_2 is

$$-44.24 \pm t_{0.025,21}\,(44.14) \quad \text{i.e.} \quad (-136.1, 47.6).$$

The corner group has an *estimated* mean 44.2 seconds below that for the row group (for the same value of EFT), but the population value could be any value in the above wide interval, including zero.

2.8 Strategies for model simplification

The simplification of the model in the above sequence—interaction, then main effects—is the most common and useful strategy for model simplification. High-order interactions are usually small, and their retention in a model makes the interpretation of the model much more difficult. Successively removing interactions from the model as far as possible, starting with the most complex, is therefore appealing.

This is, however, not the only possible strategy. Another possible approach is to examine the lower-order interactions or main effects and to simplify them, while retaining the high-order interactions in a simplified form. The data of Section 2.3 provide a good example. The parameter estimates and their standard errors (in parentheses) from the interaction model are

$$\underset{(63.1)}{226.1} + \underset{(0.997)}{3.249}\ \text{EFT} + \underset{(83.91)}{70.45}\ \text{GROU(2)} - \underset{(1.303)}{2.067}\ \text{EFT.GROU(2)}$$

As we have already seen, the interaction term can be omitted from the model. However, examination of the regression coefficients for EFT in the two instruction groups reveals that in the row group, the slope is large: 3.249, while in the corner group it is small: 3.249–2.067=1.182. This suggests that a different simplification of the model is possible: we could set the slope of the EFT regression to zero in the corner group, but keep it non-zero in the row group. Since the slope is $(\beta_1+\beta_3)$ in the corner group, this constraint is achieved by setting $\beta_1+\beta_3=0$, or $\beta_3=-\beta_1$. The interaction model (model 5) then becomes *Model 6*:

$$\begin{aligned}\mu &= \beta_0+\beta_1 x_1+\beta_2 x_2-\beta_1 x_1 x_2\\ &= \beta_0+\beta_1 x_1(1-x_2)+\beta_2 x_2\end{aligned}$$

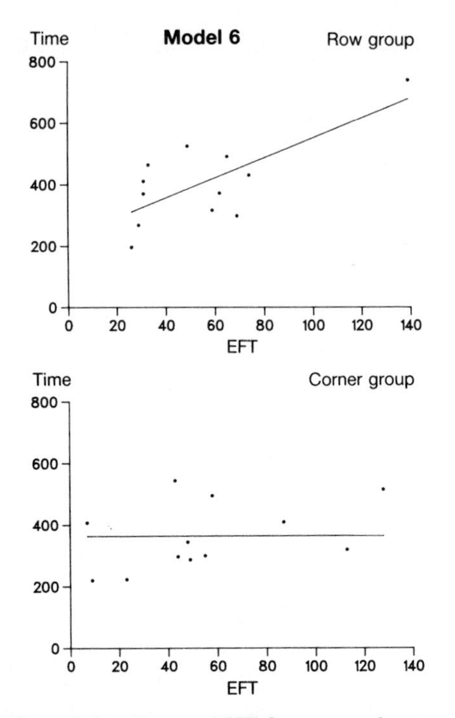

Fig. 2.2. Completion time vs EFT for row and corner groups

This model specifies a linear regression

$$\mu = \beta_0 + \beta_1 x_1$$

for the row group, and a constant mean

$$\mu = \beta_0 + \beta_2$$

for the corner group (see Fig. 2.2). It is important to note that β_2 does not now represent a constant difference in means between the groups for children with the same EFT score: this difference is now $\beta_2 - \beta_1 x_1$.

The model can be fitted by constructing a new variable $x_1(1-x_2)$ and fitting it with GROUP, as follows:

```
$CALC G = %EQ(GROUP,2) : XX = EFT*(1-G) $
$FIT XX + GROUP $D E $
```

The residual sum of squares is 239766, well inside the 15% critical value of 283499, so this model is adequate. The parameter estimates and standard errors are

$$\begin{array}{ccc} 226.1 & +\ 3.249\ \text{XX} + & 135.9\,\text{GROU(2)} \\ (64.6) & (1.021) & (71.6) \end{array}$$

The slope of the regression on EFT for the row group is the same as in the interaction model, though its standard error is slightly different since the model is different.

If we omit XX from this model we obtain Model 4 which we know is not adequate. The only other possibility is to omit GROUP. Omission of this variable is equivalent to fitting the model

$$\mu = \beta_0 + \beta_1 x_1$$

for the row group, and

$$\mu = \beta_0$$

for the corner group. This implies that the group regression lines intersect at $x_1 = 0$, so that the mean completion time for a corner group child is the same as that for a row group child with an EFT score of zero. Since there are no such children, this model has a strong property which is not verifiable from the observable data. In the absence of compelling experimental reasons for this strong *mathematical* property, we do not proceed further with this model, but return to model 6.

The interpretation of model 6 is quite different from that of model 3. Mean completion time is affected by field dependence under row instructions, but not under corner instructions. The mean time taken under row instructions is less than that under corner instructions for low field dependence, but greater for high field dependence.

These conclusions are quite complicated. The interpretation of model 3 was much simpler: no difference in mean time between instructions, and a common effect of field dependence for both instruction groups. But both models are consistent with the data, and we have no *statistical* criterion for choosing between them other than parsimony: model 3 has only one variable, model 6 has two.

A problem occurring frequently with models with non-parallel regressions is that of determining the values of the explanatory variable for which a significant difference between means can be asserted. In model 6 the difference between the mean completion times between the corner and row groups, for a given EFT score x, is $\beta_2 - \beta_1 x$. For what values of x can we assert that this is non-zero? The fitted regression lines cross when $x = \hat{\beta}_2 / \hat{\beta}_1 = 41.8$, so around this value the population means are indistinguishable. For any x the variable $\hat{\beta}_2 - \hat{\beta}_1 x$ is normally distributed with mean $\beta_2 - \beta_1 x$ and variance $(v_{22} - 2v_{12}x + v_{11}x^2)$, where $V = (v_{jk})$ is the covariance matrix of the parameter estimates, and is estimated by $\hat{V} = s^2(X'X)^{-1}$. The hypothesis $\beta_2 - \beta_1 x = 0$ can be rejected by a simultaneous test of level α valid for all x if

$$t^2(x)=\frac{(\hat{\beta}_2-\hat{\beta}_1x)^2}{\hat{v}_{22}-2\hat{v}_{12}x+\hat{v}_{11}x^2}>2\mathrm{F}_{\alpha,2,n-p-1}.$$

This procedure was first proposed by Johnson and Neyman (1936) and is often called the Johnson–Neyman technique; the simultaneous test formulation was given by Potthoff (1964). The above inequality is equivalent to

$$Q(x)=(\hat{\beta}_1^2-c\hat{v}_{11})x^2-2(\hat{\beta}_1\hat{\beta}_2-c\hat{v}_{12})x+(\hat{\beta}_2^2-c\hat{v}_{22})>0$$

where $c=2\mathrm{F}_{\alpha,2,n-p-1}$. Thus if $Q(x)=0$ has real roots $x_L<x_U$, the rejection region is $x>x_U$, $x<x_L$ if $\hat{\beta}_1^2-c\hat{v}_{11}$ is positive, or $x_L<x<x_U$ if $\hat{\beta}_1^2-c\hat{V}_{11}$ is negative. If there are no real roots, the rejection region in x is empty.

In the above example, we will take $\alpha=0.05$ for illustration, so that $c=2F_{0.05,2,21}=6.94$. The parameter estimates are given earlier in this section: we reproduce them here with the variances and covariances of the parameter estimates, obtained using $D V $:

$$\hat{\beta}_1=3.249 \qquad \hat{v}_{11}=1.041$$
$$\hat{\beta}_2=135.9 \qquad \hat{v}_{22}=5120 \qquad \hat{v}_{12}=57.89=\hat{v}_{21}.$$

The quadratic is

$$Q(x)=3.3315x^2-2(39.7825)x-17064$$

and the roots of $Q(x)=0$ are -60.6 and 84.5. The first is far outside the range of the data, but the second is well inside the data. For values of EFT 85 or greater, the mean completion time under row instructions is significantly greater than that under corner instructions. For values of EFT less than 85, the mean completion times do not differ significantly.

Given the quite different conclusions from models 3 and 6, what can we confidently say? The answer is that we are tantalizingly short of data. The conclusions from both models are strongly influenced by two of the observations, as we will see in Section 2.10. With small experimental or observational studies it will frequently happen that several different models are equally well supported by the data. In such cases all the competing models and their implications should be presented, and firm judgements suspended. This example is considered further in the next section.

2.9 Stratified, weighted and clustered samples

In most sample surveys the sample design is not simple random sampling, but some form of stratified design with unequal sampling fractions. Multi-stage *cluster* sampling is often used, and this form of sampling induces correlations between the observations in a cluster, invalidating the assumption of independence used in constructing the likelihood function in Section 2.5. Standard regression methods, as applied above to the data of Section 2.3, are invalid when applied to the observations from clustered samples. Models can be adapted for the cluster design, leading to *variance component* or *mixed* models; these are beyond the scope of this book. For a discussion of maximum likelihood estimation in such models in an educational research context, see Aitkin, Anderson and Hinde (1981), Goldstein (1986), Aitkin and Longford (1986) and Longford (1987), and for examples with binary data, see Holt, Scott and Ewings (1980) and Anderson and Aitkin (1985).

Stratified and weighted sample designs which are not clustered can be analysed by the standard methods available in GLIM, with slight modifications.

For the psychological example of Section 2.3, we stated that the sample was randomly drawn from the population of fifth-grade children in one primary school. In fact, the sample design was stratified: the population was divided into sexes, and twelve children of each sex were randomly drawn from the sex sub-populations. The twelve children of each sex were then assigned to each experimental group by restricted randomization so that each group had six children of each sex. Thus sex and experimental group are orthogonal in the analysis.

The first six children in each experimental group in Section 2.3 are girls, the last six boys. Sex is a *stratifying factor* in the design, and it is modelled as an explanatory variable like experimental group. The sex identification is not given in the data listing in Appendix 4, but we can generate it as follows:

```
$FACTOR  SEX 2 $
$CALC  SEX = %GL(2,6) $
```

Why is it necessary to model the stratifying factor? There are two reasons. First, there may be substantial parameter differences between different strata: this is often the reason for stratification in the first place. Omission of the stratum variable from the model may then substantially bias the estimates of the model parameters.

Second, the estimates of parameters based on aggregating over the strata assume that the population has the same structure over strata as the sample. Unequal sampling fractions (that is, unequal proportions of the population from different strata included in the sample) combined with large stratum differences may give quite misleading population estimates if the stratum variable is ignored. On the other hand if there are no stratum differences in parameters then unequal sampling fractions make no difference and the model can be collapsed over strata. A general discussion of weighting (and of coding of dummy variables) is given in Aitkin (1978) in connection with hierarchical and non-hierarchical approaches to model simplification.

Returning to the example, the number of possible models now increases considerably since it is possible that regressions of time on EFT are different for each sex. A sequence of models of increasing complexiy could be fitted as before. For reasons which will be made clear in the next section, we will consider only the most complex model, in which the regression of time on EFT is different for each sex/instruction group:

```
$FIT GROUP*SEX*EFT $D E $
```

The six observations for each group and the fitted regressions are shown in Fig. 2.3 (these could be obtained from GLIM by using %RE for each set of six observations; for clarity we have given a high-quality plot instead). It is immediately striking that two observations (one in the girl/corner group and one in the boy/row group) are remote from the other observations, and without these two observations there would be little evidence of any regression on EFT at all. We now turn to an examination of such features of a model.

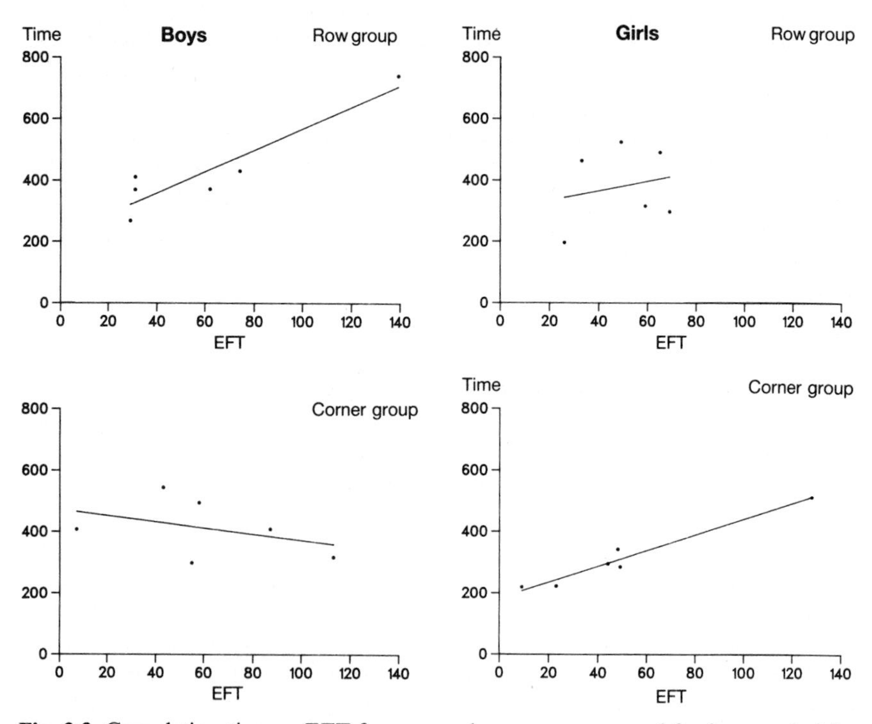

Fig. 2.3. Completion time vs EFT for row and corner groups and for boys and girls Model SEX*GROUP*EFT

2.10 Model criticism

The conclusions drawn from a statistical model depend on the validity of the model. We pointed out in Section 2.3 that models could not be *exact* representations of the population. We require only that they reproduce the main features of the population without major distortion.

A careful examination of the correspondence between data and model should be part of any statistical modelling of data. Examination of the data for failure of the model has been called *model criticism* by Box (1980, 1983), a term which we shall adopt, though the systematic examination of the model through *residuals* has a long history. Draper and Smith (1981) give a discussion of residuals in normal models, and the field is still developing rapidly (see for example Atkinson, 1982; and the recent books by Atkinson, 1985; Belsley, Kuh and Welsch, 1980; Barnett and Lewis, 1978 and Cook and Weisberg, 1982).

There are four important areas in which failures of the model may occur:

(1) mis-specification of the probability distribution for y, leading to an inappropriate likelihood function and inappropriate maximum likelihood estimates for the parameters;

(2) mis-specification of the link function;

(3) the occurrence of aberrant observations, distorting either the probability distribution or the parameter estimates from the model;

(4) mis-specification of the systematic part of the model, leading to incorrect interpretations.

Before considering these failures, we quote some standard results for the normal model.

The (raw) residual e_i for the i-th observation from the model is

$$\begin{aligned} e_i &= y_i - \hat{y}_i \\ &= y_i - \mathbf{x}_i'\hat{\boldsymbol{\beta}}. \end{aligned}$$

The GLIM directive $D R $ following the fitting of a normal model displays these residuals.

In matrix terms

$$\begin{aligned} \mathbf{e} &= \mathbf{y} - X\hat{\boldsymbol{\beta}} \\ &= (I - X(X'X)^{-1}X')\mathbf{y} \\ &= (I - H)\mathbf{y} \end{aligned}$$

where H is the projection or 'hat' matrix

$$H = X(X'X)^{-1}X'.$$

The distribution of $\mathbf{e}$ as a random variable in repeated sampling is

$$\mathbf{e} \sim N_n(\mathbf{0}, \sigma^2(I-H))$$

which is a singular n-dimensional multivariate normal distribution of rank $n-p-1$, since there are only $n-p-1$ linearly independent residuals e_i. Denote the diagonal elements of H by h_i; then

$$h_i = \mathbf{x}_i'(X'X)^{-1}\mathbf{x}_i,$$

$$\text{var } e_i = \sigma^2(1-h_i), \qquad \text{cov}(e_i, e_j) = -\sigma^2 h_{ij} \qquad i \neq j,$$

$$\text{var } \hat{y}_i = \sigma^2 h_i.$$

The values of h_i are between zero and unity with $\Sigma\, h_i = p+1$. A value of zero for h_i means the fitted value of y_i must be identically zero, while a value of unity for h_i means that the residual is identically zero, and the model must exactly reproduce the i-th observation y_i. Thus large values of h_i are an indication that the corresponding observations may be influential in determining the position of the fitted model.

The residuals have different variances, but can be converted to *standardized residuals* f_i:

$$f_i = e_i/s(1-h_i)^{\frac{1}{2}}.$$

Most of the assessment of model failure is based on the standardized residuals. Other forms of residual are sometimes useful: the *jack-knife* or cross-validatory residuals are

$$j_i = f_i/((n-p-1-f_i^2)/(n-p-2))^{\frac{1}{2}}.$$

The individual j_i have t_{n-p-2} distributions, though they are not independent; the distribution of the f_i, though it is not normal or t, does not depend on σ. Other statistics, functions of the residuals and/or the *influence* values h_i (these are also called *leverage* values), are also used by several authors.

We now consider the four areas of model failure introduced above.

2.10.1 *Mis-specification of the probability distribution*

Since the probability distribution $f(y)$ is fundamental to the likelihood function and therefore to the parameter estimates in the model, an incorrect specification of the distribution can have serious consequences for interpretation of the data. Examination of the data for failure of this assumption is through a *probability plot* (actually a *quantile* or *Q–Q* plot) of the standardized (or sometimes even the raw) residuals. The residuals f_i are ordered, and the ordered values $f_{(i)}$ plotted against the normal quantiles $z_i(a)=\Phi^{-1}\{(i-a)/(n+1-2a)\}$, where a is a suitably chosen constant $(0\leq a<1)$, and $\Phi(x)$ is the standard normal cumulative distribution function. The usual choice of a is either zero or $\frac{1}{2}$: we will use the value $a=0.3175$—for discussions of such choices, see Barnett (1975), Filliben (1975) and Draper and Smith (1981, p. 178)—since this allows the use of a test by Filliben (1975) for normality. If the probability distribution is correctly specified, the plot should be roughly a straight line. Systematic curvature, or individual observations far from the straight line, indicate failures of the probability distribution specification.

An acute difficulty in the inspection of such plots is how to decide whether the variation in the plot is too far from a straight line: it is easy to over-interpret sampling variation as real model failure. Simulation of samples from normal models gives some feeling for the behaviour of such plots in small samples; at the cost of considerable computer time, *simulation envelopes* (Atkinson, 1981, 1982, 1985) can be generated to provide a statistical test. An alternative test procedure is based on the probability plot itself (Filliben, 1975); such tests have been in common use since Shapiro and Wilk (1965) first proposed them. These test procedures use essentially the correlation between the ordered sample values and the normal quantiles or similar values (functions of the normal order statistics). This correlation should be very high, corresponding to a nearly linear plot, if the probability distribution is correctly specified, and will be reduced by systematic curvature or individual outlying observations.

A formal significance test was given by Filliben for the case of a single sample from a normal distribution. This test uses the medians of the normal order statistics, which are approximated by $z_i\,(0.3175)$ for $i=2,\ldots,n-1$; the exact values are used for $i=1$ and n. The use of the test with correlated standardized residuals affects the significance level of the test to some (unknown) extent, but the test is still useful despite this.

The residuals defined above and the influence values are not provided as part of the standard GLIM output, but they are easily calculated. The macro QPLOT gives a quantile plot and calculates the correlation between ordered sample values and normal quantiles; this correlation is used in the Filliben test.

The standard use of QPLOT assumes a previous $FIT directive, so that a y variable and fitted values must already exist.

Macro QPLOT constructs a normal deviate using

```
%ND((%GL(%NU,1)-0.3175)/(%NU+0.365))
```

except for the first and last elements, and calculates and prints the value of the Filliben test correlation coefficient. The form of residual to be plotted is specified as the argument of the macro QPLOT. The arguments RAW, STAN or JACK give the raw, standardized or jack-knife residuals respectively, RAW being the default option.

We illustrate with the example of Section 2.3. We will first ignore sex and fit the EFT*GROUP model. Plotting is usually done with the full or most complex model, since simplifications of this model may be affected by failure of the normal distribution assumption.

```
$INPUT 1 SOLV $
$YVAR TIME
$FIT EFT*GROUP $D E $
$INPUT %PLC QPLOT $
```

The subfile QPLOT contains the macro QPLOT and other macros for calculating the different forms of residuals. (For further details, see the documentation of QPLOT in the GLIM macro library.)

```
$USE QPLOT $
```

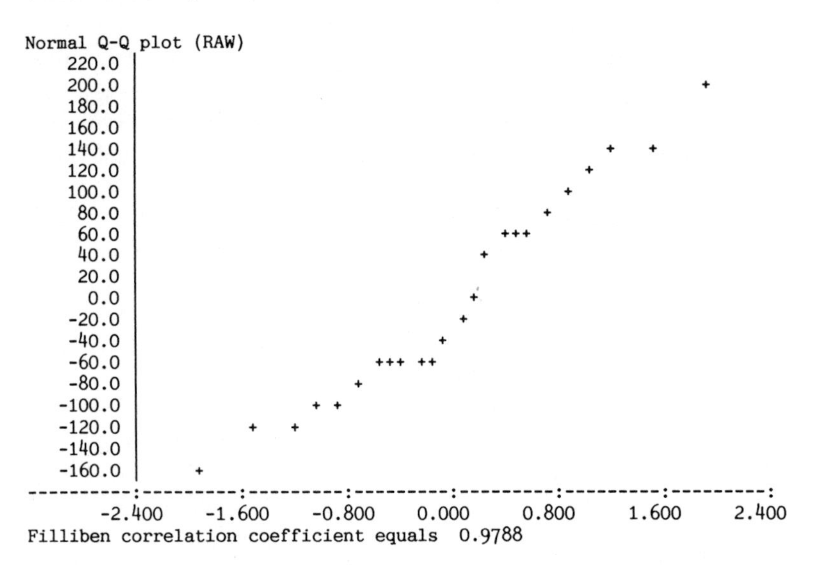

The plot has several bends, but the r-value of 0.9788 is at about the 25% point of the distribution, giving no evidence against a normal distribution. In small samples this test has low power against symmetric non-normal distributions, but high power against a skewed distribution like the lognormal.

What should we do if the Filliben test gives a significant result? If the normal distribution is not an adequate representation of the random variation, there are two possibilities: try a transformation of the variable to produce a closer agreement—usually the log transformation, or less commonly the reciprocal transformation, or use a different type of continuous probability distribution. These possibilities are examined in detail in Chapters 3 and 6.

The raw, standardized and jack-knife residuals can be explicitly calculated into named variables and used subsequently, provided that %VL, the variance of the linear predictor, is first extracted. Here the linear predictors are the same as the fitted values so the h_i are given by $\%VL/s^2$. ($EXTRACT is a directive which allows access to data structures not immediately available to the user after $FIT). For example, to obtain the standardized residuals, say SRES, we use:

```
$EXTRACT %VL $
$CALC SRES = #STAN $
```

Note that the macro STAN in the QPLOT subfile contains the GLIM expression for the standardized residuals.

A second and less common form of failure of the probability model assumption is that the nuisance parameters, assumed constant for all observations, are in fact varying systematically with the explanatory variables. A simple example of this would be if the variances were different for the two instruction groups in the example of Section 2.3. This could be assessed by plotting the residuals separately for each group against the fitted values. In general, it is necessary to plot the residuals against the fitted values and each explanatory variable: if different variability is evident then the model can be modified appropriately. Heterogeneity of variance is examined in a factorial design in Section 3.11.

2.10.2 *Mis-specification of the link function*

The choice of parameter scale on which the systematic effects are modelled is an important part of statistical modelling. When the response variable is normal it is almost universal practice to work on the scale of the normal mean μ, using the identity link: this has therefore been taken as the default option in GLIM for the link function with a normal error specification.

In non-normal distributions the choice of scale is much more open, as we shall see in later chapters. However, even for normal distribution models, the possibility of working on the log scale particularly should be borne in mind. This possibility is investigated in the examples of Chapter 3, where both the lognormal distribution for the response variable and a log link with the normal distribution are examined.

2.10.3 *The occurrence of aberrant and influential observations*

Good data analysis depends heavily on the correct recording of variable values. Misrecording of data values can result in substantial changes to the fitted model. If an observation happens to lie on the boundary of the space of the explanatory variables, misrecording of its response variable may have a powerful effect on the position of the fitted regression, but may not produce a large residual. The same effect can occur by misrecording an explanatory variable thus moving the observation out of the cluster of correctly recorded observations. The influence values h_i are particularly useful for diagnosing observations of this type.

Even when no misrecording occurs it is valuable to know whether a small number of observations is substantially influencing the position of the fitted regression. These observations can then be carefully examined and referred back to the original survey or to experimental scientists for comment. Observations with large residuals or influence values are not automatically rejected from the model: they always require evaluation.

We illustrate the use of influence values with the example from Section 2.3. For the GROUP*EFT model, we can plot the influence values against observation number using the library macro LEV:

```
$FIT GROUP*EFT $
$INPUT %PLC LEV $
$USE LEV $
```

```
Leverage plot
   0.7600 |
   0.7200 |                   +
   0.6800 |
   0.6400 |
   0.6000 |
   0.5600 |
   0.5200 |
   0.4800 |
   0.4400 |                         +
   0.4000 |
   0.3600 |
   0.3200 |  - - - - - - - - - - - - - - - - - - - - - - -
   0.2800 |                                      +
   0.2400 |                        +               +
   0.2000 |
   0.1600 |         + +           +         +
   0.1200 |   +          +  + +
   0.0800 |  +  + + +   +      +      + +  + + +
   0.0400 |
   0.0000 |
----------:---------:---------:---------:---------:---------:---------:
         0.00      5.00     10.00     15.00     20.00     25.00     30.00
```

The influence values can be printed out: they are stored in the vector LEV_. Note the underline character which is used for variables created in library macros.

```
$PRINT LEV_ $
```

The influence values are given below to two decimal places.

0.08	0.13	0.09	0.10	0.09	0.16	0.15	0.09	0.14	0.72	0.11	0.14
0.09	0.15	0.22	0.42	0.09	0.09	0.15	0.09	0.08	0.08	0.30	0.23

How should we interpret the influence values? The model specifies a linear regression in each experimental group, with different slopes for each group. For a linear regression model, it is easily shown (see e.g. Hoaglin and Welsch, 1978) that

$$\mu_i = \beta_0 + \beta_1 x_i$$

gives

$$h_i = \frac{1}{n} + (x_i - \bar{x})^2 \Big/ \sum_{j=1}^{n} (x_j - \bar{x})^2.$$

Thus h_i takes its minimum value of $1/n$ for observations at the explanatory variable mean, and its maximum value for the observation furthest from the mean: extreme values of x are the most influential. If all x_i are equidistant from the mean, $h_i = 2/n$; if all but one observation have identical values of x_i, these will have $h_i = 1/(n-1)$, and the remaining observation will have $h_i = 1$. In this case the position of the regression is determined completely by the single observation, and we could have no confidence at all in the regression.

For the general regression model with p explanatory variables, Hoaglin and Welsch proposed regarding $2(p+1)/n$ as a value of h indicating important influence. In general $(p+1)/n$ is the average value of all the h_i and also corresponds to equally influential observations. For the model GROUP*EFT with $p=3$ and $n=24$, $(p+1)/n$ is 0.167. Since the model GROUP*EFT corresponds to unrelated simple linear regressions in each instruction group of 12 children, observations near the EFT mean in each group will have influence values near $1/12 = 0.0833$.

One value stands out: 0.72 for the 10th observation in the row group. This observation has the largest EFT value of 139, and also the largest completion time of 739 seconds. It has a powerful effect on the fitted regression for any subgroup in which it appears. For the EFT+GROUP model, its influence value is reduced to 0.35, because the slope of the regression is determined by both row and column groups, and the effect of this observation is diluted by the doubled sample size.

The influence value of 0.42 for the fourth observation in the corner group (in the EFT*GROUP model) is also high. This observation has the largest EFT value of 128 in the corner group, though its completion time value of 513 seconds is not exceptionally large. Its influence value is reduced to 0.28 in the main effects model.

It is notable that neither of these observations has a large residual: in both cases the fitted regression passes close to the observed value, partly because of the influence of the observation on the position of the fitted line.

In the full SEX*GROUP*EFT model, these two observations have influence values of more than 0.8.

```
$FIT SEX*GROUP*EFT $USE LEV $
```

The reasons are clear from Fig. 2.3: the more we subdivide or stratify the observations by explanatory variables, the greater the effect these observations have on the regressions within subgroup. The regressions in the corner group for girls and the row group for boys are being very largely determined by the single largest observation in each case. We did not pursue the simplification of the SEX*GROUP*EFT model in Section 2.9 because the conclusions from this model depend so heavily on these two observations. This is a frequent problem when samples are extensively cross-classified by many explanatory factors, so that many cells have very small numbers of observations.

To examine the effect of omitting these observations, we use the weight variate LWT_ set up in the LEV macro: LWT_ is defined to be zero if the influence value $h > 2(p+1)/n$.

```
$WEIGHT LWT_ $FIT . $D E $
```

The . argument to FIT means "the previous model". The three-way interaction parameter estimate is now smaller than its standard error, and successive elimination of terms shows that all the EFT terms can be omitted from the model. Notice that the two-way interaction GROUP.SEX is quite large (as is its standard error). This interaction can be identified as due to one discrepant cell: girls in the corner group have a mean completion time of 272.6 seconds, while boys in both groups and girls in the row group have similar means, with a common mean completion time of 388.7 seconds.

These conclusions are markedly different from those obtained by retaining the two influential observations: in the latter case the SEX*GROUP*EFT model can be reduced to the EFT model 3 of Section 2.7, with no group difference and a common regression on EFT.

What action, if any, should we take? The original data should be checked with the experimenter: are the EFT and time values correct for these two observations? If they are not, the correct values can be entered. If they are, then the experiment can be interpreted as before. The difficulty, as previously mentioned, is that because the sample sizes are small several different conflicting interpretations of the experiment are possible.

It may not be possible to check the correctness of the original data. In this case an analysis of the data might be carried out as above by including and excluding the two highly influential observations, and comparing the results. It is, however, not good statistical practice to set aside routinely or exclude observations with large residuals and/or influence simply because these values are large.

2.10.4 *Mis-specification of the systematic part of the model*

This form of model failure is different from the others because it is rectified by further modelling, though diagnostic plotting of the standardized residuals against other variables may be useful.

When the explanatory variables are continuous, one form of model failure is the reliance on linear regressions when the relationships are non-linear. Inclusion of quadratic or higher-order polynomial terms in the regression is a common solution; however, this may lead to unreasonable features in the model, for example the fitted values may reach a maximum and then decrease while the observed values approach an asymptote. Transformations of the variables or of the scale of μ (a different link function) may be preferable.

Interactions between variables may also be necessary. In fitting such terms in complex models a careful eye should be kept on the influence values, because in small samples the possibility of single observations producing apparent interactions is considerable, as in the above example.

The need for non-linear terms or interactions is usually assessed graphically by plotting the raw or standardized residuals against the explanatory variables and the fitted values. Non-random scatter patterns suggest the need for model changes. We do not discuss this further here as the examples presented in later chapters are analysed in considerable detail.

2.11 The modelling of binary response data

In this section we apply the general theory of this chapter to a simple contingency table model with a binary (two-category) response variable and binary explanatory variables, arising from an observational study. Binary data are discussed in detail in Chapter 4, but we introduce the basic ideas here.

The data (Bishop, Fienberg and Holland, 1975 p. 41) come from a study of two pre-natal clinics in a large US city. Mothers attended one of the pre-natal clinics for varying periods before the birth of the baby. The length of time the mother attended the clinic has been categorized for analysis into less than one month or more than one month.

The binary response variable of interest is the infant mortality outcome: infant died within one month, or survived at least one month. There are two explanatory variables: amount of pre-natal care, and clinic (A or B); the data are given below.

Three-way contingency table

Mother's attendance	Survival experience of child: Died within one month	Survived at least one month	Total number	Mortality rate (%)
Clinic A				
Less than one month	3	176	179	1.68
More than one month	4	293	297	1.35
Clinic B				
Less than one month	17	197	214	7.94
More than one month	2	23	25	8.00

Reproduced from: Y. M. M. Bishop, 'Full contingency tables, logits, and split contingency tables.' *Biometrics* **25,** 383–99 (1969). With permission from the Biometric Society.

An additional column has been added giving the mortality rates for each group. Clinic B has a considerably higher mortality rate than Clinic A, but the duration of mother's attendance seems to be unrelated to mortality.

If the data for the two clinics are combined, a quite different picture appears.

Both clinics				
Less than one month	20	373	393	5.09
More than one month	6	316	322	1.86

Now mothers attending for less than one month seem to experience a higher mortality rate for their infants.

This example illustrates clearly two of the hazards of inference from observational data: the danger of aggregating data over important variables, and the inappropriate assumption of causality—that duration of pre-natal care causes or is responsible for the change in infant mortality. Such a conclusion is insupportable in a non-randomized study: it could be supported only if mothers were randomly assigned to the two classes of pre-natal care. In fact there is a systematic biasing factor visible from the complete data: of the mothers attending Clinic B, only ten per cent attended for more than one month, while this proportion was sixty-two per cent for mothers attending Clinic A. Since Clinic B has a much higher mortality rate, when the clinic identification is suppressed the variation in mortality rate appears to be associated with duration of pre-natal care.

It is equally invalid, however, to conclude from the full data that the difference in mortality rates is caused by different quality of prenatal advice and treatment in the two clinics. There is no randomization of mothers to clinics: each clinic serves the mothers in a local area of the city. The different mortality rates probably reflect different infant mortality rates in the sub-populations of the city served by each clinic. Without randomized assignment in such studies it is impossible to draw strong causal conclusions.

The analysis of the data through statistical models can, however, avoid the invalid conclusions produced by aggregating the data for both clinics. We now consider the statistical modelling and maximum likelihood analysis of binary data.

2.12 The Bernoulli distribution for binary data

The response in this example is a binary variable, while the explanatory variables are also binary, defined in GLIM by a $FACTOR directive as for group in Section 2.3. We will fit regression models of the form

$$\eta = \beta_0 + \beta_1 x_1 + \beta_2 x_2$$

as before, where x_1 and x_2 are the clinic and mother's attendance dummy variables. Here η is some function of the parameter of the distribution of the binary response variable. Define

$$\begin{aligned} y &= 1 \quad \text{if the child dies during the first month—``failure''} \\ &= 0 \quad \text{if the child survives at least one month—``success''} \end{aligned}$$

Survival beyond the first month for any child is an uncertain event: 3.6% of all the infants died during the first month. Let p be the probability that a randomly chosen child does not survive beyond the first month. The distribution of y is called the *Bernoulli distribution*, with

$$\begin{aligned} \Pr(y = 1) &= p \\ \Pr(y = 0) &= 1 - p \end{aligned}$$

which can be written

$$\Pr(y) = p^y (1-p)^{1-y} \qquad y = 0,1.$$

The distribution has mean p and variance $p(1-p)$.
The Bernoulli distribution is a special case of the binomial distribution

$$b(r|n,p) = \binom{n}{r} \; p^r (1-p)^{n-r},$$

with $n = 1$ and $r = y$.

How should the linear predictor η be related to p? In Chapter 4 we present and discuss several link functions useful for probabilities; here we will use, without further comment, the *logit* link or transformation, defined by

$$\eta = \text{logit}\, p = \log \{p/(1-p)\},$$

so that

$$p = e^{\eta}/(1+e^{\eta}).$$

This has the property that the range of p, $[0,1]$, is mapped onto $(-\infty,\infty)$, the range of logit p.

2.13 Maximum likelihood for the Bernoulli and binomial distributions

The data in Section 2.11 represented by four mortality rates actually come from 715 children, though the individual binary observations are no longer available. Does this matter? To see that it does not, consider just the 179 children in the first row, whose mothers attended Clinic A and received less than one month pre-natal care. The model specifies a constant probability, say p_1, for these children. The likelihood function for this set of observations is

$$L(p_1) = \prod_{i=1}^{179} p_1^{y_i}(1-p_1)^{1-y_i}$$

$$= p_1^{\Sigma y_i}(1-p_1)^{179-\Sigma y_i}.$$

Apart from the irrelevant binomial coefficient, this is just the likelihood function for $r_1 = \Sigma_1^{179} y_i$ "failures" in $n_1 = 179$ binomial trials with failure probability p_1. In the same way the observations in the other cells also lead to binomial likelihood functions. The likelihood function for the complete set of observations can then be written

$$L(\boldsymbol{\beta}) = \prod_{j=1}^{4} p_j^{r_j}(1-p_j)^{n_j-r_j}$$

where r_j is the number of deaths in the j-th row of the table, n_j is the number of children in this row, and p_j is the death probability.

The complete specification of the statistical model for the mortality data in GLIM is then:

(1) the data consist of the four observations (r_j, n_j) for each clinic/attendance combination:

j	r_j	n_j	Clinic	Attendance
1	3	179	A	<1 month
2	4	297	A	≥ 1 month
3	17	214	B	<1 month
4	2	25	B	≥ 1 month

(2) the response variable is r, the number of deaths, which has a binomial distribution $b(r|n,p)$ ("binomial error" in GLIM);

(3) the linear predictor $\eta = \boldsymbol{\beta}'\mathbf{x}$ is one of the models containing clinic and mother's attendance factors;

(4) the link function is the logit:

$$\eta = \log \{p/(1-p)\}.$$

Before describing the fitting of the model in GLIM, we consider the general properties of maximum likelihood estimation, since the deviance and parameter standard errors have different interpretations in non-normal models.

2.14 Maximum likelihood fitting of the regression model

For the general model $\eta = \boldsymbol{\beta}'\mathbf{x}$, the likelihood function of Section 2.13 is

$$L(\boldsymbol{\beta}) = \prod_j e^{\eta_j r_j}/(1 + e^{\eta_j})^{n_j},$$

$$\ell(\boldsymbol{\beta}) = \sum_j \{\eta_j r_j - n_j \log (1 + e^{\eta_j})\}$$

where

$$\eta_j = \boldsymbol{\beta}'\mathbf{x}_j = \log \{p_j/(1-p_j)\}.$$

The derivatives of the log likelihood are

$$\frac{\partial \ell}{\partial \boldsymbol{\beta}} = \sum_j (r_j - n_j e^{\eta_j}/(1 + e^{\eta_j}))\mathbf{x}_j$$

$$= \sum_j (r_j - n_j p_j)\mathbf{x}_j$$

$$\frac{\partial^2 \ell}{\partial \boldsymbol{\beta} \partial \boldsymbol{\beta}'} = -\sum_j n_j (e^{\eta_j}/(1 + e^{\eta_j})^2)\mathbf{x}_j \mathbf{x}_j'$$

$$= -\sum_j n_j p_j (1-p_j)\mathbf{x}_j \mathbf{x}_j'.$$

The system of equations in $\boldsymbol{\beta}$ given by $\partial\ell/\partial\boldsymbol{\beta} = 0$ is non-linear in $\boldsymbol{\beta}$, and can be solved using an iterative weighted least squares algorithm, the computational core of GLIM. Details are given in Appendix 1. The estimate $\hat{\boldsymbol{\beta}}$ then gives linear predictors $\hat{\eta}_j$ and probabilities $\hat{p}_j$. The observed and estimated expected information matrices are

$$\begin{aligned} \mathrm{I}(\hat{\boldsymbol{\beta}}) = \mathscr{I}(\hat{\boldsymbol{\beta}}) &= \sum_j n_j\hat{p}_j(1-\hat{p}_j)\mathbf{x}_j\mathbf{x}_j' \\ &= X'WX \end{aligned}$$

where W is a diagonal matrix of weights

$$w_j = n_j\hat{p}_j(1-\hat{p}_j).$$

and X' is $[\mathbf{x}_1 \ldots \mathbf{x}_n]$.

The maximized log likelihood is

$$\ell(\hat{\boldsymbol{\beta}}) = \sum_j \{\hat{\eta}_j r_j - n_j \log(1 + e^{\hat{\eta}_j})\}.$$

Model comparison proceeds exactly as in Section 2.7 through the likelihood ratio test. For a reduced model $\eta = \boldsymbol{\beta}_r'\mathbf{u}$, the likelihood ratio test statistic is

$$\lambda = -2\{\ell(\hat{\boldsymbol{\beta}}_r) - \ell(\hat{\boldsymbol{\beta}})\}$$

which has an asymptotic χ^2 distribution if the omitted terms from the model actually have zero regression coefficients.

The convention in GLIM for printing the maximized log likelihood function is different for the normal distribution and for the Poisson and binomial distributions. For the former, the deviance is simply the residual sum of squares for the model being fitted. For the latter, the deviance printed out by GLIM is the value of the *likelihood ratio test statistic* for the model fitted *compared to the saturated model with a parameter for every observation*. That is,

$$\text{Deviance}_{\text{model}} = -2\{\ell(\text{model}) - \ell(\text{saturated model})\}.$$

This difference between the GLIM meanings of "deviance" often causes confusion (though it is logical in another way: the deviance is in both cases proportional to a χ^2 variable if the model fitted is an adequate representation of the data). In comparing two models, the difference between their deviances is (asymptotically) χ^2, so model comparison is facilitated. In this book we will often use the term "deviance" for a model to refer to $-2\ l(\hat{\boldsymbol{\beta}})$ *without* the comparison with the saturated model. This usually causes no confusion in model simplification since differences between deviances are unaffected by the subtraction of a common constant from each deviance. With this convention, the deviance $-2l(\hat{\boldsymbol{\beta}})$ is not a measure of goodness-of-fit of the model; only differences of such deviances are appropriate as likelihood ratio test statitstics. To fit the model in GLIM, we first enter the data through the keyboard (they are not stored on file):

```
$UNIT 4
$DATA DEATH TOTAL $
$READ
3 179 4 297 17 214 2 25
$FACTOR CLINIC 2 ATTEND 2 $
$CALC CLINIC = %GL(2,2) : ATTEND = %GL(2,1) $
```

Now we declare DEATH to be the response variable, the probability distribution to be binomial with TOTAL the denominator of the death rate, (that is, the sample sizes on which the sample death rates are based), the link function to be the logit (G), and a sequence of regression models, the null model first:

```
$YVAR DEATH $ERROR B TOTAL $LINK G
$FIT : +ATTEND : CLINIC : +ATTEND : +CLINIC.ATTEND $
```

The deviances are 17.83, 12.22, 0.08, 0.04, 0.00. The last deviance is zero because the saturated interaction model with four parameters exactly reproduces the data. Successive differencing of the deviances as in Section 2.7 leads to the following tables.

Model	Residual deviance	Source	deviance	df
1	17.83			
ATTEND	12.22	ATTEND	5.61	1
+CLINIC	0.04	CLINIC	12.18	1
+ATTEND.CLINIC	0.00	Interaction	0.04	1

Model	Residual deviance	Source	deviance	df
1	17.83			
CLINIC	0.08	CLINIC	17.75	1
+ATTEND	0.04	ATTEND	0.04	1
+ATTEND.CLINIC	0.00	Interaction	0.04	1

There is a striking interchange of deviance between Clinic and Attendance when their fitting order is reversed, reflecting the substantial correlation between the variables noted in Section 2.11. It is obvious from the table that the Clinic term is essential in the model, and that once it is included, the Attendance term is unnecessary. Formal verification of this follows from the simultaneous test procedure of Section 2.7, here applied to the approximate χ^2 distribution for the likelihood ratio test.

A size α simultaneous test of the adequacy of any reduced model rejects the hypothesis of adequacy if

$$\lambda = -2\{\ell(\text{reduced}) - \ell(\text{full})\} > \chi^2_{\alpha,p}$$

The test statistic on the left of this inequality is simply the deviance for the reduced model. The size α of the test is set at about $1-(0.95)^p$, which is 0.143 in this case since $p = 3$. Choosing $\alpha = 0.15$, the critical value $\chi^2_{0.15,3}$ is 5.32. Then the CLINIC model is minimal adequate (0.15), and the ATTENDance model is not adequate. We refit the CLINIC model and calculate the fitted and observed probabilities from the model:

```
$FIT CLINIC $D E $
$CALC FP = %FV/TOTAL : P = %YV/TOTAL $
$PR FP : P $
```

The fitted death rates are 0.0147 for clinic A and 0.0795 for clinic B, which are just the death rates for each clinic pooled over the attendance classification.

Thus statistical modelling of the three-way table has shown that the table can be collapsed over Attendance to give a simple representation, but it can *not* be collapsed over Clinic without serious distortion. While the simultaneous test assumes random sampling, the reduction of the model also applies when the data are a complete population, if we accept that it is reasonable to smooth the population to the same extent as for a sample of the same size.

2.15 Likelihood-based confidence intervals

The interpretation of standard errors for parameter estimates in non-normal models requires some care. The usual practice of constructing confidence intervals for parameters θ_j based on $\hat{\theta}_j \pm \lambda v_{jj}{}^{\frac{1}{2}}$, described in Section 2.6, can give seriously misleading answers, as seen in the following example.

Suppose we observe r successes in n binomial trials with success probability p. The likelihood function is (omitting the binomial coefficient)

$$L(p) = p^r(1-p)^{n-r}.$$

Then

$$\ell(p) = r\log p + (n-r)\log(1-p)$$

$$\frac{d\ell}{dp} = \ell'(p) = r/p - (n-r)/(1-p)$$

$$\frac{d^2\ell}{dp^2} = \ell''(p) = -r/p^2 - (n-r)/(1-p)^2.$$

The likelihood function is maximized at $\hat{p} = r/n$, and

$$I(\hat{p}) = \mathscr{I}(\hat{p}) = n/[\hat{p}(1-\hat{p})]$$

so that

$$V(\hat{p}) = \hat{p}(1-\hat{p})/n,$$

the usual estimated variance for the sample proportion. An approximate 95% confidence interval for p is then $\hat{p} \pm 1.96[\hat{p}(1-\hat{p})/n]^{\frac{1}{2}}$.

Now consider two cases when $n=10$. Suppose first that $r=1$. Then $\hat{p}=0.1$ and the 95% interval is 0.1 ± 0.186, i.e. $(-0.086, 0.286)$. Since p must be *positive* (as we have observed one success), this interval is unsatisfactory, and truncating the interval to $(0, 0.286)$ does not solve the problem.

Suppose now that $r=0$. Then $\hat{p}=0$ and $V(\hat{p})=0$ also, and no sensible interval results.

The first problem can be ameliorated by working on the logit scale. By transforming to an unbounded parameter scale, we guarantee that any interval on this scale will transform to a proper interval for p. On the logit scale, writing $\theta = \text{logit } p$,

$$\ell(\theta) = r\theta - n \log(1 + e^{\theta})$$

$$\ell'(\theta) = r - ne^{\theta}/(1 + e^{\theta})$$

$$\ell''(\theta) = -ne^{\theta}/(1 + e^{\theta})^2.$$

The likelihood is maximized at $e^{\hat{\theta}}/(1+e^{\hat{\theta}}) = r/n = \hat{p}$ or $\hat{\theta} = \text{logit } \hat{p}$, and

$$I(\hat{\theta}) = \mathscr{I}(\hat{\theta}) = ne^{\hat{\theta}}/(1+e^{\hat{\theta}})^2 = n\hat{p}(1-\hat{p})$$

so that

$$V(\hat{\theta}) = [n\hat{p}(1-\hat{p})]^{-1}.$$

An approximate 95% confidence interval for θ is then $\hat{\theta} \pm 1.96\,[n\hat{p}(1-\hat{p})]^{-\frac{1}{2}}$, and this can be transformed back to a confidence interval for p.

For the case $r=1$, $n=10$, we have $\hat{\theta} = \text{logit } 0.1 = -2.197$, and $V(\hat{\theta}) = 1.111$. The 95% interval for θ is then -2.197 ± 2.066, or $(-4.263, -0.131)$. Transforming to the p-scale gives $(0.014, 0.467)$, which is very different from the first interval for p. For the case $r=0$, this method still fails since $\hat{\theta}$ and $V(\hat{\theta})$ both tend to infinity as $\hat{p} \to 0$.

Confidence intervals for p may be obtained directly from the likelihood function by inverting the likelihood ratio test. These intervals do not possess any of the above difficulties, and will be used consistently throughout this book. Intervals based on the likelihood function are called *likelihood-based confidence intervals* (for a discussion see Kalbfleisch, 1979, pp. 22 and 205).

Consider the likelihood ratio test of the hypothesis $p=p_0$, a specified value. Applying the general procedure of Section 2.7, the maximized log likelihood from the full model where p is unspecified is

$$\ell(\hat{p})=r \log \hat{p}+(n-r) \log(1-\hat{p})$$

where $\hat{p}=r/n$. The log likelihood from the reduced model with $p=p_0$ is

$$\ell(p_0)=r \log p_0+(n-r) \log (1-p_0).$$

The likelihood ratio test statistic for the hypothesis $p=p_0$ is

$$\lambda=-2[\ell(p_0)-\ell(\hat{p})]$$

and the hypothesis is rejected at level α (in large samples) if

$$\lambda>\chi^2_{\alpha,1},$$

which is equivalent to

$$\ell(p_0)-\ell(\hat{p})<-\tfrac{1}{2}\chi^2_{\alpha,1}.$$

or

$$\ell(p_0)<\ell(\hat{p})-\tfrac{1}{2}\chi^2_{\alpha,1}.$$

Thus the hypothesis $p=p_0$ is *not* rejected if the log likelihood evaluated at p_0 is not more than $\frac{1}{2}\chi^2_{\alpha,1}$ units less than the maximum of the log likelihood at $\hat{p}$. The values of p_0 satisfying this requirement are a $100(1-\alpha)\%$ *likelihood-based confidence interval* for p.

The endpoints of the interval cannot easily be obtained analytically, but it is a simple matter to compute the log-likelihood function over a grid of values of p, and to obtain the interval by inspection. To simplify this procedure, the log *relative* likelihood function is computed, where the relative likelihood function is defined by

$$R(p)=L(p)/L(\hat{p})$$

so that

$$\log R(p)=\ell(p)-\ell(\hat{p}).$$

The relative likelihood gives the shape of the likelihood function and a good indication of the tail behaviour for values of p away from the maximum likelihood estimate. The log relative likelihood gives the values of the likelihood ratio statistic for the hypotheses $p=p_0$ for all values of p_0.

The macro BINREL in the subfile BINOMIAL computes and prints the relative and log relative likelihood functions (given in GLIM vectors R_ and LR_ respectively) over a grid of values of p and produces plots of each of these functions using the plotting symbol +. The macro also calculates the approximating quadratic relative likelihood and log likelihood functions, plotting these using the plotting symbol *. These approximating functions are defined later in this section. The macro requires the values of %R $=r$ and %N $=n$, the initial and final values %A and %Z of p, and the increment %I for the grid of values of p.

```
$INPUT 1 BINOMIAL $
$CALC %R=1 : %N=10 : %A=0.025 : %Z=0.975 : %I=0.025 $
$USE BINREL $
```

The grid interval omits $p=0$ and $p=1$, where the log relative likelihood would be infinite negative: since log of zero is set to zero, the plotted points for $p=0$ or 1 would be incorrect. From the tabulated values of the log relative likelihood it is clear that it reaches -1.92 $(-\frac{1}{2}\chi^2_{0.05,1})$ well below $p=0.025$ and near $p=0.375$: we reduce the grid size and recalculate the function around these values:

```
$CALC %A=0.003 : %Z=0.007 : %I = 0.0001  $USE BINREL $
$CALC %A=0.370 : %Z=0.380 : %I = 0.0002  $USE BINREL $
```

The values of p to four decimal places at which the log relative likelihood is -1.92 are 0.0060 and 0.3716: these are the endpoints of the 95% likelihood-based confidence interval. The computation of log relative likelihoods in this way can be extended to regression models using *profile likelihoods*, discussed in Section 3.1 and Chapter 4.

Why should the likelihood-based confidence interval be preferred to one based on the information matrix for the parameter? The main reason is that the likelihood function itself shows us which values of the parameter are plausible, given the data. The parameter value maximizing the likelihood function is taken as the best estimate, and the goodness of other values of the parameter is assessed by their likelihood values relative to the maximum, that is, by their relative likelihoods. An account of this use of the likelihood function is given by Edwards (1972); he uses the term support (function) for the log likelihood function. The cut-off value of the log relative likelihood corresponding to an (approximate) $100(1-\alpha)\%$ confidence interval is then $-\frac{1}{2}\chi^2_{\alpha,1}$.

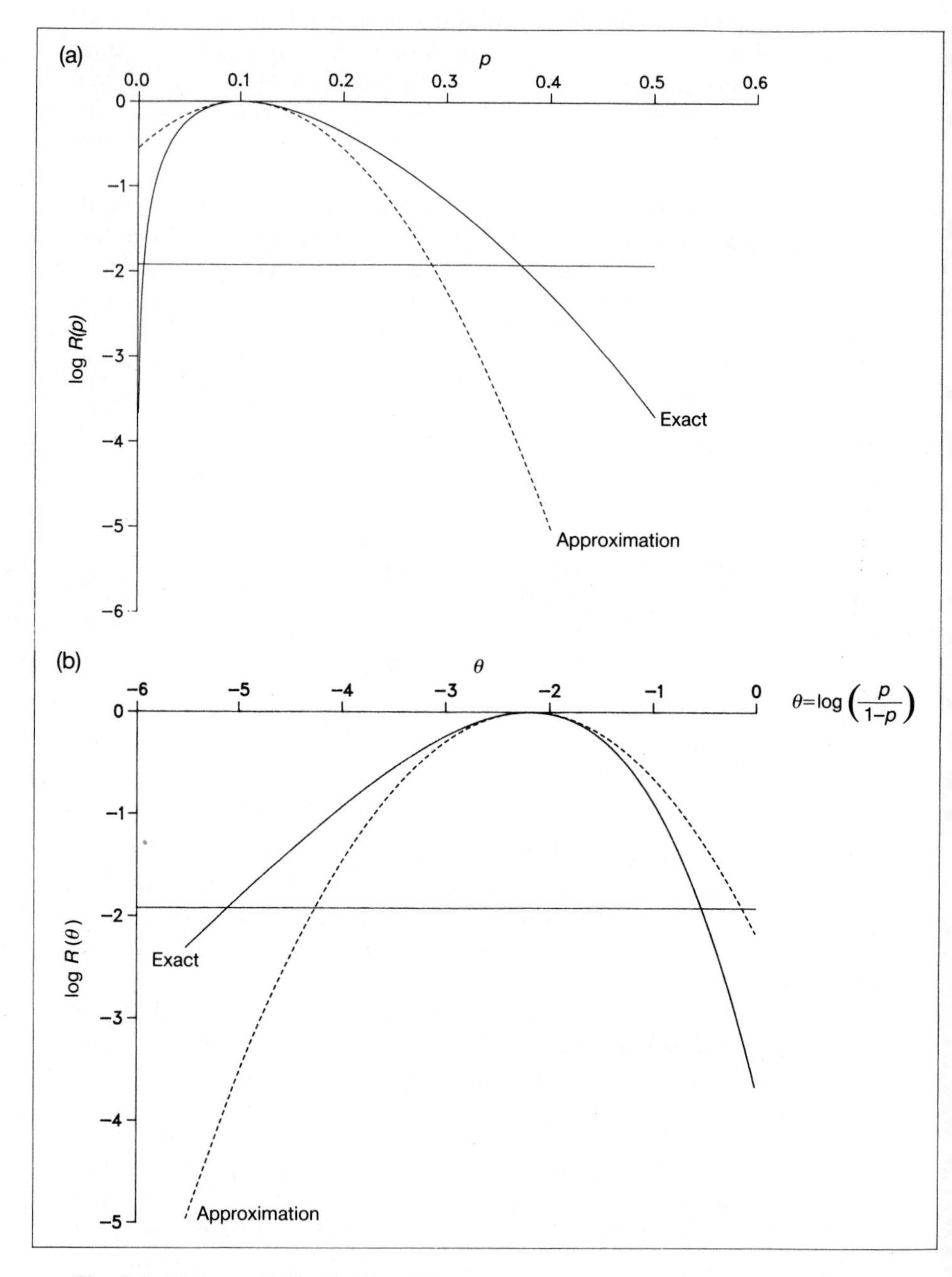

Fig. 2.4. (a) Log relative likelihood in (a) p, (b) θ, for $r=1$, $n=10$, with quadratic approximation

If the log likelihood function is *quadratic* in the parameter, then the confidence intervals based on the information matrix are identical to the likelihood-based confidence intervals. Thus the use of the standard errors from the information matrix to construct the usual confidence intervals is equivalent to the assumption that the log likelihood is quadratic about the maximum likelihood estimate. This is easily checked visually by plotting, since in general

$$\begin{aligned} \ell(\theta) &\simeq \ell(\hat{\theta}) + \tfrac{1}{2}(\theta-\hat{\theta})^2 \ell''(\hat{\theta}) \\ &= \ell(\hat{\theta}) - \tfrac{1}{2}(\theta-\hat{\theta})^2 I(\hat{\theta}), \\ \log \mathrm{R}(\theta) &\simeq -\tfrac{1}{2}(\theta-\theta)^2 I(\hat{\theta}), \end{aligned}$$

and the right-hand side is an approximation to the true log relative likelihood function. Thus, for the above example, when p is the parameter, the approximation to the log relative likelihood is

$$\begin{aligned} \log R(p) &\simeq -\tfrac{1}{2}n(p-\hat{p})^2/\{\hat{p}(1-\hat{p})\} \\ &= -5(p-0.1)^2/0.09 \\ &= -55.56(p-0.1)^2. \end{aligned}$$

The approximating quadratic relative likelihood and log likelihood are also computed and plotted in BINREL using the plotting character *. The exact and approximating quadratic log relative likelihoods in p are shown in Fig. 2.4a. The agreement is extremely poor.

If we make a parameter transformation to $\theta = \text{logit } p$, then the quadratic approximation to the log relative likelihood of θ is

$$\begin{aligned} \log \mathrm{R}(\theta) &\simeq -\tfrac{1}{2}n\hat{p}(1-\hat{p})(\theta-\hat{\theta})^2 \\ &= -0{\cdot}45(\theta + 2.197)^2. \end{aligned}$$

The exact and approximating log relative likelihoods in θ are shown in Fig. 2.4b. These log relative likelihoods have to be calculated explicitly, as BINREL only calculates likelihoods for the untransformed p scale. The agreement is better, but still not good. Note that the skewness in p is reversed in θ. The effect of parameter transformations on the likelihood function, and the choice of transformation to produce a quadratic log likelihood function, were considered by Anscombe (1964). Further discussion is given in Appendix 1.

For the case $r=0$, $n=10$, the log likelihood in p is shown in Fig. 2.5. The maximum occurs on the boundary at $p=0$. Since in this case

$$\mathrm{L}(p) = (1-p)^{10}$$

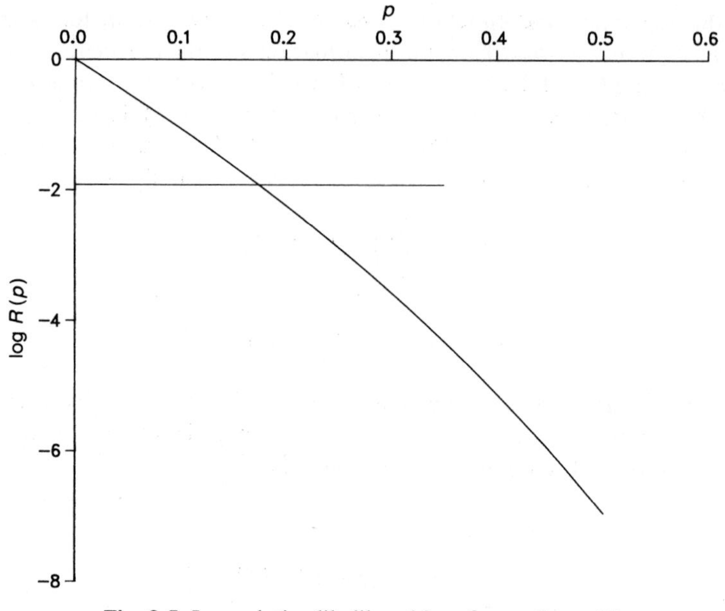

Fig. 2.5. Log relative likelihood in p for $r=0$, $n=10$

it is easy to find analytically the value of p for which $\ell(p)$ is -1.92 units below its maximum: $\ell(0)=0$, and

$$\log R(p)=\ell(p)=10 \log (1-p)=-1.92$$

gives $p=0.1747$, so the 95% likelihood based confidence interval is (0,0.1747).

3
Normal regression and analysis of variance

3.1 The normal distribution and the Box–Cox transformation family

The normal distribution was discussed in Chapter 2 and the assumption of a normal distribution examined by normal plotting and the Filliben correlation coefficient test. For the example of Section 2.3, no evidence of non-normality was found in Section 2.10, but what should we do if the test shows a significant departure from normality? The use of a different family of distributions is considered in Chapter 6; here we extend the usefulness of the normal distribution by embedding it in a larger family of distributions: the Box–Cox transformation family (Box and Cox, 1964).

The motivation for this general family is the frequent occurrence of *skewed* data in which a log transformation, or less commonly a reciprocal transformation, of the response variable produces a nearly normal distribution. This should be carefully distinguished from the use of a log or reciprocal *link function*: in the first case the observed data have a lognormal or reciprocal normal distribution, in the second they have a normal distribution. We will illustrate both cases with examples in Sections 3.3 and 3.4.

The general form of the transformation can be represented by

$$\begin{aligned} y(\lambda) &= (y^{\lambda}-1)/\lambda \qquad && \lambda \neq 0 \\ &= \log y && \lambda = 0 \end{aligned}$$

where λ is the transformation parameter and y the response variable. Here y must be *positive*. For $\lambda=1$, $y(1)=y-1$, and as $\lambda \to 0$, $y(\lambda) \to \log y$, so that $y(\lambda)$ is a continuous function of λ. For $\lambda=-1$, $y(-1)=1-y^{-1}$.

Given data $(y_i, \mathbf{x}_i)$ we assume that there is some value of λ for which $y_i(\lambda)$ has a normal distribution with mean $\boldsymbol{\beta}'\mathbf{x}_i$ and variance σ^2. We want to estimate λ, and decide the appropriate scale of y on which to fit the regression model. The choice of the scale for y, that is, of the value of λ, is determined both by the plausible values of λ from the fitted model and by the interpretability of the scale. Because the transformation is defined by a parameter, we can proceed by maximum likelihood. For each fixed value of λ, we estimate $\boldsymbol{\beta}$ and σ by maximum likelihood, and substitute these values in the likelihood. The resulting function of λ only is called the *profile likelihood in* λ; its log is the *profile log likelihood.* It is used in the same way as the one-parameter likelihood in Section 2.15 to construct likelihood-based confidence intervals for λ.

The probability density function of $y^{(\lambda)}$ is normal (μ^*,σ^{*2}) and hence that of y is, for $\lambda \neq 0$,

$$f(y|\lambda,\mu^*,\sigma^*) = \frac{1}{\sigma^*\sqrt{2\pi}} y^{\lambda-1} \exp[-\{(y^\lambda-1)/\lambda-\mu^*\}^2/2\sigma^{*2}]$$

and for $\lambda = 0$,

$$f(y|0,\mu^*,\sigma^*) = \frac{1}{\sigma^*\sqrt{2\pi}} y^{-1} \exp\{-(\log y-\mu^*)^2/2\sigma^{*2}\}.$$

In general $\mu^* = \boldsymbol{\beta}^{*\prime}\mathbf{x}$; for $\lambda \neq 0$ define $\beta_0 = 1 + \lambda\beta_0^*$, $\beta_j = \lambda\beta_j^*$, $\sigma = |\lambda|\sigma^*$. Then the density for $\lambda \neq 0$ can be written

$$f(y\,|\,\lambda,\boldsymbol{\beta},\sigma) = \frac{1}{\sigma\sqrt{2\pi}}|\lambda|y^{\lambda-1} \exp\{-(y^\lambda-\boldsymbol{\beta}'\mathbf{x})^2/2\sigma^2\}.$$

For $\lambda = 0$, define $\beta_j = \beta_j^*$, $\sigma = \sigma^*$, and the density for $\lambda = 0$ can be written

$$f(y|0,\boldsymbol{\beta},\sigma) = \frac{1}{\sigma\sqrt{2\pi}} y^{-1} \exp\{-(\log y-\boldsymbol{\beta}'\,\mathbf{x})^2/2\sigma^2\}.$$

For the given observations $(y_i,\mathbf{x}_i)$, the log likelihood function (ignoring the constant $\sqrt{2\pi}$) is, for $\lambda \neq 0$,

$$\ell(\lambda,\boldsymbol{\beta},\sigma) = -n\log\sigma + n\log|\lambda| + (\lambda-1)\Sigma\log y_i - \Sigma(y_i^\lambda-\boldsymbol{\beta}'\mathbf{x}_i)^2/2\sigma^2$$

and for $\lambda = 0$,

$$\ell(0,\boldsymbol{\beta},\sigma) = -n\log\sigma - \Sigma\log y_i - \Sigma(\log y_i-\boldsymbol{\beta}'\mathbf{x}_i)^2/2\sigma^2.$$

For fixed λ, the partial derivatives with respect to $\boldsymbol{\beta}$ and σ are then

$$\frac{\partial\ell}{\partial\boldsymbol{\beta}} = \Sigma\mathbf{x}_i(y_i^\lambda-\mathbf{x}_i'\boldsymbol{\beta})/\sigma^2$$

$$\frac{\partial\ell}{\partial\sigma} = -n/\sigma + \Sigma(y_i^\lambda-\mathbf{x}_i'\boldsymbol{\beta})^2/\sigma^3$$

where y_i^λ is to be taken as $\log y_i$ if $\lambda = 0$. Denote the solutions of the equations

$\partial \ell / \partial \boldsymbol{\beta} = 0$, $\partial \ell / \partial \sigma = 0$ by $\hat{\boldsymbol{\beta}}(\lambda)$, $\hat{\sigma}(\lambda)$. These are the same as in Section 2.7, with y_i^λ replacing y_i:

$$\hat{\boldsymbol{\beta}}(\lambda) = (X'X)^{-1}X'\mathbf{y}^\lambda$$
$$\hat{\sigma}^2(\lambda) = \Sigma\{y_i^\lambda - \mathbf{x}_i'\hat{\boldsymbol{\beta}}(\lambda)\}^2/n = \text{RSS}(\lambda)/n$$

where RSS(λ) is the residual sum of squares for the given λ. Note that RSS(λ) is discontinuous at $\lambda = 0$, due to the reparametrization above.

Substituting into the log likelihood function, we obtain the *profile log likelihood function* (omitting constants):

$$p\ell(\lambda) = -\tfrac{1}{2}n \log \text{RSS}(\lambda) + n \log |\lambda| + (\lambda - 1)\Sigma \log y_i \qquad \lambda \neq 0$$
$$p\ell(0) = -\tfrac{1}{2}n \log \text{RSS}(0) - \Sigma \log y_i.$$

Note that, though RSS(λ) is discontinuous at $\lambda = 0$, $p\ell(\lambda)$ is not.

An (approximate) $100(1-\alpha)\%$ confidence interval for λ consists of those values of λ for which $p\ell(\lambda)$ is within $\frac{1}{2}\chi^2_{\alpha,1}$ units of its maximum: the interval is most simply found by tabulation of $p\ell(\lambda)$ over a grid of values of λ. This interval is used to identify the values of λ giving an interpretable scale.

The subfile BOXCOX contains several macros for the construction and plotting of the profile log likelihood. The macro BOXCOX constructs a plot of $-2p\ell(\lambda)$ rather than $p\ell(\lambda)$ itself, so that differences from the *minimum* value of $-2p\ell(\lambda)$ are compared with $\chi^2_{\alpha,1}$. Illustration of its use is postponed to the next section.

An important feature of response-variable transformations is that on the transformed scale the model represents variation in the mean of the (normal) transformed variable, but on the original scale the variation is in the median of the variable.

This is most simply seen for the log transformation. Suppose $\log y \sim N(\boldsymbol{\beta}'\mathbf{x}, \sigma^2)$. Then y has a lognormal distribution, and

$$\text{median}\,(y) = \exp\,(\boldsymbol{\beta}'\mathbf{x})$$
$$E\,(y) = \exp\,(\boldsymbol{\beta}'\mathbf{x} + \tfrac{1}{2}\sigma^2)$$
$$\text{var}\,(y) = \{\exp\,(\sigma^2) - 1\}\exp\,(2\boldsymbol{\beta}'\mathbf{x} + \sigma^2).$$

Thus the (additive) regression model for the mean of $\log y$ is a multiplicative model for the median of y, and also for the mean of y, though the intercept is changed by $\frac{1}{2}\sigma^2$, and the variance of y is not constant. Though the model is fitted on the log scale, we usually want to interpret the model on the original scale: if fitted values from the model are transformed by %EXP(%FV) to the original scale, these are fitted values for the median response, not for the mean:

for the mean the fitted values are %EXP(%FV + %SC/2), where here %SC is %DV/%DF, the unbiased estimate s^2 of σ^2.

For transformations y^λ with $\lambda \neq 0$, if μ is the mean of y^λ then

$$\text{median}\,(y) = \mu^{1/\lambda}$$
$$\text{E}\,(y) \approx \mu^{1/\lambda}\{1 + \sigma^2(1-\lambda)/(2\lambda^2\mu^2)\}$$
$$\text{var}\,(y) \approx \mu^{2/\lambda}\sigma^2/(\lambda^2\mu^2)$$

Again the apparent discontinuity between $\lambda = 0$ and $\lambda \neq 0$ is caused by the use of y^λ rather than $(y^\lambda - 1)/\lambda$.

3.2 Modelling and background information

The most useful models are those which use background information or theory from the field of application. However, theory may be incorrect, or speculative, or there may be no adequate theory and the modelling may be meant to assist the development of an adequate theory.

We illustrate with a set of data on tree volumes taken from the *Minitab Handbook*, (Ryan, Joiner and Ryan, 1976) and discussed at length by Atkinson (1982). The volume of usable wood V in cubic feet (1 foot = 30.48 cm) is given for each of a sample of 31 black cherry trees, and the height H in feet and the diameter D in inches (1 inch = 2.54 cm) at a height 4.5 feet above the ground. We want to develop a model which will predict the usable wood volume from the easily measured height and diameter.

The data are in subfile TREES.

```
$INPUT 1 TREES $INPUT %PLC QPLOT NORMAC LEV $
```

We begin by plotting V against D and H.

```
$PLOT V D : V H $
```

Clearly accurate prediction of V can be made from D: the variation about a smooth curve is quite small. The curvature is evident: a straight line relationship will not be adequate. We will use the *multiple correlation R* (or more usually its square R^2) as a measure of the predictability of the response from the explanatory variables: R is just the correlation between the response and the fitted values from the model. Equivalently,

$$R^2 = 1 - \text{RSS}/\text{TSS}$$

where RSS is the residual sum of squares from the given model and TSS is the residual sum of squares from the null model: $TSS = \Sigma(y_i - \bar{y})^2$ is usually called the "total sum of squares".

The value of R^2 is not calculated by GLIM, but it can be easily obtained after any fit using the macro RSQ in the subfile NORMAC.

We first try the model using H and D.

```
$YVAR V  $ERROR N
$FIT H+D $D E $USE RSQ $
```

For this model R^2 is 0.948, a very high value implying close agreement between observed and fitted volumes. We now examine the residuals and influence values.

```
$USE LEV $
$PRINT LEV_ $
```

The 20th and the last observations have influence values of 0.21 and 0.23, just larger than $6/31 = 0.194$. The last tree is the largest, the 20th has a large diameter for its height.

```
$CALC RRES =#RAW $
$PLOT RRES H : RRES D $
```

The plot against H looks fairly random but that against D shows a marked dip in the middle, suggesting that the model is mis-specified.

```
$USE QPLOT $
```

The Filliben correlation is very high (0.9901); the test does not pick up here the non-random behaviour of the residuals. The curvature in the V versus D plot suggests that we may be working on the wrong scale. To investigate this we construct a Box–Cox plot. We first input the subfile BOXCOX:

```
$INPUT %PLC BOXCOX $
```

To use the BOXCOX macro, we first need to specify the untransformed y-variate and the model as the contents of two macros called by BOXCOX:

```
$MACRO YVAR V $ENDMAC
$MACRO MODEL H+D $ENDMAC
$USE BOXCOX $
```

The macro prompts the user for the range of values of λ over which the values of minus twice the log likelihood are to be plotted. It is seldom necessary to choose the range wider than -2 to $+2$; initially steps of 0.5 are sufficient.

The user will notice a considerable pause in computer response at this point before plotting, since nine $FIT directives have to be executed, plus the calculation of the log likelihood after each fit.

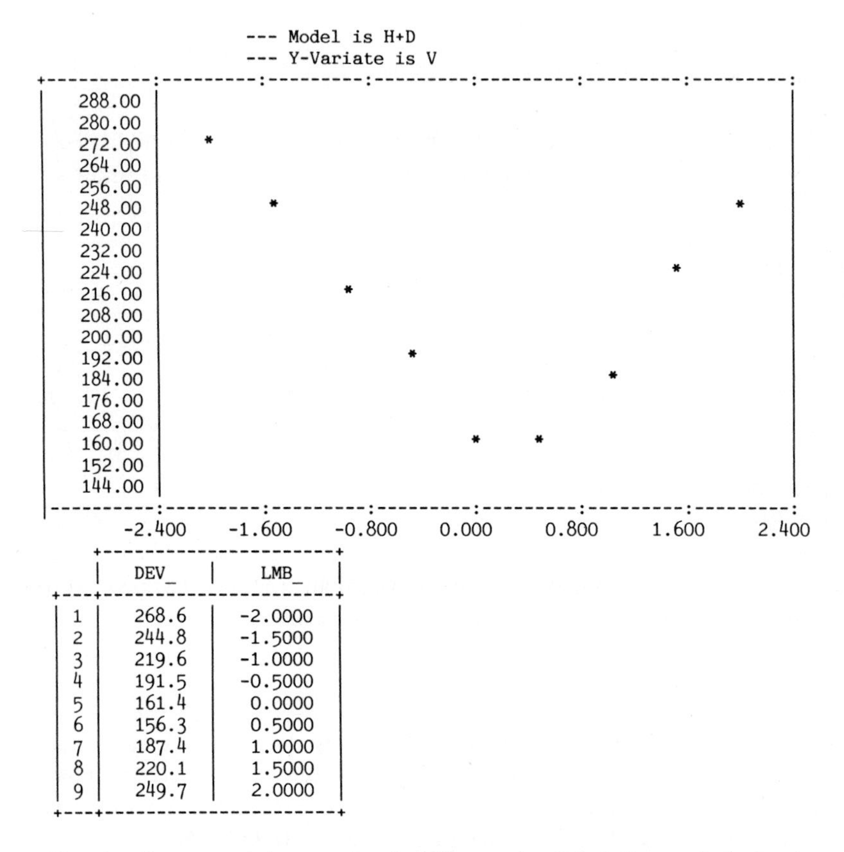

	DEV_	LMB_
1	268.6	-2.0000
2	244.8	-1.5000
3	219.6	-1.0000
4	191.5	-0.5000
5	161.4	0.0000
6	156.3	0.5000
7	187.4	1.0000
8	220.1	1.5000
9	249.7	2.0000

The plot shows a minimum of $-2p\ell(\lambda)$ near $\lambda = 0.5$. A finer tabulation in steps of 0.1 over the range 0 to 1 is required.

```
$USE BOXCOX $
```

The minimum of $-2p\ell(\lambda)$ occurs near $\lambda = 0.3$, with values of 161.4 at $\lambda = 0$, 152.2 at $\lambda = 0.3$ and 187.4 at $\lambda = 1$. The approximate 95% confidence interval, based on a value of 152.2 ± 3.8, is (0.12, 0.49), which is a quite narrow interval. How should we interpret this result?

The nature of the problem throws light on this question. Volume is measured in cubic feet, but height and diameter in feet (or inches). We are attempting to predict a volumetric measurement from a linear measurement. This suggests that the "side of the equivalent cube" might be more appropriate as a response, the cube root of volume. The profile likelihood in λ points very closely to this value. To see why, we calculate $V^{1/3}$ and plot it against D and H:

```
$CALC VTH = V**(1/3) $
$PLOT VTH D : VTH H $
```

The plot against D is now very closely linear: the curvature has been removed by the cube root transformation. We now refit the model:

```
$YVAR VTH
$FIT #MODEL $D E $USE RSQ $USE QPLOT $
```

Note here the use of the text substitution symbol #: the contents of the macro MODEL are used as the argument of $FIT. The value of R^2 is 0.9777, a substantial increase over that for $\lambda = 1$. The fitted model for VTH is, with standard errors in parentheses,

$$\hat{\mu} = \underset{(0.184)}{-0.0854} + \underset{(0.0028)}{0.0145}\,H + \underset{(0.0056)}{0.1515}\,D$$

with an estimated error variance of $s^2 = 0.00686$. Residual plots against D and H show a random scatter, though the residuals seem to increase noticeably in magnitude with H, and a quantile plot gives a closely linear fit with a Filliben correlation of 0.9895.

This model can also be fitted using the macro BOXFIT, which prompts for a specified value of λ:

```
$USE BOXFIT $
--- Model is H+D
--- Y-Variate is V
-- model changed
Value of lambda?   0.3333
```

The value of $-2 \log p\ell(\lambda)$ is 152.2, the same (to one decimal place) as for $\lambda = 0.3$.

The fitted median values of V from the model are $\hat{\mu}^3$, and the fitted mean values of V are approximately $\hat{\mu}^3(1 + 3s^2/\hat{\mu}^2)$. These can be compared with the observed values, for example

```
$CALC FV = %FV**3 $
$PRINT FV : V $
$PLOT V FV '+' $
```

Fitted median and mean values both agree closely with the observed values. We have a closely fitting model to which we are led by physical considerations of dimensionality, and the Box–Cox plot. Other physical considerations, however, lead to a different model. The curvature which we noted in the V/D plot can also be removed by a log transformation:

```
$CALC LV = %LOG(V) : LD = %LOG(D) : LH = %LOG(H) $
$PLOT LV LD : LV LH $
```

This suggests a regression of log V on the logged explanatory variables.

```
$YVAR LV  $FIT LH + LD $D E $USE RSQ $USE QPLOT $USE LEV $
```

The value of R^2 is 0.9777, exactly the same (to four decimal places) as that for the $V^{1/3}$ model. Residual plots show little unusual apart from three rather large negative residuals, though the same pattern of residuals increasing with H is visible as with the cube root transformation. The quantile plot of the standardized residuals shows some variation but the Filliben correlation coefficient of 0.983 is high. The influence values are much the same as before, though that for the 20th observation is larger (0.24) and that for the last is smaller (0.18).

Should we be working on the log scale of V? What is the effect of the log transformations of H and D on the Box–Cox transformation of V? We specify a new model for the BOXCOX macro by overwriting the contents of the macro MODEL.

```
$MACRO MODEL LH + LD $ENDMAC
$USE BOXCOX $
```

Using the grid $-2(0.5)2$ for λ shows a minimum near $\lambda=0$. A finer tabulation in steps of 0.05 shows a minimum of $-2p\ell(\lambda)$ of 150.1 at $\lambda=-0.05$, and a 95% confidence interval of $(-0.24, 0.11)$, which includes zero but excludes $\lambda=1/3$. Thus if we log transform H and D, the cube root transformation of V is not appropriate, but the log transformation is. The value of $-2p\ell(\lambda)$ at $\lambda=0$ of 150.7 is slightly smaller than for $V^{1/3}$, H and D, which was 152.2.

We can compare these by a likelihood ratio test, since both models can be expressed in the form

$$V^* \sim \mathrm{N}(\beta_0+\beta_1 H^*+\beta_2 D^*\ , \sigma^2)$$

with

$$V^*=(V^\lambda-1)/\lambda_V$$

$$H^*=(H^\lambda-1)/\lambda_H$$

$$D^*=(D^\lambda-1)/\lambda_D.$$

The comparison is of the values $\lambda_V = 1/3$, $\lambda_H = \lambda_D = 1$ with $\lambda_V = \lambda_H = \lambda_D = 0$, the other parameters being nuisance parameters in each case, the likelihood being maximized over them. The difference of 1.5 is not significant for χ^2_3, so we cannot choose convincingly between the two models. (Each set of parameter values could be compared with the maximum likelihood estimates $\hat{\lambda}_V, \hat{\lambda}_H, \hat{\lambda}_D$ by the usual χ^2_3 likelihood ratio test, but our interest here is choosing between two specified sets.)

The fitted model for log V is

$$\hat{\mu} = \underset{(0.800)}{-6.632} + \underset{(0.204)}{1.117}\ \mathrm{LH} + \underset{(0.075)}{1.983}\ \mathrm{LD}$$

with $s^2 = 0.00662$. We note that the coefficient of LH is close to 1, and that of LD close to 2. Again the nature of the problem throws light on this result. The tree may be idealized as a regular solid figure, like a cylinder or a cone. The volume V of a cylinder of length ℓ and diameter d is $\pi d^2 \ell/4$; of that of a cone of height ℓ and base diameter d is $\pi d^2 \ell/12$. In either case

$$\log V = \mathrm{c} + \log \ell + 2 \log d,$$

where c is log $\pi/4$ for the cylinder and log $\pi/12$ for the cone. These volume formulae assume that ℓ and d are measured in the same units. In our example H is in feet but D is in inches. We first convert D to feet:

```
$CALC DF = D/12 : LDF = %LOG(DF) $
```

We redefine the y-variate and refit the model.

```
$YVAR LV
$FIT + LDF - LD $D E $
```

The intercept is now -1.705. Now we fit the model in which the coefficients are fixed at 1 and 2. The model is then

$$\mu = \beta_0 + x_1 + 2x_2$$

in which only β_0 is to be estimated. Here $(x_1 + 2x_2)$ is treated as a single variable whose coefficient is to be fixed at 1.0. This is achieved by declaring this variable to be an *OFFSET* and fitting the null model in which only β_0 is estimated.

```
$CALC Z = LH + 2*LDF $
$OFFSET Z
$FIT $D E $USE RSQ $
```

The value of R^2 is 0.9774, a trivial change from the full model. The intercept is now -1.199 with a standard error of 0.014, and $s^2=0.00626$. Does the value of β_0 correspond to a cylinder or a cone or neither?

```
$CALC %LOG(%PI/4) : %LOG(%PI/12) $
```

The values are -0.242 and -1.340. The tree is closer to a cone than a cylinder, but it has a greater volume than the cone, which seems reasonable.

The fitted *median* values of V from the model are $\exp(\hat{\mu})$, which can be compared with the observed values:

```
$CALC FV = %EXP(%FV) $
$PRINT FV : V $
$PLOT FV V '+' $
```

The fitted *mean* values of V are $\exp(\hat{\mu}+\frac{1}{2}s^2)$, which are only 0.3% greater than the median values.

The two fitted models agree closely with the observed values (a maximum error of 5 cubic feet for the largest tree), with the error standard deviation of 0.079 in the log model corresponding to a percentage error of 8% in the fitted volume. Here the data are insufficient to discriminate between models based on different physical arguments.

3.3 Link functions and transformations

We noted in Chapter 2 that even in normal models the choice of the *link function* relating the parameter of the probability distribution to the linear predictor is not always obvious. The preceding example provides a good illustration. For the first model relating tree volume to diameter and height, a cube root transformation of V linearized the V/D plot. It is not, however, necessary to assume that $V^{1/3}$ has a normal distribution with mean θ equal to the linear predictor $\beta_0+\beta_1 H+\beta_2 D$, in order to fit such a model. We can assume instead that V is *normal*, but the link function relating V to the linear predictor is the *cube root*, that is, if μ is the mean tree volume,

$$\eta=g(\mu)=\mu^{1/3}, \quad \text{or} \quad \mu=\eta^3=(\beta_0+\beta_1 H+\beta_2 D)^3,$$

while $V\sim N(\mu,\sigma^2)$.

The fitting of this model by maximum likelihood is equivalent to fitting by non-linear least squares, since the model for μ is now non-linear in the parameters. A simple change in the link function to the Exponent link is all that is needed; the offset also needs to be removed:

```
$YVAR V $LINK E 0.3333 $OFFSET $FIT H+D $D E $USE RSQ $
```

The value of R^2 is 0.9773, very slightly less than for the $V^{1/3}$ model. The fitted model, with standard errors in parentheses, is almost the same as in Section 3.2:

$$\hat{\mu}^{1/3} = \underset{(0.224)}{-0.0510} \underset{(0.0033)}{+0.0143\,H} \underset{(0.0058)}{+0.1503\,D}$$

The residual sum of squares for this model is 184.2. Residual plots show the same feature as for the cube root transformation of V, the residuals again increasing in absolute size with H, though not with D.

```
$USE QPLOT $
```

The quantile plot looks linear and the Filliben correlation is 0.9844. This model also seems satisfactory.

We can repeat this procedure using the log link instead of the lognormal distribution for V:

```
$LINK L  $FIT LH + LD $D E $USE RSQ $
```

The value of R^2 is 0.9778, and the fitted model is again very similar to that in Section 3.2:

$$\widehat{\log \mu} = \underset{(0.944)}{-6.537} \underset{(0.242)}{+1.088\,LH} \underset{(0.082)}{+1.997\,LD}$$

and the residual sum of squares is 179.7, a slightly better fit. The residual plot against H shows the same feature of rapidly increasing residuals, though the Filliben correlation is again high (0.9871).

Now we have four possible models. Can we choose between response variable and link transformations? Extending the argument of Section 3.2, we can write the competing models in terms of *four* λ parameters:

$$V^* \sim N(\eta^*, \sigma^2)$$

with

$$\eta^* = (\eta^{\lambda_L} - 1)/\lambda_L$$

and

$$\eta = \beta_0 + \beta_1 H^* + \beta_2 D^*$$

as in Section 3.2. Here λ_L is the link function parameter. The two models of Section 3.2 had $\lambda_L = 1$ with $\lambda_V = 1/3$ and $\lambda_V = 0$; the two models above have $\lambda_L = 1/3$ and $\lambda_L = 0$ with $\lambda_V = 1$. The comparable values of $-2 \log p\ell(\lambda)$ for the latter models are obtained directly as $n \log \text{RSS}$; they are 161.7 for the cube root link and 160.9 for the log link. Compared with the value of 150.7 for the lognormal model, both the link function transformation models are significantly worse. We therefore conclude that the log or cube root variable transformation models are better, and we prefer the log model because of its "solid body" interpretation. Thus the likelihood ratio test enables us to choose between transforming the distribution and transforming the link function.

A detailed discussion of transformations of the explanatory variables is given by Box and Tidwell (1962). Scallan, Gilchrist and Green (1984) extend the idea of the Box–Cox transformation to "parametric link functions" in which $\eta = g(\mu) = (\mu^\lambda - 1)/\lambda$ and consider the estimation of λ.

3.4 Regression models for prediction

In Chapter 2 we noted two uses of regression models: for simplified descriptions of populations, and for prediction of values of the response variable for new observations.

The tree volume example provides a good illustration. The trees were felled and the volume of usable wood measured so that the volume of usable wood could be estimated for stands or forests of similar trees. The real purpose of the model is not for simple description—though this is an essential part of the modelling—but for the prediction of wood volume of similar trees.

This prediction is easily obtained from the fitted model. Given $y_i \sim$ N $(\boldsymbol{\beta}'\mathbf{x}_i, \sigma^2)$ independently, $i = 1, \ldots, n$, and a new observation $\mathbf{x}_{n+1}$ on $\mathbf{x}$, we to predict the corresponding value y_{n+1} of y. We assume that the same distributional model applies for y_{n+1}. For simplicity of notation we write

$$y = y_{n+1}, \mathbf{x} = \mathbf{x}_{n+1}.$$

Then

$$y \sim N(\boldsymbol{\beta}'\mathbf{x}, \sigma^2)$$

and

$$\hat{\boldsymbol{\beta}}'\mathbf{x} \sim \mathrm{N}(\boldsymbol{\beta}'\mathbf{x}, \sigma^2\mathbf{x}'(X'X)^{-1}\mathbf{x}),$$

$$\mathrm{RSS} \sim \sigma^2\chi^2_{n-p-1}$$

independently of y. It follows immediately that

$$y - \hat{\boldsymbol{\beta}}'\mathbf{x} \sim N(0, \sigma^2(1+h))$$

$$(y - \hat{\boldsymbol{\beta}}'\mathbf{x})/s(1+h)^{\frac{1}{2}} \sim t_{n-p-1}$$

where

$$s^2 = \mathrm{RSS}/(n-p-1)$$

and

$$h = \mathbf{x}'(X'X)^{-1}\mathbf{x}$$

is the variance of the estimated linear predictor $\hat{\boldsymbol{\beta}}'\mathbf{x}$, divided by σ^2.

Thus probability statements may be made about the random variable y from the t-distribution, for example

$$\Pr(|y-\hat{\boldsymbol{\beta}}'\mathbf{x}| < t_{\alpha/2,n-p-1}\, s(1+h)^{\frac{1}{2}}) = 1-\alpha$$

is a $100(1-\alpha)\%$ *prediction interval* for y. It is not a *confidence interval* since y is a random variable, not a parameter. Confidence interval statements can be made about the mean of y, $\boldsymbol{\beta}'\mathbf{x}$, as this is a parametric function of $\boldsymbol{\beta}$. Since

$$\hat{\boldsymbol{\beta}}'\mathbf{x} \sim \mathrm{N}(\boldsymbol{\beta}'\mathbf{x},\sigma^2 h),$$

a $100(1-\alpha)\%$ confidence interval for $\boldsymbol{\beta}'\mathbf{x}$ is

$$|\boldsymbol{\beta}'\mathbf{x}-\hat{\boldsymbol{\beta}}'\mathbf{x}| < t_{\alpha/2,n-p-1}\, sh^{\frac{1}{2}}.$$

This is shorter than that for y as the additional variation in the random variable does not have to be allowed for.

The same results for y can be obtained by treating y as a formal unknown parameter in the model for $y_1,\ldots,y_n, y$ and constructing a profile likelihood for y by maximizing the likelihood of $y_1,\ldots,y_n, y$ over $\boldsymbol{\beta}$ and σ for a fixed value of y. Details are given in Appendix 1. This device is very useful in non-normal models, and is used in subsequent chapters.

Construction of the prediction interval depends on the predicted value $y=\hat{\boldsymbol{\beta}}'\mathbf{x}$ and the variance s^2h of the linear predictor $\hat{\boldsymbol{\beta}}'\mathbf{x}$. These are easily obtained in GLIM by the device of adjoining an additional observation $(1,\mathbf{x})$ to the calibration data $(y_i,\mathbf{x}_i)$ $i=1,\ldots,n$. (More than one new observation $\mathbf{x}$ can be added in this way, but if there are many it is simpler to use a data file with all observations, complete and incomplete.)

We illustrate with the trees data. Suppose in addition to the 31 observations we have a new tree with height H = 80 feet and diameter D = 15 inches. What predictive statement can be made about this tree's usable wood volume?

We extend the length of the data to 32 by defining new variables of this length, and copying the original variables into them. We use the directive $ASSIGN to do this.

```
$ASSIGN VX  =  V,1  :
        HX  =  H,80  :
        DX  =  D,15  $
```

The values of VX will be the same as those of V, except for the additional last element which is set to unity (arbitrarily: any other value could be used, since this observation is to be weighted out in the fit). The last elements of HX and DX contain the specified values for the new tree.

We now fit the required model—we will use the log model for illustration. The standard length of vectors has first to be reset to 32.

```
$CALC LVX=%LOG(VX) : LHX=%LOG(HX) : LDX=%LOG(DX) $
$UNIT 32
$CALC W = 1 : W(32) = 0 $
$YVAR LVX $WEIGHT W  $FIT LHX+LDX $D E $
```

The parameter estimates are identical to those in Section 3.2. The predicted value for the last observation is now the last component of %FV, and the variance of the linear predictor $s^2\mathbf{x}'(X'X)^{-1}\mathbf{x}$ is in the last component of %VL, which has to be extracted:

```
$EXTRACT %VL $
$ACC 8
$LOOK %NU %FV %VL $
```

The fitted value is 3.633, the variance 0.000333 and the residual mean square estimate s^2 is 0.006624. A 95% prediction interval for the tree log-volume is then

$$3.633 \pm t_{0.025,\,28}(0.000333+0.006624)^{\frac{1}{2}}$$

or

$$(3.462,\ 3.804).$$

The corresponding interval for predicted tree volume is (31.9, 44.9).

3.5 Model choice and mean square prediction error

We noted in Chapter 2 that regression models are used both for smooth representation and for prediction. The model selection procedure described in Chapter 2 is intended to provide the model which gives the simplest representation of the population consistent with the data. It does not necessarily follow that this model is the best one for prediction.

Since the outcome of the prediction process is a predicted value (and an interval for the true value), it seems reasonable to choose the model which gives the closest prediction. This can be defined in several ways; we will adopt the usual convention that the *squared prediction error*

$$\sum_{i \in N} (y_i - \hat{y}_i)^2$$

is to be minimized, where the minimization is over the set N of future observations. Since this is usually unknown, we are forced to characterize the set of future observations by reference to the calibration set of observations already available: the new observations $\mathbf{x}$ are assumed to be similar to the existing observations $\mathbf{x}_i$. (If they were not, then we should have reservations about extrapolating the fitted regression into regions of the explanatory variable space not represented in the data.)

This similarity is expressed in one of two ways: the existing $\mathbf{x}_i$ are thought of as fixed, and the new observations $\mathbf{x}$ have a uniform distribution over the $\mathbf{x}_i$; or the existing $\mathbf{x}_i$ are thought of as values of a random variable $\mathbf{X}$ (apart from the constant 1), and the new $\mathbf{x}$ are independent values of this random variable. Minimization of the *expected* or *mean* square prediction error (MSPE) is then taken to be the aim of the model selection procedure, where the expectation is over the uniform distribution of the $\mathbf{x}_i$, or the common multivariate distribution of $\boldsymbol{X}$. Because of the difficulty of specifying the latter distribution (a multivariate normal distribution is a common, but unrealistic, specification) we consider only the first case.

For the normal model $\mathbf{y} \sim N(\boldsymbol{\beta}'\mathbf{x},\sigma^2)$, the mean square prediction error (e.g. Aitkin, 1974) is

$$E_f = (n+p+1)\sigma^2/n.$$

For the subset model $\boldsymbol{\beta}_r'\mathbf{u}$, omitting the component $\boldsymbol{\beta}_s'\mathbf{v}$ from $\boldsymbol{\beta}'\mathbf{x}$, the MSPE is (using the notation of Section 2.7)

$$E_r = (n+p_1+1)\sigma^2/n + \boldsymbol{\beta}_s' S_{v.u}\boldsymbol{\beta}_s/n$$

where

$$S_{v.u} = V'V - V'U(U'U)^{-1}U'V.$$

Thus prediction is improved (MSPE decreases) if

$$E_f - E_r > 0$$

i.e. if

$$p_2\sigma^2 > \boldsymbol{\beta}_s' S_{v.u}\boldsymbol{\beta}_s/n.$$

Thus if $\boldsymbol{\beta}_s$ is small, prediction will be improved by omitting $\mathbf{v}$. It is possible to test formally whether this inequality is violated for any subset $\mathbf{u}$, thus giving a simultaneous test for the non-reduction of the MSPE (Aitkin, 1974) for all possible subsets. This test is identical to the simultaneous test in Chapter 2 for $\boldsymbol{\beta}_s = \mathbf{0}$, except for a different critical value.

However this test does not identify the model with the *smallest* MSPE, only those models with MSPEs *not significantly larger* than that of the full model. Replacing parameter values in E_f and E_r by sample estimates from each model allows a comparison of the *estimated* MSPEs, though a formal test is not available.

For the full model the difference between the estimated MSPEs is

$$\begin{aligned} E_f - E_r &= 2(\mathrm{RSS}_f/(n-p-1) - \mathrm{RSS}_r/(n-p_1-1)) \\ &= 2(s_f^2 - s_r^2) \end{aligned}$$

where s_f^2 is the residual mean square estimate of σ^2 from the full model and s_r^2 that from the reduced model.

Thus we might choose as the best model the one with the *smallest residual mean square estimate of* σ^2. As noted above, this approach ignores the sampling variation in these estimates: it is quite possible for the best model to have a sample variance s^2 larger, through sample fluctuations, than that of another model. The use of this criterion is considered in an example in Section 3.7, but we first examine the use of *cross-validation*.

3.6 Prediction model selection through cross-validation

Considerable experience with the use of regression models for prediction has shown that when applied to new observations or samples, the model predicts much less well than in the calibration sample. This phenomenon is called "shrinkage on cross-validation", the shrinkage referred to being that of R^2. Stone (1974) gives some historical examples.

In regression studies with large samples it is possible to divide the sample (randomly) into two halves. The model is fitted on the first half, and used for the prediction of the values of y in the second half. This *cross-validation* of the model on the second independent sample gives a more realistic assessment of its predictive value, and of the value of reduced models. A recent example with small samples is given by Copas (1983).

With small samples cross-validation can be achieved by omitting each observation in turn from the data, fitting the model (or models) to the remaining observations, predicting the value of y for the omitted observation, and comparing the prediction with the observed value. Let $\hat{y}_{(i)}$ be the predicted value of y_i when the i-th observation is omitted from the data. Then

$$\mathrm{CVE} = \sum_{i=1}^{n} (y_i - \hat{y}_{(i)})^2/n$$

is the *cross-validation estimate* of the MSPE; it is an unbiased estimate of σ^2. The sum $n\mathrm{CVE} = \Sigma(y_i - \hat{y}_{(i)})^2$ was called PRESS (Prediction Sum of Squares) by Allen (1971) who gave a more general definition.

The computation of PRESS for any model is easily achieved, without omitting observations, from the standard output of a FIT directive.

Let $X_{(-i)}$ by the matrix of explanatory variables with $\mathbf{x}_i$ deleted and $\mathbf{y}_{(-i)}$ the vector of response values with y_i deleted, and let $\hat{\boldsymbol{\beta}}_{(-i)}$ be the corresponding estimate of $\boldsymbol{\beta}$. Then

$$\hat{y}_{(i)} = \mathbf{x}_i'\hat{\boldsymbol{\beta}}_{(-i)}$$

and

$$X'X = X'_{(-i)}X_{(-i)} + \mathbf{x}_i\mathbf{x}_i'$$

$$X'\mathbf{y} = X'_{(-i)}\mathbf{y}_{(-i)} + \mathbf{x}_i y_i.$$

Then

$$(X'_{(-i)}X_{(-i)})^{-1} = (X'X - \mathbf{x}_i\mathbf{x}_i')^{-1}$$
$$= (X'X)^{-1} + (X'X)^{-1}\mathbf{x}_i^{-1}(1 - \mathbf{x}_i'(X'X)^{-1}\mathbf{x}_i)^{-1}\mathbf{x}_i'(X'X)^{-1}$$

$$X'_{(-i)}\mathbf{y}_{(-i)} = X'\mathbf{y} - \mathbf{x}_i y_i$$

$$\hat{\boldsymbol{\beta}}_{(-i)} = (X'_{(-i)}X_{(-i)})^{-1}X'_{(-i)}\mathbf{y}_{(-i)}$$

and

$$\hat{y}_{(i)} = \mathbf{x}_i'\boldsymbol{\beta}_{(-i)} = (\hat{y}_i - h_i y_i)/(1 - h_i)$$

$$e_{(i)} = y_i - \hat{y}_{(i)} = (y_i - \hat{y}_i)/(1 - h_i) = e_i/(1 - h_i)$$

where

$$h_i = \mathbf{x}_i(X'X)^{-1}\mathbf{x}$$

is the influence of the i-th observation in the complete sample. The PRESS criterion is thus

$$\text{PRESS} = \sum_{i=1}^{n} e_{(i)}^2 = \sum_{i=1}^{n} e_i^2/(1 - h_i)^2$$

which is easily calculated using the library macro PRESS available in the subfile PRESS; this macro also provides the cross-validation estimate CVE of σ^2, and the cross-validation $R^2{}_{\text{CV}}$ (the squared correlation between the y_i and $\hat{y}_{(i)}$).

Again, we might choose for prediction the model with the smallest value of PRESS, or more realistically, one of the models with a small value of this criterion.

We examine in the next section the use of these criteria for prediction model choice on a complex example. We conclude this section with a simple illustration on the trees data, using the log volume model. The coefficients of log height and log diameter have simple physical interpretations: should we use the model 2*LD+LH for prediction, or the model with estimated coefficients?

```
$INPUT 1 TREES $INPUT %PLC NORMAC PRESS $
$CALC LV=%LOG(V) : LD=%LOG(D) : LH=%LOG(H) $
$YVAR LV $FIT LD+LH $D E$USE RSQ $USE PRESS $
```

For the full model we have

$$\text{RSS}=0.1855,\ s^2=0.006624,\ \text{PRESS}=0.2186$$

and

$$\text{CVE}=0.007050$$

which is 6% larger than s^2. The cross-validation R^2_{CV} is 0.9737, very little less than the value 0.9777 for R^2 itself.

```
$CALC Z = 2*LD+LH $OFFSET Z
$FIT $D E$USE PRESS $
```

For the reduced model

$$\text{RSS}=0.1877,\ s^2=0.006256,\ \text{PRESS}=0.2004,\ R^2=0.9774$$

and

$$\text{CVE}=0.006465$$

is now only 3% larger than s^2. The cross-validation R^2_{CV} is 0.9759, *greater* than R^2_{CV} for the full model. Both s^2 and PRESS decrease in the reduced model, by 6–8% of their values in the full model. This shows that prediction can be based on the reduced model.

```
$END
```

3.7 The reduction of complex regression models

The value of modelling and model simplification becomes clear when we are dealing with complex data sets with many possible explanatory variables. A good example is given by Henderson and Velleman (1981), also discussed by Aitkin and Francis (1982). The data are in subfile CAR1, and are quarter-mile acceleration time in seconds (QMT) and fuel consumption in miles per (US) gallon (MPG) for 32 cars tested by the US. *Motor Trend* magazine in 1974. Nine explanatory variables are given: shape of engine S (straight = 1, vee = 0), number of cylinders C, transmission type T (automatic = 0, manual = 1), number of gears G, engine displacement in cubic inches DISP, horsepower HP, number of carburettor barrels CB, final drive ratio DRAT, and weight of the car in thousands of pounds WT. In the original analyses referenced above, quarter-mile time is taken as an explanatory variable for MPG, but it is not a basic design variable and is therefore omitted in this analysis of fuel consumption. A separate analysis of QMT as a response variable is given in Section 3.8.

Our object is to obtain a simple model relating MPG to the explanatory variables. The 32 cars are not a random sample of the car population to which this model can be generalized: the *Motor Trend* sample is heavily weighted to European and US high-performance sports cars and luxury cars.

We begin by plotting MPG against each of the explanatory variables. This can help identify outlying observations, suggest the important explanatory variables, and give a general feel for the data.

```
$INPUT 1 CAR1 $INPUT %PLC NORMAC LEV $
$PLOT MPG S : MPG C : MPG T : MPG G : MPG DISP $
```

Curvature in the plot against DISP is very noticeable, suggesting that a scale change is required. We try the log scale:

```
$CALC LMPG = %LOG(MPG) : LDIS = %LOG(DISP)
$PLOT LMPG LDIS $
```

The plot is now much more nearly linear. The plots against HP and WT also show curvature and these are similarly linearized by a log transformation:

```
$CALC LHP = %LOG(HP) : LWT = %LOG(WT)
$PLOT  LMPG LHP : LMPG LWT $
```

The plots do not show any marked outliers, and it is clear that HP, WT and DISP are important explanatory variables, as one might expect.

We now proceed to fit a model using all the explanatory variables, with HP, WT and DISP replaced by their log transformation. Should the model be fitted to MPG or log MPG, or perhaps 1/MPG–gallons per mile, the European standard for fuel consumption (litres per km) used by Henderson and Velleman (1981)? The Box–Cox family suggests a clear answer.

```
$INPUT %PLC BOXCOX $
$MACRO YVAR MPG $ENDMAC
$MACRO MODEL S+C+T+G+LDIS+LHP+CB+DRAT+LWT $ENDMAC
$USE BOXCOX $
```

Specifying the range -2 to 2 in steps of 0.5 in response to the prompts, we obtain the values of $-2p\ell(\lambda)$ reproduced below.

λ	-2	-1.5	-1	-0.5	0	0.5	1	1.5	2
$-2p\ell(\lambda)$	195.7	181.7	168.5	157.0	149.2	147.6	152.5	162.1	174.0

The minimum of 147.3 occurs near $\lambda = 0.4$, and the approximate 95% confidence interval for λ includes $\lambda = 0$ but neither $\lambda = -1$ or $\lambda = 1$. Thus the log scale for MPG is indicated.

Does the transformation of DISP, HP and WT to their logs affect the transformation of MPG? To check, we can repeat the Box–Cox transformation without transforming these variables.

```
$MACRO MODEL S+C+T+G+DISP+HP+CB+DRAT+WT $ENDMAC
$USE BOXCOX $
```

The values of $-2p\ell(\lambda)$ are now:

λ	-2	-1.5	-1	-0.5	0	0.5	1	1.5	2
$-2p\ell(\lambda)$	184.8	172.6	162.2	154.9	152.2	154.8	161.7	171.5	182.8

The minimum now occurs almost at $\lambda = 0$, and again neither $\lambda = -1$ nor $\lambda = 1$ is included in the approximate 95% confidence interval. The transformation of the explanatory variables has had very little effect on the estimation of λ, though $-2p\ell(\lambda)$ is reduced by 3 at $\lambda = 0$ by the log transformations, showing a slightly better fit.

We proceed then with an examination of the first model above, using LDIS, LHP and LWT. We first examine the residuals and influence values from the model.

```
$MACRO MODEL S+C+T+G+LDIS+LHP+CB+DRAT+LWT $ENDMAC
$USE BOXFIT $
```

Specifying $\lambda = 0$ gives the parameter estimates and the quantile plot for the model residuals, which is closely linear, apart from two rather large negative residuals. A plot of residuals against observation number shows that these large values are for the Cadillac and Continental, the luxury American cars. Which observations are influential?

```
$USE LEV $PRINT LEV_$
```

Here $p = 9$, $n = 32$ and $2(p+1)/n = 0.625$. Only two values exceed this, 0.6254 for observation 19 and 0.677 for observation 29. Observation 27 might also be considered influential with a value of 0.6175. What is unusual about these cars? The Ford Pantera (29) has the second largest HP, but has low gearing for its power—the third highest value of DRAT. Most of the powerful cars have low values of DRAT. The Honda Civic (19) has the lowest HP, the highest DRAT, and the second lowest DISP and WT. The Porsche 914 (27) has low

DISP and HP, the second highest DRAT and low WT. These three cars all lie on the edge of the HP/DRAT scatter, as can be seen from a plot of HP against DRAT. The Cadillac and Continental, which had the largest negative residuals, by contrast have very little influence, with values of 0.186 and 0.180.

What happens if we exclude the three observations with high influence? We return to this point below, but for the moment assume that no action is needed. We proceed to simplify the full regression model, using the simultaneous test procedure of Section 2.7. We specify a simultaneous test size α of about $1-(0.95)^p=0.370$, rounded to $\alpha=0.35$. Then any model using a subset of variables is adequate at level $\alpha=0.35$ if its residual sum of squares RSS_r does not exceed $(1+pF_{\alpha,p,n-p-1}/(n-p-1))RSS_f$, where RSS_f is the residual sum of squares from the full model, which is 0.2861, while $F_{0.35,9,22}=1.19$, so that the limit of RSS_r for adequate subsets is 0.4252.

A convenient model simplification procedure in GLIM is *backward elimination* (see Draper and Smith 1981 Ch.6 for a discussion of this and other methods), in which at each step the least important variable is dropped from the current model. Importance is assessed by the t-ratio $\hat{\beta}_j/\text{s.e.}(\hat{\beta}_j)$; these are not provided as part of the standard output in GLIM, but they may be obtained using the macro TVAL in the subfile NORMAC. Elimination of variables continues as long as the RSS from the model does not exceed the upper bound for an adequate subset.

Elimination can equivalently be based on the squared multiple correlation R^2. Any submodel is adequate if its R^2 does not fall below the value

$$R_0^2=1-(1-R^2_f)\{1+pF_{\alpha,p,n-p-1}/(n-p-1)\},$$

where R^2_f is the value of R^2 for the full model.

To fit the full model we have to declare the y-variate to be LMPG, since the BOXFIT macro resets the y-variate to that specified in the macro YVAR.

```
$YVAR LMPG
$FIT #MODEL $D E $
$USE RSQ $USE TVAL $USE PRESS $
```

We find $R^2=0.8959$, $R^2_{CV}=0.8081$. Here $R=0.9468$ is the correlation between the observed LMPG values and the fitted values from the full regression model, which is very high: the full model gives a close reproduction of the actual LMPG values. The estimate of σ^2 from the full model is 0.0130 with $s=0.114$. The lower limit R_0^2 is then 0.8453.

Backward elimination now proceeds from the full model in the following sequence, using at each step \$FIT – (VARIABLE) \$D E \$ \$USE RSQ \$ \$USE PRESS \$USE TVAL \$. At each step the smallest t-value in magnitude is listed and the corresponding variable, RSS, R^2, R^2_{CV} and s^2 values, and the cross-validation estimate CVE of σ^2:

Step	RSS	R^2	R^2_{CV}	s^2	CVE	Variable be to omitted	t
0	0.2861	0.8959	0.8081	0.0130	0.0171	C	-0.082
1	0.2862	0.8959	0.8298	0.0124	0.0150	DRAT	-0.320
2	0.2875	0.8954	0.8337	0.0120	0.0146	S	-0.340
3	0.2888	0.8949	0.8422	0.0116	0.0137	T	-0.461
4	0.2913	0.8940	0.8476	0.0112	0.0132	LDIS	-0.723
5	0.2971	0.8919	0.8529	0.0110	0.0127	CB	-1.070
6	0.3097	0.8873	0.8569	0.0111	0.0123	G	1.052
7	0.3220	0.8829	0.8549	0.0111	0.0125	LHP	-4.372
	0.5342	0.8057	0.7766	0.0178	0.0192		

At this point elimination ceases as the omission of LHP results in an inadequate model. The minimal adequate model found by backward elimination uses LHP and LWT. The fitted model for LMPG, with parameter standard errors, is

$$\hat{\mu} = \underset{(0.53)}{4.836} - \underset{(0.058)}{0.255}\ \mathrm{LHP} - \underset{(0.087)}{0.562}\ \mathrm{LWT}$$

while $s = 0.105$. This model could also be used for prediction, if this were appropriate: its values of s^2 and CVE are very slightly greater than the minimum values. It is notable that the cross-validation R^2_{CV} is initially substantially less than R^2, showing shrinkage of R^2 from the full model on cross-validation, but as the model is progressively simplified R^2_{CV} *increases*, though its maximum value of 0.8569 is still well below R^2 for the full model.

We noted above that three observations had substantial influence on the model. If these are deleted by being given weight zero in a weighted fit, we find that backward elimination leads to the same model with very similar estimated coefficients

$$\hat{\mu} = 4.928 - 0.268\ \mathrm{LHP} - 0.587\ \mathrm{LWT}$$

with slightly larger standard errors, reflecting the smaller sample size. Thus despite their high influence values these observations are not discrepant from the rest, and should not be excluded. We retain the preceding model as the final model.

It is tempting to interpret the final model causally (as did Aitkin and Francis, 1982): if horsepower could be increased by 10% (an increase of 0.095 in LHP) keeping weight constant, then predicted LMPG would decrease by

0.024, giving a 2.5% decrease in MPG. Whether this is possible or not, no such experimental manipulations occurred in the data. It is more realistic to think of the final model as a simple, parsimonious and accurate representation of the fuel consumption results for this set of cars. It may not generalize to other cars, or to the same cars tested in other years, because of the selective sample of cars on which the model has been based. Consequently, this model is inappropriate for prediction.

It is important to stress that backward elimination is only one way of proceeding to simplify the model. *Forward selection* and *stepwise* procedures (Draper and Smith, 1981 Chapter 6) are in more common use, but since GLIM does not have a look-ahead facility for determining the best variable to add to a model, these procedures are computationally tedious in GLIM. Generating all possible regressions (2^p models with p explanatory variables) is feasible if p is not too large; since many of these models will not be adequate the amount of computation can be substantially reduced by careful bookkeeping. We have followed the practice in this book of not searching extensively for additional models; it sometimes happens however that several parsimonious and adequate models exist using quite different subsets of variables. Causal interpretations in such cases are extremely hazardous.

We conclude by noting that the model predicts median MPG by

$$\exp(\hat{\mu}) = 126.0\ HP^{-0.255}\ WT^{-0.562},$$

the prediction of median GPM is just the reciprocal of this.

3.8 Sensitivity of the Box–Cox transformation to outliers

Regression diagnostics allow us to identify influential observations which may substantially affect the position of the fitted model. The extension of the model to include the Box–Cox transformation parameter raises the possibility that the estimate of this parameter may also be strongly affected by influential observations. The identification of observations influencing the transformation parameter is discussed by Atkinson (1982); he describes the use of an additional "constructed variable" in the model. The need for caution in the use of the Box–Cox transformation is clearly shown in the next example, the modelling of acceleration time QMT in the CAR1 data.

```
$INPUT 1 CAR1 $INPUT %PLC BOXCOX $
$MACRO YVAR QMT $ENDMAC
$MACRO MODEL S+C+T+G+DISP+HP+CB+DRAT+WT $ENDMAC
$USE BOXCOX $
```

We first use the explanatory variables DISP, HP and WT without transformation. Tabulating over the usual range -2 to $+2$, we find that $-2p\ell(\lambda)$ decreases monotonically with λ, and is still decreasing rapidly at $\lambda=-2$. Extending the grid, the minimum is found to occur at about $\lambda=-3.5$. (We should note that it is possible for numerical instability to occur for large negative values of λ. This problem, however, is easily resolved by scaling, e.g. by dividing y by 10.)

How should we interpret this unusual value? A normal quantile plot on this scale ($USE BOXFIT with argument -3.5) shows a good straight-line fit, with no marked outliers, but the unusual estimate of λ makes us suspect an error in the data. Refitting the model on the log scale with $\lambda=0$ gives a quantile plot with a large negative residual for observation 9, the Mercedes 230. Background knowledge of the scientific field is helpful, as in the tree example. The Mercedes has the slowest of all acceleration times, and is substantially slower than the diesel Mercedes 240D, which is much less powerful, has higher gearing, and is slightly heavier. The body shapes are identical, so this slow acceleration is very hard to understand. One possible explanation is that the acceleration times of these two cars have been reversed at some point in their publication, or in transcription from *Motor Trend* magazine.

We now exclude the Mercedes 230 from the analysis.

```
$CALC W = 1 : W(9) = 0 $ WEIGHT W
$USE BOXCOX $
```

The Box–Cox macros allow for the weighting-out of observations, so the likelihood calculations and plotting are still correct. The minimum of $-2p\ell(\lambda)$ of 43.95 is now very close to $\lambda=-1.5$, while the values at $\lambda=-1$, 0 and 1 are 44.19, 45.98 and 49.21 respectively. Both the reciprocal and log transformations of QMT are appropriate but QMT itself is not. Thus the omission of a single observation has considerably changed the conclusion about the appropriate scale of y. The large outlier is only accommodated in a normal model by an extreme transformation of scale.

The reciprocal transformation corresponds to using average speed over the quarter-mile as the response variable. On the reciprocal scale the fitted full model gives $s^2=2.056\times10^{-6}$ and an R^2 of 0.9468, and backward elimination as in Section 3.5 leads to the minimal adequate (0.35) model below for $1/QMT$:

$$\begin{array}{ccccc} \hat{\mu}=0.052 & -\ 0.00536\ \mathrm{S} & +0.00178\ \mathrm{G} & +(5.389\times10^{-5})\ \mathrm{HP} & -0.00231\ \mathrm{WT} \\ (0.003) & (0.00078) & (0.00050) & (0.686\times10^{-5}) & (0.00050) \end{array}$$

with $s^2=2.169\times10^{-6}$ and an R^2 of 0.9305. The interpretation of the coefficients is hindered by their small size; if we note that SP$=900/$QMT is the average speed in miles per hour (1 mph; 1.6 kph) over the distance, the corresponding estimates and standard errors for the model for SP are just 900 times the above values:

$$\begin{aligned}\hat{\mu}_{\mathrm{SP}} &= 900\ \hat{\mu}\\ &= \underset{(2.5)}{46{\cdot}9} - \underset{(0.70)}{4.82\ \mathrm{S}} + \underset{(0.45)}{1.60\ \mathrm{G}} + \underset{(0.0062)}{0.0485\ \mathrm{HP}} - \underset{(0.45)}{2.08\ \mathrm{WT}}\end{aligned}$$

Now we examine the effect of log transforming HP, DISP and WT as in Section 3.5

```
$CALC LHP = %LOG(HP) : LWT = %LOG(WT)
: LDIS = %LOG(DISP) $
$MACRO MODEL S+C+T+G+LDIS+LHP+CB+DRAT+LWT $ENDMAC
$USE BOXCOX $
```

The minimum of $-2p\ell(\lambda)$ now occurs between $\lambda=0$ and $\lambda=0.5$, with values at $\lambda=-1, 0$ and 1 of 43.08, 41.20 and 41.79 respectively. All three values are now plausible for λ, though $\lambda=0$ is better supported than $\lambda=-1$. The effect of log transforming the explanatory variables has been to reduce $-2p\ell(\lambda)$ at $\lambda=-1$, 0 and 1 by 1.19, 4.78 and 7.42 respectively. Thus if log QMT or QMT itself are to be used, the explanatory variables should be transformed; if 1/QMT is to be used, the transformation of these variables has little effect.

It can be verified that backward elimination using log QMT or QMT leads to minimal adequate (0.35) models using the same four variables plus DRAT. Omission of DRAT gives a model in each case which is not minimal adequate (0.35). The simpler model obtained with 1/QMT gives a slight preference for this transformation.

```
$END
```

3.9 The use of regression models for calibration

In some applications of regression, we want to estimate a new value x_{n+1} of a single *explanatory* variable, given data (y_i, x_i) and a new observed *response* value y_{n+1}. This *calibration problem* arises, for example, when x is a precise but expensive measurement, and y a cheap but imprecise measurement of the same quantity. The calibration sample (y_i, x_i) is chosen to cover the range of values of x of interest to the experimenter.

A likelihood treatment of the problem is given by Minder and Whitney (1975); Brown (1982) gives a recent discussion and references. Confidence intervals for x_{n+1} are easily obtained. For simplicity of notation we write $y=y_{n+1}$, $x=x_{n+1}$. Then

$$y-\hat{\alpha}-\hat{\beta}x \sim N(0, \sigma^2+v_{11}+2v_{12}x+v_{22}x^2)$$

where

$$V=(v_{jk})=\sigma^2(X'X)^{-1}.$$

Then

$$(y-\hat{\alpha}-\hat{\beta}x)/(s^2+\hat{v}_{11}+2\hat{v}_{12}x+\hat{v}_{22}x^2)^{\frac{1}{2}}\sim t_{n-2}$$

and therefore with probability $1-\alpha$,

$$|y-\hat{\alpha}-\hat{\beta}|<t_{\alpha/2,n-2}\,(s^2+\hat{v}_{11}+2\hat{v}_{12}x+v_{22}x^2)^{\frac{1}{2}}.$$

This is equivalent to the quadratic inequality

$$Q(x)=x^2(\hat{\beta}^2-c\hat{v}_{22})-2x(\hat{\beta}(y-\hat{\alpha})+c\hat{v}_{12})+(y-\hat{\alpha})^2-c(s^2+\hat{v}_{11})<0$$

where

$$c={t^2}_{\alpha/2,n-2}.$$

V is the estimated covariance matrix of the parameter estimates, available from GLIM using $D V $. The confidence interval for x is defined by the roots of $Q(x)=0$.

As a numerical illustration, for the data in Chapter 2 on completion times for the Block Design test, we obtained a minimal adequate model using only EFT (model 3). Suppose we have an additional observation with completion time equal to 400 seconds. What can we say about EFT for this observation?

```
$INPUT 1 SOLV $
$YVAR TIME $FIT EFT  $ACC 8 $D E V $
```

Increased accuracy is required to calculate the roots of the quadratic: the four digits for $\hat{\beta}$ and $\hat{v}_{11}$ are not sufficient. The estimates are

$$\begin{array}{lll} \hat{\alpha}=271.128 & \hat{v}_{11}=1845.44 & \\ \hat{\beta}=2.0405 & \hat{v}_{12}=-24.490 & \hat{v}_{22}=0.441594 \\ y=400 & s^2=11694.3 & \end{array}$$

and for a 95% confidence interval, $t_{0.025,22}=2.074$, so $c=4.301$. The quadratic in x is

$$Q(x)=2.264x^2-2(157.6)x-41626.4$$

and the equation $Q(x)=0$ has roots -82.8 and 222.0. For $Q(x)$ to be negative, x must be between the roots so the 95% interval for x is $(-82.8, 222.0)$.

Negative values are impossible from the nature of EFT, which is an unsatisfying feature of the confidence interval construction procedure. (This difficulty can be avoided by constructing a profile likelihood for x over the range of possible values of x. See Appendix 1. Minder and Whitney construct a marginal likelihood for x.) The obvious estimate for x is $(y-\hat{\alpha})/\hat{\beta}$ which is 63.2; the interval is extremely wide here because of the large residual variation about the regression line. For regressions which are not well determined, it is possible for $Q(x)$ to have no real roots, with $Q(x)<0$ for *all* x. Then *all* values of x are consistent with the observed y. This may seem strange, but the special case $\hat{\beta}=0$ makes the result obvious: if $\hat{\beta}$ is actually zero, then y can provide no information about x, and a value of $\hat{\beta}$ near zero (relative to its standard error) will lead to the same result.

```
$END
```

3.10 Measurement error in the explanatory variables

We noted in Chapter 2 that an important assumption of the model was that the explanatory variables are measured without error. In survey studies in the social sciences, this is frequently not the case: many variables contain measurement error, in the sense that repeated measurements of the same individual under the same circumstances give different values of the variable. In psychometric testing it is standard practice to quote the reliability of a test, which is the correlation of two parallel measurements on the same individual under identical conditions. For example, the quoted reliability for the Block Design Subtest of the WISC is 0.87 for ten-year-old (American) children (Wechsler 1949). How should measurement error be allowed for in the model and the analysis? It is not possible to allow for measurement error when modelling with GLIM except when replicate or "parallel" measurements on the unreliable explanatory variables are available: we give therefore only a short discussion of the problem.

Consider the simple linear model with one explanatory variable. Let t_i be the true value of the explanatory variable for the i-th observation, assuming it could be measured without error, and suppose that the observed value x_i is related to t_i by

$$x_i = t_i + \xi_i$$

where ξ_i is the measurement error, which we will assume to be normally distributed $N(0,\sigma_m^2)$, independently of the random component of the model for y.

One possibility, perhaps the most commonly used, is to ignore measurement error. Suppose y is a student achievement test at the end of an academic year, and x is a measure of initial ability, for example Stanford–Binet IQ or Verbal Reasoning Quotient. We might argue that the model

$$y_i|x_i \sim N(\beta_0 + \beta_1 x_i, \sigma^2)$$

is still appropriate despite unreliability in x, because x is the only estimate we have of the student's initial achievement, and the regression has to be based on observable data. For a descriptive use of the model, this may be reasonable, but if the observations come from an experimental design (as in the first example of Chapter 2) from which causal conclusions are to be drawn, the ignoring of measurement error may give quite misleading conclusions about experimental variables in the model.

To proceed further we need to make an assumption about the distribution of t_i, the unobserved true value of x_i. We will assume they are normally distributed $N(\mu,\sigma_t^2)$. Then the complete model is

$$y_i|t_i \sim N(\beta_0 + \beta_1 t_i, \sigma^2)$$
$$x_i|t_i \sim N(t_i, \sigma_m^2)$$
$$t_i \sim N(\mu, \sigma_t^2).$$

Write $\sigma_x^2 = \sigma_t^2 + \sigma_m^2$. Then the marginal distribution of x_i *is* $N(\mu,\sigma_x^2)$, and the conditional distribution of t_i given x_i is

$$t_i|x_i \sim N(\rho x_i + (1-\rho)\mu,\ \rho(1-\rho)\sigma_x^2)$$

where

$$\rho = \sigma_t^2/\sigma_x^2$$

is the *reliability* of the measurement x_i. The conditional distribution of y_i given x_i is then

$$y_i|x_i \sim N(\beta_0 + \beta_1(\rho x_i + (1-\rho)\mu),\ \sigma^2 + \beta_1{}^2\rho(1-\rho)\sigma_x^2)$$
$$= N(\beta_0{}^* + \rho\beta_1 x_i,\ \sigma^2 + \beta_1{}^2\rho(1-\rho)\sigma_x^2)$$

where

$$\beta_0{}^* = \beta_0 + (1-\rho)\beta_1\mu.$$

Thus the slope of the regression of y_i on x_i is not β_1 but $\rho\beta_1$, and the standard least squares estimate of β_1 systematically underestimates the true parameter value (i.e. is inconsistent). The estimates of any other parameters in a regression function will also be inconsistent. Further, the variance of the residuals, though constant, also depends on β_1, so that maximum likelihood estimation of β_1 *even if ρ is known* is not equivalent to least squares estimation.

A consistent estimate of the regression coefficient β_1 can be obtained by dividing the least squares estimate by ρ; the estimate obtained however will not be efficient since the additional information about β_1 in the variance is neglected; further, the other parameters in the model are not correctly estimated.

An extensive discussion of consistent estimation in the general regression model with measurement error in *all* the explanatory variables is given by Fuller and Hidiroglou (1978). Maximum likelihood estimation in such models can be achieved in several cases: when reliabilities or error variances are known; when the ratio of error variances to residual variance in y is known; or when 'parallel' measurements of the unreliable explanatory variables are available (from parallel forms or split-half measures). Quite general models with measurement error in the explanatory variables can be fitted by maximum likelihood using LISREL (Jöreskog and Sörbom, 1981).

3.11 Factorial designs

We noted in Chapter 2 that the choice of an appropriate model is greatly simplified in *orthogonal factorial designs* because the analysis of variance table is unique. Orthogonal factorial designs have many other well-known advantages, which we will not discuss here. In many sample surveys with categorical explanatory variables, formally similar *cross-classifications* result, but the sample sizes in the cells of the table are generally unequal, and sometimes show marked associations with the explanatory variables. In such cases great care is necessary in model simplification and interpretation. A simple example was discussed in Sections 2.11–2.14.

We conclude this chapter with two examples: an orthogonal two-factor design discussed by Box and Cox (1964) and an unbalanced four-way cross-classification discussed by Aitkin (1978).

Subfile POISON contains the survival times of rats in units of 10 hours after poisoning. The design is a completely randomized 3×4 factorial with three replicates, with three types of poisons and four treatments.

```
$INPUT 1 POISON $INPUT %PLC BOXCOX QPLOT $
$YVAR TIME $FIT : +TYPE : +TREAT : +TYPE.TREAT $D E $
```

The analysis of variance table is

Source	SS	df	MS	F
TYPE	1.0330	2	0.5165	23.27
TREAT	0.9212	3	0.3071	16.71
Interaction	0.2501	6	0.0417	1.88
Residual	0.8007	36	0.0222	

We do not need to use the simultaneous test of Section 2.7 because the sums of squares of the main effects are invariant under reordering. Each mean square of the main effects are invariant under reordering. Each mean square can be compared directly with the residual mean square by the usual F-test. (If a large number of separate tests is to be carried out in this way, some control over the simultaneous Type I error rate is desirable, and can be simply achieved by dividing the overall test size γ by the number of tests—Bonferroni testing.) The main effects are highly significant, the interaction not significant at the 10% level ($F_{0.10,6,36}=1.945$), though the interaction parameter for TYPE(3).TREAT(2) is rather large (-0.3425 with s.e. 0.1491).

Inspection of the mean survival times is possible, but not convenient, from the parameter estimates for the interaction model. A table of means can be constructed using the $TABULATE directive (not available in GLIM3.12).

```
$TABULATE THE TIME MEAN FOR TYPE ; TREAT $
```

The means are arranged by rows for TYPE and columns for TREAT.

	1	2	3	4
1	0.4125	0.8800	0.5675	0.6100
2	0.3200	0.8150	0.3750	0.6675
3	0.2100	0.3350	0.2350	0.3250

Since the design is balanced, the main effects can be inspected for simple one-way tabulations of means:

```
$TABULATE THE TIME MEAN FOR TYPE $
$TABULATE THE TIME MEAN FOR TREAT $
```

The TYPE means are 0.6175, 0.5444, 0.2763 and those for TREAT are 0.3142, 0.6767, 0.3925 and 0.5342.

Tabulation of the variances is always useful in cross-classifications:

```
$TABULATE THE TIME VARIANCE FOR TYPE ; TREAT $
```

	1	2	3	4
1	0.0048250	0.0258667	0.0245583	0.0127333
2	0.0056667	0.1131000	0.0032333	0.0734250
3	0.0004667	0.0021667	0.0001667	0.0007000

There are remarkable disparities in variance, with $s^2_{max}/s^2_{min}=0.1131/0.000167=677.2$. This is very large compared with $F_{max,\ 0.01,12,3}=361$. (An ordinary $F_{3,3}$-test is not appropriate because this value is the *largest* of all the possible variance ratios; the F_{max} test due to Hartley (1950) must be used. Critical values can be found in the *Biometrika Tables* by Pearson and Hartley

(1966).) Further, the small variances are associated with small means, as is easily seen from a plot of the data. Stronger evidence comes from a χ_3^2 quantile plot of the ordered variances, shown in Fig. 3.1a. The plotting points are $(q_i, s^2_{(i)})$ where $s^2_{(i)}$ is the i-th smallest variance, and we take

$$q_i = F^{-1}((i-0.3175)/(n+0.365))$$

as for normal quantile plotting, but F is now the c.d.f. of the χ_3^2 distribution (given for example in Lindley and Scott, 1984). Also plotted is the line through the origin $s^2 = q\tilde{\sigma}^2/3$, where $\tilde{\sigma}^2$ is the residual mean square from the full model. The points should fall close to the line since each variance should be distributed as $(\sigma^2\chi_3^2)/3$ if the model is correct. (This plot cannot be constructed in GLIM as the inverse χ^2 c.d.f. is not available.)

The assumption of constant variance is clearly violated, and the mean–variance association suggests the scale of the response is wrong. Further evidence comes from a normal quantile plot of the full model residuals.

```
$USE QPLOT $
```

The plot shows pronounced curvature, and the Filliben correlation is 0.9471, far below the 1% point. We look for an appropriate scale.

```
$MACRO YVAR TIME $ENDMAC
$MACRO MODEL TYPE*TREAT $ENDMAC
$USE BOXCOX $
```

The value of $-2p\ell(\lambda)$ changes rapidly over the range -2 to $+2$, and a finer tabulation is necessary over the range -1.5 to 0.5. The minimum occurs close to $\lambda = -0.8$, and the 95% confidence interval is approximately $(-1.29, -0.34)$ which includes $\lambda = -1$ but not $\lambda = 0$ or 1. The reciprocal scale is clearly indicated; the response variable on this scale can be interpreted as the rate of dying.

```
$CALC RATE = 1/TIME   $YVAR RATE
$FIT : +TYPE  : +TREAT : +TYPE.TREAT $
```

The analysis of variance table is

Source	SS	df	MS	F
TYPE	34.877	2	17.439	72.46
TREAT	20.414	3	6.805	28.35
Interaction	1.571	6	0.262	1.09
Residual	8.643	36	0.240	

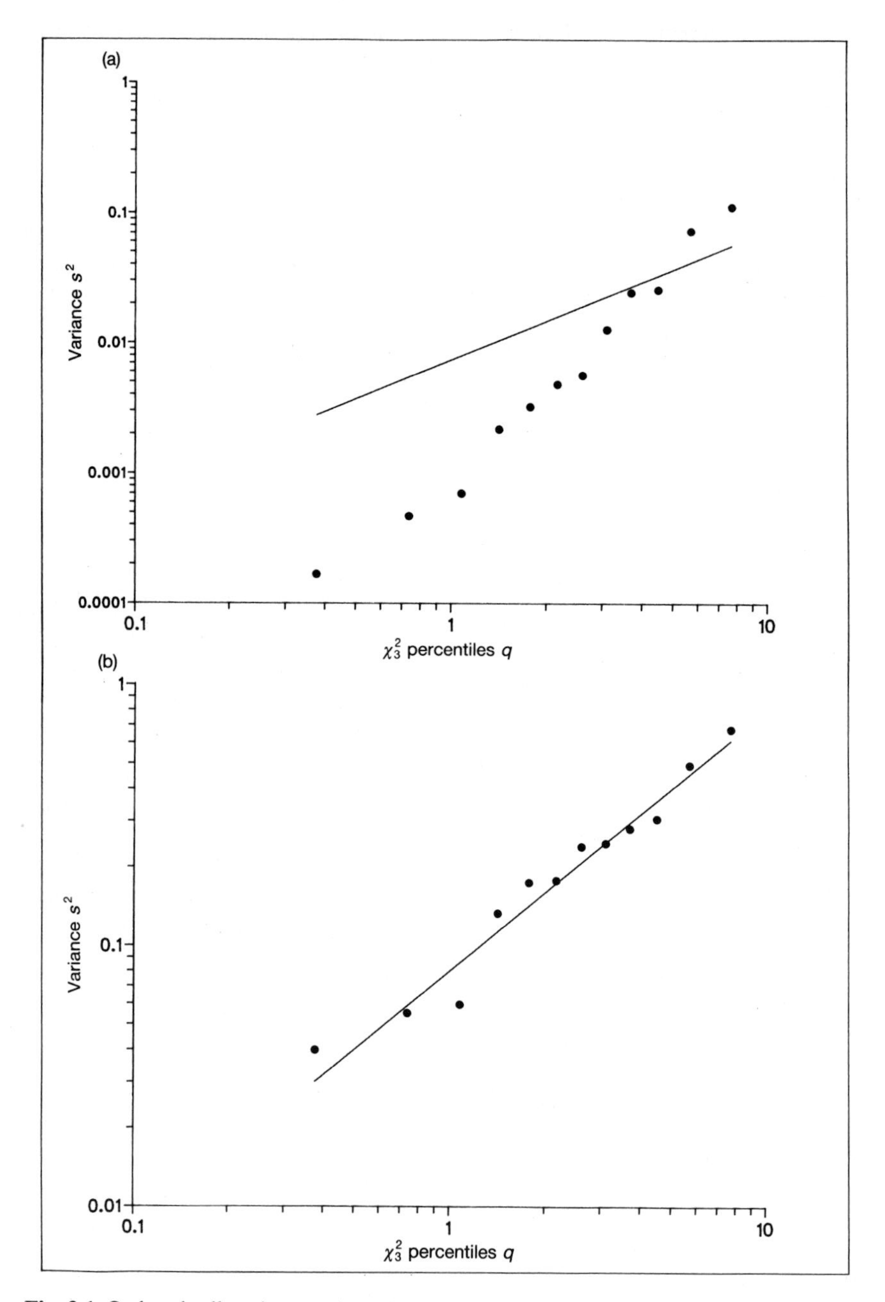

Fig. 3.1. Ordered cell variances plotted against percentiles of χ_3^2; (a) for time, (b) for rate

The F-values for the main effects are now much larger than they were before, but that for interaction is smaller: the interaction mean square is almost the same size as the residual mean square.

```
$TABULATE THE RATE MEAN FOR TYPE ; TREAT $

      1      2      3      4
1  2.487  1.163  1.863  1.690
2  3.268  1.393  2.714  1.702
3  4.803  3.029  4.265  3.092

$TABULATE THE RATE VARIANCE FOR TYPE ; TREAT $

      1         2         3         4
1  0.24667  0.03980  0.23949  0.13302
2  0.67622  0.30602  0.17432  0.49267
3  0.28051  0.17761  0.05514  0.05956
```

The variance heterogeneity is greatly reduced, with $s^2_{max}/s^2_{min} = 16.99$, much less than $F_{max\,0.05,12,3} = 124$. More importantly, there is no association between cell mean and variance. A χ^2_3 quantile plot of the variances is shown in Fig. 3.1b, with the theoretical straight line which now fits closely to the plotted points. A normal quantile plot of the model residuals also shows close linearity, with a Filliben correlation of 0.9855, which is at about the lower 25% point. The reciprocal transformation therefore seems satisfactory.

The rate means for TYPE are 1.801, 2.269 and 3.797, and those for TREAT are 3.519, 1.862, 2.947 and 2.161. Transforming back the means, the fitted *median* survival times are 0.555, 0.441 and 0.236 for the poison types, and 0.284, 0.537, 0.339 and 0.463 for the treatments.

The effect of the reciprocal transformation on the precision of the analysis in this example is important. The assumption of constant variance on the original scale was invalid, and the pooled variance estimate on that scale was inflated by the large cell variance estimates for the cells with large means. The reciprocal transformation stabilizes the variance estimates, increasing the precision of the analysis.

We saw earlier in this chapter that the estimate of λ on the Box–Cox transformation could be affected by a change in model. What happens if we fit the main effects model instead of the interaction model?

```
$MACRO MODEL TYPE+TREAT $ENDMAC
$USE BOXCOX $
```

The minimum of $-2p\ell(\lambda)$ now occurs at -0.75, and the 95% confidence interval is $(-1.14, -0.36)$. The model change has had very little effect.

Box and Cox (1964) point out that the transformation in this case is trying simultaneously to achieve three ends: normality of the residuals, additivity of the means and homogeneity of variance, on the transformed scale. By including the interaction, we require the transformation to attempt to satisfy only two of these requirements. We could relax the requirements further by fitting the interaction model with a different variance term for each cell, thus removing the requirement of homogeneity of variance. Box and Cox discuss this; we do not pursue it here because the model with non-homogeneous variances cannot easily be specified in GLIM. The requirement of homogeneity of variance is a strong one; without it, any value of λ from below -1 to about 2.5 is plausible (Box and Cox 1964, p. 236).

```
$END
```

3.12 Cross-classifications

Subfile QUINE contains data on absence from school, from a sociological study of Australian Aboriginal and white children by Quine (1975). Quine sampled children from four age groups (final grade in primary schools and first, second and third form in secondary schools), with nearly equal numbers of the two sexes and equal numbers from the two cultural groups. The children in each group were classified as slow or average learners; in the third form group there were no slow learners.

The response variable of interest was the number of DAYS absent from school during the school year. Children who had suffered a serious illness during the year are excluded from the analysis, so the original balance in cultural group, and near balance in sex are lost. The structure of the data is an unequally replicated (unbalanced) $4 \times 2 \times 2 \times 2$ cross-classification with four empty cells, and 146 observations. The cross-classifying factors are defined as age A, sex S, culture C and learner L. The factor coding is given with the data.

We first inspect the data by tabulating the cell means and variances.

```
$INPUT 1 QUINE $
$TABULATE THE DAYS MEAN FOR C;S;A;L $
$TABULATE THE DAYS VARIANCE FOR C;S;A;L $
$TABULATE FOR C;S;A;L $
```

The last directive gives a count of the number of observations in each cell. The second table shows great heterogeneity of variance, with $s^2_{\max}/s^2_{\min} = 2114$. The $F_{\max}$ test is not appropriate because the degrees of freedom are not the same for all variances, but it is clear that a transformation is needed: variance increases systematically with mean as in the previous example. The χ^2 quantile plot cannot be used because the cell variances have different numbers of degrees of freedom. A normal quantile plot of the model residuals shows a clear need for a transformation.

```
$INPUT %PLC QPLOT $
$YVAR DAYS  $FIT C*S*A*L $USE QPLOT $
```

The Filliben correlation is 0.9790, which is highly significant (although the table in Filliben extends only to $n=100$, the value 0.979 is at the 0.005 point for this sample size, and will be much further below for larger sample sizes).

In applying the Box–Cox transformation, however, we encounter a difficulty: zero values in the data. The log and negative power transformations of these values would be infinite, and would be set to zero in GLIM with a warning message. Further, the likelihood in Section 3.1 involves the term $(\lambda-1)\Sigma \log y_i$. The BOXCOX macro can only be used with a positive random variable. What should we do?

We consider five possibilities:

(1) try different link functions, as in the tree volume example of Section 3.2;

(2) use a two-parameter Box–Cox transformation (Box and Cox, 1964) in which $(Y+\lambda_2)^{\lambda_1}$ is assumed normal for suitable $\lambda_2>0$ and λ_1;

(3) add some small constant to the zeroes, and apply the standard Box–Cox transformation;

(4) estimate the transformation parameter from a variance–mean plot;

(5) treat days absent as a discrete rather than a continuous response variable, and fit a Poisson model. Discussion of this possibility is postponed to Chapter 5.

The first possibility can be quickly disposed of. A different link function will affect the fitted values and residuals from the model, but it will not much affect the variance heterogeneity, which is defined by the observed response values in each cell. This can be affected only by a response variable transformation.

The second possibility is discussed at length by Box and Cox. The positive parameter λ_2 added to Y ensures that the Box–Cox transformation is being applied to a positive variable; the profile likelihood function in λ_1 can then be tabulated over a grid of values of λ_2 to identify both the joint maximum likelihood estimate of (λ_1,λ_2), and a confidence region for the two parameters.

This is easily achieved. We begin (arbitrarily) with $\lambda_2=0.1$.

```
$INPUT %PLC BOXCOX $
$MACRO MODEL C*S*A*L $ENDMAC
$CALC D = DAYS + 0.1 $
$MACRO YVAR D $ENDMAC
$USE BOXCOX $
```

Initial tabulation over the usual grid for λ_1 shows that the likelihood is highly concentrated near $\lambda_1=0.5$; a tabulation in steps of 0.025 over (0.25,0.5) is necessary. The minimum of $-2p\ell(\lambda)$ of 1348.9 occurs at $\hat{\lambda}_1=0.35$, when $\lambda_2=0.1$.

```
$CALC D = DAYS + 0.2  $USE BOXCOX $
```

Repeating with $\lambda_2=0.2$, the larger minimum of 1352.7 occurs at $\hat{\lambda}_1=0.325$. Reducing the value of λ_2 to 0.05 gives a minimum of 1344.4 at $\hat{\lambda}_1=0.375$, and reducing it to $\lambda_2=0.01$ gives 1331.8 at $\hat{\lambda}_1=0.375$. These values are summarized below.

λ_2	$\hat{\lambda}_1$	Deviance
0.2	0.325	1352.7
0.1	0.35	1348.9
0.05	0.375	1344.4
0.01	0.375	1331.8

Thus the smaller the value of $\lambda_2>0$, the greater the likelihood. The reason for this result is that for $y_i=0$, as $\lambda_2\to 0$ the terms $(\lambda_1-1)\log(y_i+\lambda_2)$ are increasing rapidly, and are "driving" the value of the likelihood and the location of its maximum. Thus a value of λ_1 around 0.35, though strongly indicated by the two-parameter Box–Cox transformation, may be misleading.

The same effect occurs with the third possibility: if some small number δ (like 0.01) is added to the zero values only, then $\log y$ for these values will again be very large, while the values of $\log y$ and $\log(y+\lambda_2)$ for $y>0$ and small λ_2 will be almost identical.

The fourth possibility follows from the relation between the mean and variance of y in Section 3.1. To a first degree of approximation, if σ is small compared with μ, then when $y^\lambda \sim N(\mu,\sigma^2)$,

$$E(y) \approx \mu^{1/\lambda}$$

$$\text{var}(y) \approx \mu^{2/\lambda-2}\sigma^2/\lambda^2$$

so that the standard deviation of y, $SD(Y)$, is given by

$$SD(y) \approx \mu^{1/\lambda-1}\sigma/\lambda$$

$$= \{E(y)\}^{1-\lambda}\sigma/\lambda$$

and hence

$$\log SD(y) \approx (1-\lambda)\log E(y) + \text{const},$$

so a plot of log $SD(y)$ against log $E(y)$ should have slope approximately $1-\lambda$. Thus an estimate of λ is {1 minus the slope of the regression of log $SD(y)$ on log $E(y)$}.

The slope can be estimated by fitting the regression of log s on log $\bar{y}$ for each cell of the cross-classification. This procedure could have been used on the POISON data discussed in Section 3.11, where we noted the great heterogeneity of variance. Fitting the model of log (mean) to the response variable log (SD) for that example, the estimated slope $1-\lambda$ is 1.977, with standard error 0.263. Thus the estimate of λ is -0.977, clearly pointing to the reciprocal transformation. The approximate 95% confidence interval for λ based on $\hat{\lambda}$ plus or minus two standard errors is of doubtful accuracy here because the values of log (SD) are not normally distributed (though in large samples they are close to normality). Even so the interval (-0.977 ± 0.526), i.e. $(-1.50, -0.45)$ agrees reasonably with the approximate 95% confidence interval $(-1.29, -0.34)$ from the Box–Cox transformation, despite the use of cell means and variances rather than the complete sample.

The standard deviation/mean plot can be constructed conveniently using the $TABULATE directive by saving the table means, standard deviations and sample sizes in vectors. We illustrate with the Quine data.

```
$TABULATE THE DAYS MEAN FOR C;S;A;L
    INTO TMEAN USING TN $
$TABULATE THE DAYS DEVIATION FOR C;S;A;L
    INTO TSD $
$CALC LSD = %LOG(TSD) : LM = %LOG(TMEAN)
 : W = %GT(TN,1) $
$UNIT 32
$YVAR LSD $WEIGHT W $FIT LM $D E $
```

Note that the sample sizes are calculated explicitly as weights in the first tabulate directive and are saved in the vector TN. Cells for which TN = 0 or 1 are weighted out in the fit. The $UNIT length has to be reset before variates of non-standard length can be used as arguments to $YVAR or $FIT. The estimated slope is $1-\lambda = 1.013$, with s.e. 0.136; an approximate 95% interval for λ is $(-0.285, 0.259)$. The log transformation is clearly indicated. This has the reasonable interpretation of multiplicative effects of the factors on absence.

The problem remains of what adjustment to make to the zero values, or all the data, in order to use the transformation. A small constant added to the zero values will, as we have seen, have a considerable effect on the fitted model; we decide, rather arbitrarily, to add 1.0, the next smallest value, to all the observations.

```
$UNIT 146
$CALC D1 = DAYS + 1 $
$MACRO YVAR D1  $ENDMAC
$MACRO MODEL  C*S*A*L  $ENDMAC
$USE BOXCOX $
```

The minimum of $-2p\ell(\lambda)$ of 1339.52 occurs at $\lambda=0.2$, while at $\lambda=0$ the value is 1369.84, an increase of 10.3. The log transformation is thus not well supported, but for the reasons given above we proceed with it.

```
$USE BOXFIT $USE QPLOT $
```

Specifying $\lambda \neq 0$, the quantile plot gives a Filliben correlation of 0.9883, which is rather low: for $n=100$ this value is at the lower 7% point of the distribution, and will be lower for larger n. A noticeable curvature is present in the plot as well. The transformation does not look altogether satisfactory.

```
$CALC LD1 = %LOG(D1) $
```

Repeating the sequence of tabulate directives above for the variable LD1, we find that the variance heterogeneity is greatly reduced, with $s^2_{\max}/s^2_{\min}=78.1$. A plot of log (SD) against log (mean) shows no clear association, the slope of the regression of log (SD) on log (mean) being 0.392, with s.e. 0.309. Thus the log transformation has achieved homogeneity of variances, though the curvature in the quantile plot shows that normality has not been adequately achieved.

We proceed then with simplification of the model. Elimination of terms proceeds by backward elimination much as in the general regression model discussed in Sections 2.7, 3.7 but with some changes.

First, the natural ordering of main effects, two-way interactions, three-way interactions, . . . means that elimination of terms from the model respects their hierarchical order: main effects are not eliminated when their interactions are retained (except in certain cases illustrated in this example).

Second, the detailed examination of individual parameter estimates in the model can often be avoided by an initial sequence of $FIT+ directives from the null model to the saturated model, giving an analysis of variance table for the particular sequence of terms used. Inspection of this table will often suggest the important terms in the model.

Third, terms with multiple degrees of freedom (like age in this example) may need to be split into individual degree-of-freedom components, since a moderate sum of squares for several pooled degrees of freedom may contain one large single component which should not be dropped from the model. The parameter estimates for effects with multiple degrees of freedom should therefore be carefully examined.

The Quine data provide a good illustration of these points. In the full model for DAYS there are 28 estimable parameters. Four interactions—A(4).L(2) and its interactions with C and S—are not estimable since there are no slow learners in the last age group. These are *extrinsically aliased* in GLIM: their non-estimability is detected and they are set equal to zero: that is, effectively deleted from the model. The value of $1-0.95^{27}$ is 0.750; such a large type I error rate implies substantial overfitting of the model. The value of $1-0.975^{27}$ is 0.495 which is still high, but we will choose $\alpha=0.50$ as the level of the simultaneous test for the complete model, bearing in mind that further terms may possibly be omitted. The residual sum of squares for the full model is 97.32, with $s^2=0.825$; the upper limit of the residual sum of squares for an adequate (0.50) model is 97.32 $(1+27F_{0.50,27,118}/118)$ which is 119.18.

We begin with a complete sequence of fits, in an order which is essentially arbitrary.

```
$YVAR LD1
$FIT : +C : +S : +A : +L : +C.S : +C.A : +C.L : +S.A : +S.L
  : +A.L : +C.S.A : +C.S.L : +C.A.L : +S.A.L : +C.S.A.L $
```

The analysis of variance table is

Source	SS	df	MS
C	16.14	1	16.14
S	0.58	1	0.58
A	4.39	3	1.46
L	0.73	1	0.73
C.S	0.16	1	0.16
C.A	8.12	3	2.71
C.L	0.02	1	0.02
S.A	12.27	3	4.09
S.L	0.89	1	0.89
A.L	3.71	2	1.86
C.S.A	3.14	3	1.05
C.S.L	5.38	1	5.38
C.A.L	3.34	2	1.67
S.A.L	2.31	2	1.16
C.S.A.L	0.94	2	0.47
Residual	97.32	118	0.8248

For this sequence of terms, the residual SS is below 119.18 for all models below C.L. However the minimal model in this sequence cannot be regarded as satisfactory since it excludes C.S.L, a large interaction (in the order fitted). We note large mean squares for S.A and C.A as well, and a very large mean square for C.

The sum of squares for C.S.A.L with 2 df only just exceeds the residual mean square, so it is clear that C.S.A.L can be dropped from the model. Similarly S.A.L. can be omitted since even if its sum of squares is concentrated in one df it will still not be large compared with the residual mean square. The SS for C.A.L. is rather larger, however, and so its parameter estimates need to be examined. We note also that the SS for C.S.L. is large: this term looks important.

```
$FIT - S.A.L - C.S.A.L $D E $
```

Further simplification of the model is now based on comparisons of interaction parameter estimates with their standard errors: parameters with small estimates relative to their standard errors are candidates for omission from the model. Inspection of the parameter estimates shows that those for C.S.A are smaller relative to their standard errors than those for C.S.L or C.A.L: this suggests that C.S.A can be omitted.

```
$FIT - C.S.A $D E $
```

The RSS increases to 102.33, an increase of only 1.75 on 3 df. Even if this were concentrated in 1 df, it would only be twice the residual mean square; thus C.S.A can certainly be omitted. The remaining three-way interaction parameters are listed below.

1.559	0.671	C(2).S(2).L(2)
1.895	0.994	C(2).A(2).L(2)
1.579	0.998	C(2).A(3).L(2)
0.000	aliased	C(2).A(4).L(2)

The C.S.L interaction is 2.5 times its standard error and needs to be retained. The C.A.L interaction terms are rather large: we try omitting them.

```
$FIT - C.A.L $D E $
```

The RSS increases by 3.18 on 2 df to 105.51. This is nearly four times the residual mean square and might correspond to an important single df. If the two estimable parameters for C.A.L are equated, most of this SS is associated with the common parameter, but to simplify the model interpretation, we decide to omit the interaction completely. Since C.S.L is retained in the model, its marginal two-way interactions C.S, C.L and S.L, and the mean effects C, S and L are also retained. Inspection of the parameter estimates for A.L suggest that this term can be omitted:

```
 0.406    0.519     A(2).L(2)
-0.731    0.530     A(3).L(2)
 0.000    aliased   A(4).L(2)
```

```
$FIT - A.L $D E $
```

The RSS increases by 4.36 on 2 df to 109.87, again a fairly large change, though no single df contains most of the SS. The estimates for C.A suggest that this could also be omitted:

```
-0.671    0.481    C(2).A(2)
-0.898    0.483    C(2).A(3)
 0.001    0.499    C(2).A(4)
```

```
$FIT - C.A  $D E $
```

The RSS increases by 4.13 on 3 df to 114.00, well inside the value of 119.18 for an adequate (0.50) model. The parameter estimates for S.A are large in two cases and this term needs to be retained.

```
 0.013    0.506    S(2).A(2)
-1.195    0.513    S(2).A(3)
-1.529    0.507    S(2).A(4)
```

Since S and A have also to be retained, no further simplification of the model is possible by the omission of complete terms. The model is now S*A+C*S*L. Since sex is involved in two different interactions with A and C.L, interpretation of the fitted model is greatly assisted by considering the sexes separately. We fit the model A+C*L for each sex separately, and examine further possible simplifications.

```
$CALC GIRL = %EQ(S,1) $
$WEIGHT GIRL
$FIT A + C*L  $D E $
```

The RSS is 53.11, and the parameter estimates for C*L are shown below.

```
-0.100    0.407    C(2)
-0.086    0.382    L(2)
-0.854    0.499    C(2).L(2)
```

The C.L interaction is only 1.7 times its standard error and can therefore be omitted.

```
$FIT - C.L  $D E $
```

The RSS increases by 2.63 on 1 df to 55.74, the C estimate is −0.666 (0.240) and the L estimate is −0.530 (0.285), so that L can also be omitted.

```
$FIT - L  $D E$
```

The RSS increases by 3.20 on 1 df to 58.94.

2.468	0.271	1
−0.241	0.356	A(2)
0.668	0.321	A(3)
0.908	0.355	A(4)
−0.664	0.244	C(2)

The estimate for A(2) is less than its standard error, while those for A(3) and A(4) differ by less than one standard error. We therefore equate the first and second levels, and the third and fourth levels, of A into a reduced factor A34.

```
$CALC A34 = 1 + %GE(A,3) $
$FACTOR A34 2 $
$FIT - A + A34  $D E $
```

The RSS increases by 0.93 on 2 df to 59.87.

2.365	0.224	1
0.872	0.242	A34(2)
−0.674	0.242	C(2)

The nearly equal magnitudes of the two parameters suggests that they could be equated: the effect on absence of being in the first culture group rather than the second is nearly equivalent to the effect of being in the second A34 group rather than the first.

```
$CALC D2 = %EQ(A34,2) - %EQ(C,2) $
$FIT D2  $D E $
```

The RSS increases by 0.29 on 1 df to 60.16.

2.470	0.119	1
0.773	0.162	D2

How do we interpret this model for girls? We work backwards through the dummy variable definitions. We defined D2 = %EQ(A34,2) − %EQ(C,2). The first component is 1 for girls in the third and fourth age groups, and zero otherwise, while the second component is 1 for white girls, and zero for Aboriginals. Thus D2 can take the values −1, 0 and 1, as follows:

C \ A	1	2	3	4
1	0	0	1	1
2	−1	−1	0	0

Note that the model does not contain an L effect: there is no need to distinguish slow learners from average learners for girls. There is a culture effect of 0.773: Aboriginal girls (C1) are away more often than white girls (C2) by an estimated factor of $e^{0.773} = 2.17$, or nearly 2. For simplicity of presentation we round this value to 2, by fitting the model with fixed coefficient log (2) (=0.693) for D2:

```
$CALC Z1 = %LOG(2)*D2 $
$OFFSET Z1
$FIT $D E $
```

The RSS increases by 0.23 on 1 df to 60.39, and the intercept estimate is unchanged at 2.470 (s.e. 0.119).

Finally, there is an age effect of the same magnitude as the culture effect: girls in the second and third forms of high school are away twice as often as those in first form or in primary school. The onset of menstruation would be a logical explanation of this effect. We give below a table of observed and [fitted] mean days absent (based on $\exp(\hat{\mu} + \frac{1}{2}s^2) - 1$) together with the sample sizes (n) for the girl classification, with the culture order reversed so that absence increases from the top left corner to bottom right corner.

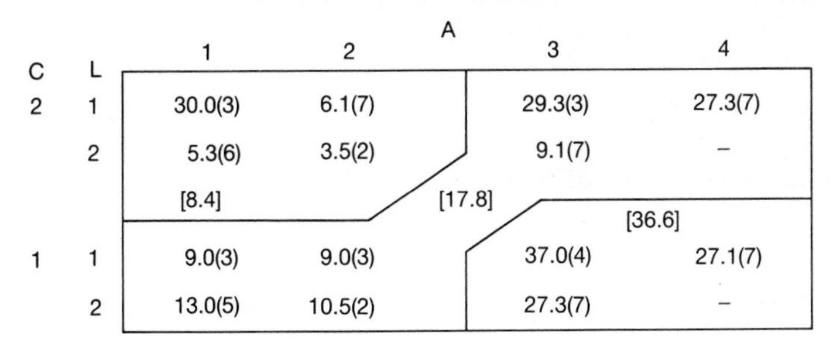

C	L	A = 1	A = 2		A = 3	A = 4
2	1	30.0(3)	6.1(7)		29.3(3)	27.3(7)
	2	5.3(6)	3.5(2)		9.1(7)	–
		[8.4]		[17.8]		
					[36.6]	
1	1	9.0(3)	9.0(3)		37.0(4)	27.1(7)
	2	13.0(5)	10.5(2)		27.3(7)	–

Observed and [fitted] mean days absent for girls, and (sample sizes)

Now we turn to the boys.

```
$OFFSET $
$CALC BOY = %EQ(S,2) $
$WEIGHT BOY $FIT A + C*L $D E $
```

The RSS is 60.89, and the C.L interaction is large:

−1.198	0.289	C(2)
−0.774	0.354	L(2)
1.225	0.409	C(2).L(2)

The C.L interaction, i.e. the additional C effect at L(2), almost cancels out the C effect at L(1), since $-1.198+1.225 \simeq 0$. Thus for normal learners, there is no culture difference in absence, but there *is* a culture difference for slow learners. We simplify the interaction model by setting equal the C(2) and C(2).L(2) coefficients. The model

$$\beta_1 C(2)+\beta_2 L(2)-\beta_1 C(2).L(2)$$

is equivalent to

$$\beta_1 C(2)\,(1-L(2))+\beta_2 L(2).$$

```
$CALC C2 = %EQ(C,2) : L2 = %EQ(L,2) : D2 = C2*(1-L2) $
$FIT - C*L + L + D2 $D E $
```

The RSS increases by 0.01 to 60.90.

−0.760	0.316	L(2)
−1.198	0.287	D2

The L and D2 coefficients differ by about 0.4, or slightly more than one standard deviation of L(2). We try equating them, that is equating the culture effect for slow learners to the learner effect for Aboriginals.

```
$CALC D3 = D2 + L2 $
$FIT - D2 - L + D3 $D E $
```

The RSS increases by 1.64 on 1 df to 62.54.

−0.665	0.335	A(2)
−0.759	0.366	A(3)
−0.378	0.358	A(4)
−1.019	0.259	D3

The coefficients of A(2) and A(3) are nearly equal, and that of A(4) differs from them by less than one standard error. It also differs from zero, that is from A(1), by one standard error so it could be set to zero. The grouping of primary and third form boys seems less reasonable than that of all high school boys, so we choose the latter grouping. We collapse the age classification into two categories by equating the last three levels.

```
$CALC A234 = 1 - %EQ(A,1) $
$FIT - A + A234 $D E $
```

The RSS increases by 1.34 on 1 df to 63.88. The coefficients of A234 and D3 differ by about one standard error, so we try equating them.

```
$CALC D4 = D3 + A234 $
$FIT - D3 - A234 + D4 $D E $
```

The RSS increases by 0.57 on 1 df to 64.45.

3.640	0.344	1
−0.790	0.201	D4

Note that the estimate for the boys of −0.790 is very close in magnitude to that (0.773) for the girls, and is correspondingly close to $-\log 2 = -0.693$. As for the girls, we round it to this value:

```
$CALC Z2 = -%LOG(2)*D4 $
$OFFSET Z2
$FIT $D E $
```

The RSS increases by only 0.19 to 64.64 and the intercept is 3.482 (s.e. 0.101).

What is the interpretation of this model? By definition,

$$\begin{aligned} D4 &= D3 + A234 \\ &= D2 + L2 + A234 \\ &= C2(1-L2) + L2 + A234. \end{aligned}$$

Again D4 is a score which can take values from 0 to 2, as follows:

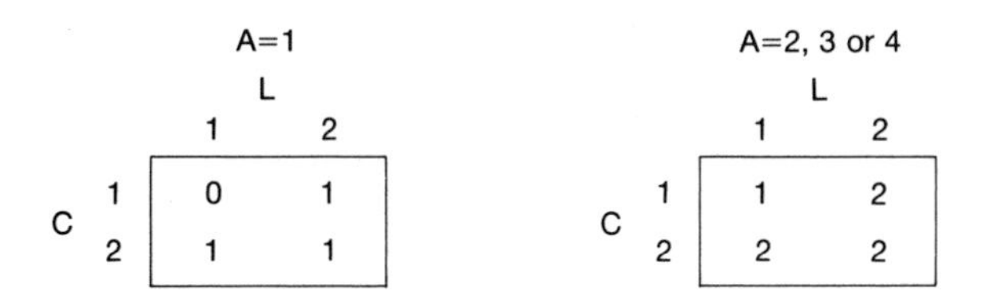

A=1

C \ L	1	2
1	0	1
2	1	1

A=2, 3 or 4

C \ L	1	2
1	1	2
2	2	2

Thus boys in high school (A234) are away half as often as boys in primary school (since the coefficient of D4 is *negative*) uniformly for culture and slow/average learners, while Aboriginal slow learners (C1,L1) are away twice as often as the other three culture/learning groups. This pattern can be seen in the tabulation of observed and [fitted] means below with (sample sizes), where we change the order of A and C so that absence increases from the top left-hand corner to the bottom right-hand corner.

		A			
C	L	4	3	2	1
	2	13.5 (10)	1.0 (1)	11.0 (6)	18.5 (4)
2			[11.2]		
	1	–	6.2 (9)	6.0 (11)	25.0 (1)
	2	14.6 (9)	2.0 (1)	11.4 (5)	21.3 (4)
1				[23.5]	
	1	–	36.4 (8)	22.6 (10)	3.0 (1) [48.0]

Observed and [fitted] mean days absent for boys, and (sample sizes)

The observation of 3 days absent for the one slow learning primary Aboriginal boy is badly fitted by the model. This is a consequence of the omission of the C.A.L interaction in the process of model simplification.

For the two final models for each sex, the total RSS is $60.39 + 64.64 = 125.03$. This exceeds the limit of 119.18 for an adequate (0.50) model: the model is not adequate at this level. The value of 0.50 is, however, fairly arbitrary: the value of 125.03 would correspond to an adequate model if α were taken as 0.35 (say). As always there is a balance between smoothing to obtain the simplest structure and losing important detail. The choice of α governs the degree of smoothing: we have obtained a simple structure for each sex by ignoring rather large interactions at an early stage of the modelling. The poor fit of the model for slow learning Aboriginal boys is an unsatisfactory feature of the model; the reader is encouraged, in this and in other examples, to try other models. The Poisson distribution is used in Section 5.2 for the Quine data, but is found to be unsatisfactory. The arbitrary nature of the transformation of days for the lognormal distribution is also unsatisfactory; no completely satisfying analysis of the Quine data is given in this book.

3.13 Missing data

Throughout this chapter we have assumed that complete data on response and explanatory variables are available for every sample member. The reality of survey data collection, however, is that missing values always occur, through non-response, misrecording or accident. The data matrix of individuals by variables generally has a scatter of missing values, and this considerably complicates model fitting by maximum likelihood.

The ideal case, which we consider first, is that in which the missing values are "missing at random". We may imagine a random "failure process", in which the value of any variable for each individual has a constant probability (possibly different for each variable) of failing, that is, of being missing, and failures occur independently. In this case fully efficient maximum likelihood estimates of the model parameters may be obtained using the EM algorithm (Dempster, Laird and Rubin, 1977), at the expense of assuming an explicit probability model for the variables for which values are missing. Little and Schluchter (1985) give computational details for a general model with normal or categorical variables.

We do not discuss this approach further because it can not be programmed in GLIM except in special cases, since GLIM lacks the general matrix handling facilities needed for calculating the information matrix in missing data models.

The ideal case of values missing at random may not hold. The probability of a value being missing may differ over individuals by depending on the values of other variables for the individual, or on the value that would have been observed.

As a simple example, consider a survey of monthly earnings of individuals classified by age, sex, educational level and industrial sector of employment. The object of analysis is to model average earnings by a regression on the other explanatory variables, but the value of earnings *only* is missing for some individuals. If these values are missing at random, then maximum likelihood estimation of the regression model parameters is equivalent to omitting the individuals with missing values and analysing only the complete cases.

Suppose now that the probability of an income value being missing increases with age of the respondent. This can be established by a logit regression (Section 4.2) of the missing data indicator on the explanatory variables. Then higher ages are under-represented in the complete cases, and if we intend to report the overall mean monthly earnings for the population the sample means will have to be weighted to allow for the under-representation of older people. The fitted regression of earnings on the explanatory variables is, however, not affected by the selection occurring in the sample, because the fitted model is conditional on the observed values of the explanatory variables. It is only marginal, or unconditional inferences which are affected. See Section 2.9 for a discussion of this point. The variances of the regression coefficients are however increased by the restriction of range.

Suppose finally that the probability of an income value being missing increases with decreasing income. Then low *incomes* are under-represented in the sample, and any regression will be systematically biased. No form of weighting of the observed sample means can make appropriate adjustment for the sample selection: we would have to fit a model in which the sample selection mechanism is explicitly represented.

We conclude with a discussion of approximate methods available in statistical packages.

3.14 Approximate methods for missing data

Three approximate methods are in common use. The first omits all observations with missing data on any variables, and uses only complete records. There are two serious disadvantages of this method. The first is the obvious one that the number of observations may be substantially reduced. In extreme cases there may be no complete records. The second is that serious bias in the model parameter estimates may result. When observation are missing on both response and explanatory variables, the parameter estimates from the complete cases are no longer consistent.

The second approximate method estimates the mean and variance of each variable from all the complete observations on that variable, and the covariance between each pair of variables from the complete observations on those pairs of variables. This does not avoid bias in the parameter estimates, and may result in a covariance matrix for the explanatory variables which is not positive definite.

The third approximate method fills in the missing values with imputed values, with the object of creating a complete data matrix which can be used for different analyses. In the simplest case the variable mean is used as the imputed value; this again biases the parameter estimates. More complex forms of imputation have recently been developed based on the EM algorithm in which regression estimates of the missing values, or values generated randomly from the conditional distribution of the missing values, are imputed. For a discussion of these procedures, see Panel on Incomplete Data (1983).

4
Binomial response data

4.1 Binary responses

Much social science data consist of *categorical* variables. Familiar examples are religion, nationality, residence (urban/rural), type of dwelling, level of education and social class. The categories may be unordered (religion, nationality) or ordered (degree of disablement, attitude to a social question).

The simplest categorical variable is one with just two categories. Simple Yes/No or Agree/Disagree responses to questionnaire items are examples, while in medical science Present/Absent, Susceptible/Not susceptible, Exposed/Not exposed are common examples. In Chapter 2 we considered the use of binary variables, coded 0 or 1, as explanatory variables in a regression model. Here, however, we are interested in the relationship between a binary response variable and other explanatory variables.

We begin with a simple example from Finney (1947), see also Pregibon (1981). The data in subfile VASO were obtained in a carefully controlled study of the effect of the RATE and VOLume of air inspired by human subjects on the occurrence (coded 1) or non-occurrence (coded 0) of a transient vaso-constriction RESPonse in the skin of the fingers. Three subjects were involved in the study: the first contributed 9 observations at different values of RATE and VOL, the second 8, and the third 22 observations. The experiment was designed to ensure as far as possible that successive observations obtained on each subject were *independent*: serial correlation between successive observations on the same subject in such studies is always a possibility.

Our aim is to develop and fit a statistical model relating RESP to RATE and VOL. As in Section 2.12, in general we define the response variable

$$\begin{aligned} y &= 1 \text{ “success” (occurrence of the response)} \\ &= 0 \text{ “failure” (non-occurrence)} \end{aligned}$$

and let p be the success probability for a randomly chosen individual at given values of RATE and VOL. Then y has a Bernoulli distribution

$$P(y) = p^y(1-p)^{1-y}, \quad y = 0 \text{ or } 1,$$

where p is related to RATE and VOL through a suitable regression function and link function. The random variable y has mean p and variance $p(1-p)$.

We now consider in detail the use of different link functions for probability parameters.

4.2 Transformations and link functions

The relation between RESP, the binary response variable, and RATE and VOL will be through a regression model for p in which RATE and VOL, and possibly other functions of these variables, are used as explanatory variables in the linear predictor. How is RESP to be related to the linear predictor?

In Chapter 3 the regression model was for the mean μ of the normal distribution, although as illustrated in Section 3.3 other functions of the mean could be used. Since the mean of the Bernoulli distribution is p, this suggests a linear regression model

$$p = \beta_0 + \beta_1 \text{RATE} + \beta_2 \text{VOL}$$

for the probability of the occurrence of vaso-constriction. Such *linear models for proportions* are sometimes used, but they suffer from obvious difficulties. If $\beta_1 > 0$, then for RATE sufficiently large, p will exceed 1, while for RATE sufficiently small, p may be negative.

In practical data analysis these natural bounds for p are sometimes not exceeded, so that linear models *may* give sensible answers. The model is sometimes fitted by least squares, treating the binary (0,1) response as though it were normally distributed:

```
$INPUT 1 VASO $
$YVAR RESP $ERROR N $FIT RATE+VOL $D E $PR %FV $
```

```
        estimate        s.e.      parameter
  1      -0.6126       0.2177      1
  2       0.3434       0.07837     RATE
  3       0.4011       0.08463     VOL
  scale parameter taken as   0.1490

  1.155    1.166  0.7473  0.2034  0.8072  0.8702 -0.1144  0.4124  0.0060
-0.0971 -0.0960  0.5524  0.6583  0.7491  0.9761  0.8732   1.220  0.2143
 0.4333  0.7276  0.2347  0.2355  0.3925  0.4561  0.6405  0.1432  0.6245
 0.4210  0.4758  0.1665  0.7280  0.3403  0.4571  0.5841  0.5556  0.8519
 0.4210  0.3407  0.4669
```

The parameter estimates look sensible, but three fitted values are greater than 1.0 and three are negative: the model is exceeding the natural bounds for p and is not satisfactory.

Since the variance $p(1-p)$ of RESP is not constant, this least squares fitting is not equivalent to maximum likelihood fitting. The linear model can be fitted by maximum likelihood, but we do not discuss this model further since the model fitting procedure will fail if fitted values fall outside the range (0,1), and some form of constrained maximization of the likelihood would then have to be used. Linear models for proportions or probabilities are now rarely used.

Transformations of the success probability are the standard solution to the problem of the finite range for p. Let $H(\theta)$ be a strictly increasing function of θ, where $-\infty<\theta<\infty$, such that $H(-\infty)=0$, $H(\infty)=1$. This condition is satisfied if H is a *cumulative distribution function* (cdf) for any continuous random variable defined on $(-\infty,\infty)$. Noting that for each value of p there is a value of θ with $p=H(\theta)$, we can define a transformation of p from (0,1) to $(-\infty,\infty)$ by

$$\theta=H^{-1}(p)=g(p).$$

A linear model for this transformed parameter

$$\theta=\beta_0+\beta_1\text{RATE}+\beta_2\text{VOL}$$

now has the property that any finite value of the linear predictor will give a value of p in the range (0,1).

Three such transformations are provided as standard options in GLIM, where they are defined by the link function $g(p)$, introduced in Section 2.4. They are the *logit* transformation or link

$$\theta=\text{logit } p=\log\{p/(1-p)\},\quad H(\theta)=\{e^\theta/(1+e^\theta)\},$$

for which $H(\theta)$ is the cdf for the logistic distribution, the *probit* transformation or link

$$\theta=\text{probit } p=\Phi^{-1}(p),\quad H(\theta)=\Phi(\theta),$$

for which $H(\theta)$ is the cdf for the standard normal distribution, and the *complementary log–log* (*CLL*) transformation or link

$$\theta=\log\{-\log(1-p)\},\quad H(\theta)=1-\exp(-e^\theta),$$

for which $H(\theta)$ is the cdf for the extreme value distribution (see Chapter 6).

The probit and logit links are similar, and can be made almost identical by a scale change to equate the variances of the underlying normal (1) and logistic ($\pi^2/3$) distributions. They generally give very similar fitted models. (For an example where this is *not* the case, see Section 5.7.) They are symmetric in p and $1-p$. The complementary log–log link is not symmetric, but for small p it is very close to the logit link, since in this case $\log(1-p)\approx-p$, so that

$$\log(-\log(1-p))\approx\log p,\quad \log\{p/(1-p)\}\approx\log p+p.$$

A fourth link function, the *log–log* link, can also be used though it is not given as a separate link option in GLIM. It is defined by

$$\theta = -\log(-\log p), \quad H(\theta) = \exp(-e^{-\theta}),$$

for which $H(\theta)$ is the cdf for the reversed extreme value distribution: if z has the extreme value distribution with cdf

$$H(z) = 1 - \exp(-e^{z}),$$

then $w = -z$ has the reversed extreme value distribution with

$$H(w) = \exp(-e^{-w}).$$

For p near 1, the log–log link is very close to the logit link.

As $p \to 0$ or 1, all these link functions approach $\pm\infty$. The logistic transformation is also called the *log odds*, $p/(1-p)$ being the odds in favour of $y = 1$.

To fit models to binary data in GLIM, we have to specify that the probability distribution is binomial ($ERROR B): the Bernoulli distribution is not provided as a separate distribution, since it is the special case of the binomial when the number of trials n is equal to 1 (see Section 2.11–2.14 for discussion). Since the binomial distribution depends on both n and p, the binomial denominator, or number of trials n_i on which the observed number of successes r_i is based, has to be specified as a second argument of the $ERROR directive, using a variate name following the B argument. The Y variate is specified as the number of successes r_i.

In the Bernoulli case, we have only one observation at each value of $\mathbf{x}_i$ and the denominator must be calculated explicitly to be 1. The link function is specified to be either logit (G), probit (P) or complementary log–log (C), and then the model is specified in the $FIT directive. (These, and only these, link functions are available for the binomial distribution.)

The binomial denominator N is first defined.

```
$CALC N = 1 $
$YVAR RESP $ERROR B N $LINK G $FIT RATE+VOL $D E $
```

Here the two-variable model uses the logit transformation. The fitted model is

$$\hat{\theta} = \underset{(3.224)}{-9.530} + \underset{(0.912)}{2.649}\ \text{RATE} + \underset{(1.425)}{3.882}\ \text{VOL}$$

where $\theta = \text{logit}\ p$.

The automatic output from $FIT includes the number of iterations or *cycles* required for the algorithm to converge. The scaled deviance is not as in the normal model a residual sum of squares, but minus twice the maximized value of the log-likelihood function, relative to a model with a parameter for every observation. Differences between scaled deviances for two models, one of which is a submodel of the other, are distributed in large samples like χ_r^2, if the r parameters omitted from the larger model are actually zero in the population. Thus the deviance fills the role of the residual sum of squares in the normal model, in providing a significance test for the importance of parameters omitted from the model.

The deviance for the fitted model is 29.77 with 36 df. How is this to be interpreted? The maximal model, with a parameter for every observation, will fit the data exactly and have a deviance of zero, so if we were to use the large-sample theory of the likelihood ratio test, we would treat the deviance as approximately χ_{36}^2 if the fitted model is an adequate representation of the data. For binary data, however, large-sample theory does not apply to the distribution of the residual deviance from the fitted model, because the maximal model is equivalent to n separate models, one for each single observation: we are fitting n models to n samples of size 1. Failure of the model to fit at individual observations has to be assessed by residual examination, which we now consider.

4.3 Model criticism for binary data

The $D R $ directive in GLIM when used with binary data prints the Pearson residuals

$$e_i = (y_i - \hat{p}_i)/\{\hat{p}_i(1-\hat{p}_i)\}^{\frac{1}{2}}$$

where y_i is 0 or 1. These are approximately standardized variables with mean 0 and variance approximately 1; the variance is approximate because of the estimation of $\boldsymbol{\beta}$ in the linear predictor. They are not, however, normally distributed, so a normal quantile plot is not appropriate and may be misleading. Note that e_i can take only the two possible values a_i and $-1/a_i$ for each i, where

$$a_i = -\{\hat{p}_i/(1-\hat{p}_i)\}^{\frac{1}{2}}.$$

"Deviance residuals" have been defined by McCullagh and Nelder (1983); these are the signed square roots of the individual components of the deviance: their sum of squares is the deviance. Residual analysis in this chapter will be based on the Pearson residuals. Large values of the Pearson (or other) residuals indicate failure of the model to fit at the corresponding points. The sum of squares of the Pearson residuals is the Pearson goodness-of-fit X^2, which is available in the system scalar %X2.

```
$D R$
```

For the fitted model for RESP, there are two large residuals: 3.751 for the fourth observation and 3.458 for the 18th. These two observations are "successes" at points where the fitted probability of success is very low: 0.066 for observation 4 and 0.077 for observation 18. They contribute a total of 26.03 to the Pearson X^2 of 38.85 (in large samples with binomial data the deviance and the Pearson X^2 agree closely, but they may differ substantially in small samples or with binary data).

A plot shows these points up clearly. We plot successes and failures using plotting characters F and S corresponding to the levels of a factor equal to RESP+1:

```
$CALC RF = RESP+1  $FACTOR RF 2 $
$PLOT RATE VOL 'F S' RF $
```

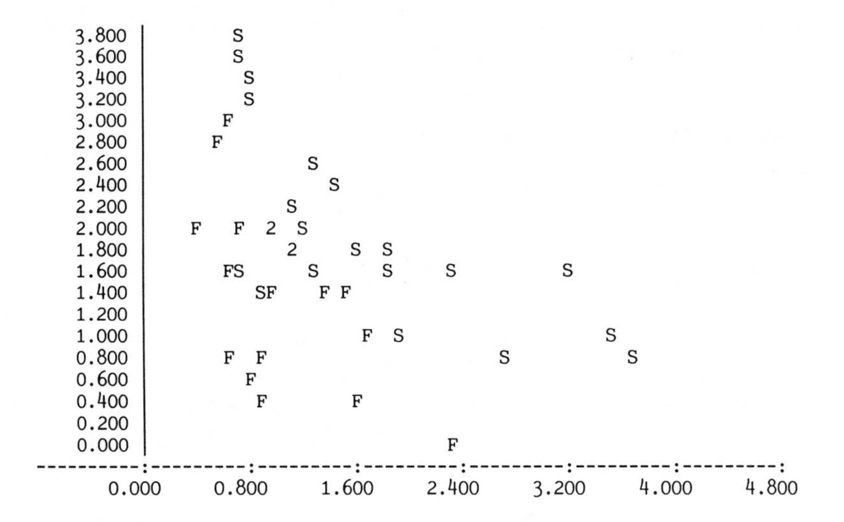

The two "success" observations with large residuals appear on the bottom left hand side of the cluster of "success" observations.

What now? As with residual examination in normal regression models, we do not automatically remove these observations. They would be checked for correctness with the experimenter; if they *are* correct, then it is the model which needs modification, or else we accept the poor fit at these points as random variation.

One possible modification is the incorporation of a subject effect. The data come from three subjects who are not identified in the model. If subjects vary in their base level vaso-constriction response, then the fitted model is mis-specified and a better fit may be obtained by the inclusion of the subject effect.

```
$FACTOR S 3 $
$CALC I = %CU(1) :
      S = 1 + %GT(I,9) + %GT(I,17) $
$FIT + S $D E R $
```

The deviance is now 26.16, a reduction of 3.61 for 2 df. There is no strong evidence of a subject effect. The residual at the fourth observation has decreased to 1.907, but that at the eighteenth has *increased* to 4.595. We cannot attribute the poor fit at these points to the absence of a subject effect.

Could the form of the model be inappropriate? The analyses in Finney (1947) and Pregibon (1981) use log RATE and log VOL rather than RATE and VOL, though there were no strong scientific reasons for choosing between them.

```
$CALC LV = %LOG(VOL) : LR = %LOG(RATE) $
$FIT LV + LR $D E R $PR %X2 $
```

The deviance is now 29.23, slightly less than for the first model. The Pearson X^2 is reduced from 38.85 to 34.15, and the residuals at the fourth and eighteenth observations are reduced to 3.518 and 2.906. The formal likelihood ratio test cannot discriminate between models with logged and unlogged explanatory variables, as Finney (1947) noted, but the smaller residuals for the logged model suggest that it is slightly preferable.

Are there influential observations which may be producing a marked effect on the fit of the model? The simple results of Chapter 3 on regression diagnostics can be generalized in various ways (Pregibon, 1981) at the expense of additional computing. We will use the diagonal elements of the hat matrix from the iteratively reweighted least squares algorithm. At the final (r-th) iteration GLIM computes the parameter estimates $\hat{\beta}$ using

$$X'W^{(r)}X\hat{\boldsymbol{\beta}} = X'W^{(r)}\mathbf{z}^{(r)}$$

where $W^{(r)}$ is the diagonal matrix of iterative weights and $\mathbf{z}^{(r)}$ the adjusted dependent variate at the final iteration. The hat matrix is then

$$H = (W^{(r)})^{\frac{1}{2}}X(X'W^{(r)}X)^{-1}X'(W^{(r)})^{\frac{1}{2}}.$$

The diagonal elements of H are obtained in GLIM as the product of the variance of the linear predictor and the iterative weight variate, since

$$\begin{aligned}\text{var}\,(X\hat{\boldsymbol{\beta}}) &= X\,\text{var}\,(\hat{\boldsymbol{\beta}})\,X' \\ &= X(X'W^{(r)}X)^{-1}X'.\end{aligned}$$

We examine the influence of observations in the LV+LR model.

```
$EXTRACT %VL $CALC H = %VL*%WT : I = %CU(1) $
$PRINT H$
$PLOT H I '+' $
```

The value of $2(p+1)/n$ is 0.154. The 13th and 29th values exceed this marginally, and the 31st considerably: it is 0.246. Is there anything unusual about the position of this observation?

```
$LOOK 31 LV LR %WT $
$PLOT LV LR '+' $
```

The 31st observation is near the edge of the main cloud of points. One other point is strikingly remote from the main cloud; it is the 32nd observation, but it has little influence:

```
$LOOK 32 H LV LR %WT $
```

So little weight is assigned to this observation that its discrepant position (due to a very small value of RATE) does not affect the regression. If the most influential observation is omitted, very little change in the fitted model occurs:

```
$CALC W = 1 : W(31) = 0 $WEIGHT W $FIT . $D E $
```

The deviance changes by 0.77 and the parameter estimates by about 1/10 of their standard errors. Omitting observation 32 instead again has little effect.

For the model RATE+VOL, the 32nd observation has an even larger value of h_i – 0.334 – but omitting this observation again has little effect, the deviance changing by 1.45 and the parameter estimates by about 30% of their standard errors.

Omission of the two observations with large residuals, on the other hand, has a very marked effect.

```
$CALC W = 1 : W(4) = W(18) = 0
$FIT  RATE + VOL  $D E $
```

Note that the weight variate is still in force though its values have changed.

The deviance is now 10.70, and the parameter estimates have become much larger:

$$\hat{\theta} = -41.99 + 10.74\ \text{RATE} + 17.49\ \text{VOL}$$
$$\quad (22.09) \quad (5.64) \qquad\qquad (9.35)$$

Warning messages are printed during the 9 cycles required for convergence that observations 1, 2 and 17 are "held at limit": the fitted probabilities for these observations are approaching zero or 1, and the corresponding fitted logits $\hat{\theta}$ are approaching $\pm\infty$. The calculation of $\exp(\hat{\theta})$ would cause numerical overflow or underflow; this is prevented by the fitted logits being held at fixed large positive and negative values. For observation 1, for example, the fitted logit from the model is 31.58, corresponding to a fitted probability of $1 - 0.2 \times 10^{-13}$. The large estimates for the parameters show that a rapid change from 0 to 1 in the response probability occurs over a small interval of each variable.

```
$FIT LR + LV $D E$
```

For the LR + LV model, 10 observations are held at the limit, and the model coefficients are even larger; the deviance is 7.36. The exclusion of the two observations has resulted in a remarkably close fit of the model to the data. This is, of course, the whole object of modelling, but when it is obtained by the removal of observations which do not fit the model, we should be very cautious indeed. If there is no experimental reason to doubt the correctness of these two observations, then the original model should be retained. As usual, more data would clarify the position.

We consider finally the simplification and interpretation of the log model. First we verify that both variables are needed in the model:

```
$WEIGHT  $FIT LV + LR  $D E $
```

The fitted model is

$$\hat{\theta} = -2.875 + 5.179\ \text{LV} + 4.562\ \text{LR}$$
$$\quad (1.319) \ (1.862) \qquad (1.835)$$

The t-values for LV and LR (see Section 2.7 for a definition) are 2.781 and 2.486. For a normal model, these large values would indicate that both variables are important. The asymptotic theory of Section 2.7 suggests that the same interpretation may hold for non-normal models, but we noted in Section 2.15 that the use of parameter standard errors obtained from the expected information matrix could be seriously misleading in small samples when the likelihood function has substantial skew. To verify the importance of each variable we omit it explicitly from the model.

```
$FIT - LV  $FIT + LV - LR $
```

When LV is omitted from the model, the change in deviance is 19.63, which is in very poor agreement with $(2.781)^2 = 7.73$; similarly when LR is omitted, the change in deviance is 17.88, in equally poor agreement with $(2.486)^2 = 6.18$. In this example the t-values for the parameters substantially understate the importance of these variables to the model, and a confidence interval based on $\hat{\beta} \pm \lambda$ s.e. $(\hat{\beta})$ gives a misleading impression of the plausible values of the parameters because the likelihood function in each parameter is not symmetric about the ML estimate.

We note that the coefficients of LV and LR are very similar, differing by only one third of a standard error. This suggests that they could be equated, as Finney (1947) noted, to give a single variable model using log (VOL∗RATE).

```
$CALC LVR = LV + LR $
$FIT LVR $D E $
```

The deviance increases by 0.29 on 1 df from 29.23 to 29.52, and the fitted model is

$$\hat{\theta} = \underset{(1.279)}{-3.031} + \underset{(1.738)}{4.901}\,\mathrm{LVR}$$

The corresponding fitted model for the odds ratio is

$$\frac{\hat{p}}{1-\hat{p}} = \exp(\hat{\theta}) = e^{-3.031}(\mathrm{VOL*RATE})^{4.901}$$

The power of VOL∗RATE is very close to 5—only 1/17 of a standard error away—and we note that

$$e^{-3.031} = (e^{-0.6062})^5 = (0.545)^5 \approx 2^{-5}.$$

A simpler model

$$\frac{\hat{p}}{1-\hat{p}} = \exp(\hat{\theta}) = (\mathrm{VOL*RATE}/2)^5$$

can therefore be fitted by specifying an offset:

```
$CALC OFS = 5*(LVR - %LOG(2)) $
$OFFSET OFS
```

and fitting the model with *no* estimated parameters:

```
$FIT -LVR -1 $
```

(In earlier releases of GLIM3, the intercept was specified by %GM: this may also be used in GLIM3.77.)

A GLIM error message is produced: GLIM does not allow the fitting of a model with no estimated parameters. We can circumvent this by fitting a model with one variable which is identically zero:

```
$CALC ZERO = 0 $
$FIT ZERO-1 $
```

The deviance is 30.17, an increase of 0.65 for the two fixed parameters. Note that the df is now 39, equal to the number of observations.

The fitted model and observed data are now easily plotted in one dimension.

```
$PLOT %FV RESP LVR '+o' $
```

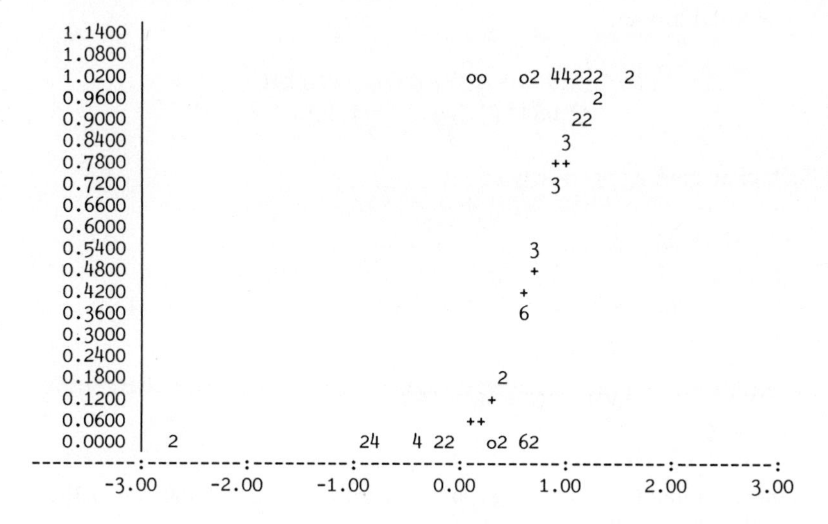

The fitted probability increases from 0 to 1 as LVR increases from about -0.1 to 1.8. The two observations 4 and 18 with large residuals are those at the top of the plot at the left-hand end of the row of 'o's. It can be seen that when they are removed from the fit, the slope of the regression will be much steeper as there is almost no overlap in LVR between the 0 and 1 responses.

Note that the value of 5 for the power of VOL∗RATE is chosen only for simplicity. No strong substantive interpretation should be drawn from it, in the absence of a physical model for an integer power. We cancel the offset setting before proceeding with further model fitting.

```
$OFFSET $
```

These analyses can be repeated with other link functions, with similar results. For example, the LV+LR model gives deviances 29.23, 29.29 and 26.62 with LINKs G, P and C respectively. To fit the log–log link, we have to define the response to be 1-RESP and use link C:

```
$CALC NORE = 1-RESP $
$YVAR NORE $LINK C $FIT LV + LR $
```

The deviance is 31.53, a worse fit. The comparison of deviances from different link functions requires some care; for a discussion see Appendix 1. We refit the model reverting to the CLL link:

```
$YVAR RESP $FIT LV + LR $D E $
```

giving the fitted model

$$\hat{\theta} = -2.968 + 4.327\ \mathrm{LV} + 3.960\ \mathrm{LR}.$$
$$\quad (1.084)\ (1.540) \qquad (1.368)$$

The coefficients are again nearly equal.

```
$FIT LVR $D E $
```

The deviance increases by only 0.19 on 1 df to 26.81, and the coefficient of LVR is 4.078, which we fix at 4:

```
$CALC OFS = 4 * LVR  $OFFSET OFS
$FIT  $D E $
```

The deviance is unchanged at 26.81, and the intercept is -2.950 (s.e. 0.301). Fixing it at -3, we refit the model:

```
$CALC OFS = OFS-3  $
$FIT ZERO-1  $D E $
```

The deviance increases to 26.84. We again plot the fitted model.

```
$PLOT %FV  RESP LVR '+o' $
```

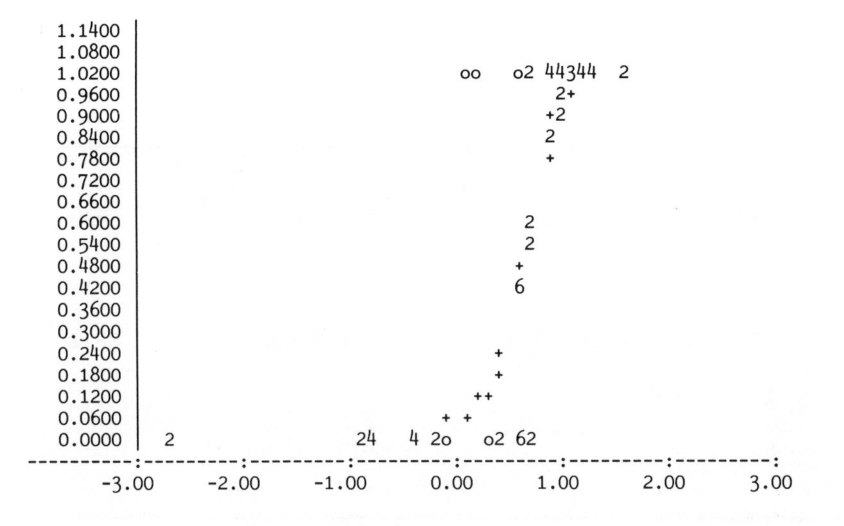

The asymmetry in the CLL transformation is visible: the fitted probability increases more slowly initially than for the logit, but it approaches its upper asymptote more rapidly just beyond LVR = 1 instead of at 1.8.

The somewhat better fit of the CLL model gives us a mild preference for this model, which can be expressed as

$$-\log(1-\hat{p})=e^{\hat{\theta}}=e^{-3}\,(\text{VOL}*\text{RATE})^{4}.$$

$END

4.4 Contingency table construction from binary data

In larger samples with binary data and continuous explanatory variables it is frequently possible to construct a contingency table by grouping the explanatory variables into ordered categories. This allows more detailed model criticism than with the original binary data, at the expense of some loss of precision in parameter estimation. We illustrate with an example given by Wrigley (1976) of data from a 1974 study of chronic bronchitis in Cardiff conducted by Jones (1975). The data consist of observations on three variables for each of 212 men in a sample of Cardiff enumeration districts (subfile BRONCHITIS in Appendix 4). The variables are CIG, the number of cigarettes smoked per day, POLL, the smoke level in the locality of the respondent's home (obtained by interpolation from 13 air pollution monitoring stations in the city), and R, an indicator variable taking the value 1 if the respondent suffered from chronic bronchitis, and 0 if he did not. The presence or absence of chronic bronchitis was determined using a special questionnaire devised by the Medical Research Council; detection of chronic bronchitis using this questionnaire had been found to be almost completely consistent with the clinical diagnosis. (The text in Wrigley refers to CIG as the total consumption of cigarettes ever smoked in hundreds, but the published data are either total consumption expressed in units of ten thousand, or are current rates of smoking in cigarettes per day. We have assumed the latter.)

Our aim is to develop and fit a statistical model relating R to CIG and POLL. We begin by plotting the data using *N* and *B* as plotting characters as in the previous example.

```
$INPUT 1 BRONCHITIS $
$CALC RF = R + 1 $FACTOR RF 2 $PLOT POLL CIG 'N B' RF $
```

```
      70.00 |
      69.00 |
      68.00 |        N   B
      67.00 BN  N     2
      66.00 2  N           B                               B
      65.00 2               N
      64.00 3N N  B  N N        B B
      63.00 5NNN2 2    2 B      B
      62.00 5 NN  B   N     B   B  N2
      61.00 52 NN    N N     B      N
      60.00 53 N   N     BB N B
      59.00 5N3N2   B  2B          B
      58.00 72  3  NB        BB     N               N
      57.00 623 2N3    BN B NN        N
      56.00 3523  N BN     3 B    N    N
      55.00 224  N              B         B
      54.00 242NN2 2     B   NB
      53.00 2732  NN  N N    2            B
      52.00 N
      51.00 |
----------:---------:---------:---------:---------:---------:---------:
         0.00      6.00      12.00     18.00     24.00     30.00     36.00
```

The pattern of *N*s and *B*s is not at all clear, and the large number of multiple points for non- and light smokers obscures the picture. Two very heavy smokers have different bronchitis outcomes.

We proceed to fit the logit model. The binomial denominator N=1 is already defined in the subfile.

```
$YVAR R $ERROR B N $LINK G $
$FIT CIG + POLL $D E $PR %X2 $
```

The deviance is 174.21 with 209 df., but this does not necessarily indicate a good fit, since as we noted in Section 4.2, for binary data the residual deviance does not follow a χ^2 distribution. The Pearson X^2 is 191.6. To display the residuals would take many screens of output; instead we display just the large residuals (greater than 2 in absolute value) using the W option for $DISP, and setting %RE=1 for the large residuals and zero for the rest:

```
$CALC E  = (%YV-%FV)/%SQRT(%FV*(1-%FV)):
      E2 = E**2 : %RE = %GT(E2,4) $
$D W $
```

There are 11 residuals greater than 2 in absolute value, and these correspond to observations 7, 48, 59, 87, 122, 132, 142, 147, 149, 171 and 195. Where in the variable space are these observations?

```
$PLOT POLL CIG 'N B' RF $
$DEL %RE $
```

Of the 10 positive residuals, six occur for bronchitis sufferers at CIG = 0 and the others for low values of CIG and POLL; the negative residual occurs for a non-occurrence for the heavy smoker, observation 147 with CIG = 24.9 and POLL = 58. The model is fitting badly for non- and light smokers.

How should we amend the model? The 'additive' model assumes that the regression of logit p on POLL is the same for smokers and non-smokers: is this reasonable? We might hypothesize that the irritating effects of air pollution might be greatly increased by heavy smoking, producing an interaction. Non-smokers might have a response to POLL quite different from smokers.

This possibility is investigated by fitting models for smokers and non-smokers separately:

```
$CALC WNS = %EQ(CIG,0)  $WEIGHT WNS
$FIT POLL $D E $
```

The slope of 0.1809 (s.e. 0.1247) is not very different from that for the two groups combined. The deviance is 39.38.

```
$CALC WS = 1 - WNS  $WEIGHT WS
$FIT CIG + POLL $D E $
```

The slope for POLL is 0.1172 (s.e. 0.0539) and that for CIG is 0.2288 (s.e. 0.0450); the deviance is 134.00. The total deviance for the two models is 173.38: compared with the deviance of 174.21 for the original model, almost nothing has been gained by a separate regression on POLL for non-smokers. The poor fit for non-smokers is thus not due to a simple interaction.

We now extend the model to a *second-degree response surface*:

```
$CALC P2 = POLL**2 : CP = CIG*POLL : C2 = CIG**2 $
$WEIGHT  $FIT + P2 + CP + C2 $D E $
```

The deviance decreases by 10.5 with 3 df, almost all of which is due to the term

C2. Reduction of the model by omitting CP and P2 gives a deviance increase of only 0.36, and C2 has a parameter estimate of −0.0130 (s.e. 0.0038), compared with the linear term estimate 0.441 (s.e. 0.083). Thus the rate of increase of bronchitis logit falls off as CIG increases, and becomes zero at CIG = 0.4410/2(0.0130) or 17.0.

This is a surprising result. Even more surprising is the model obtained if we extend the second-degree to a *third-degree response surface*:

```
$CALC P3 =POLL*P2
  : P2C = P2*CIG : PC2 = POLL*C2 : C3 = CIG*C2 $
$FIT  POLL + CIG + P2 + CP + C2 + P3 + P2C + PC2 + C3
$D E $
```

The deviance change compared to the second degree response surface model is 11.48 on 4 df. After elimination of unnecessary third degree terms, we find that C2, CP, P2 and P2C must be retained in the model. How can we interpret such a complex model?

The complexity makes us suspect some systematic failure of the model to represent the data. To investigate the appropriateness of the model, we group both CIG and POLL into classes and construct a three way contingency table (actually a two way cross-classification of R and N). The number of bronchitis sufferers in each cell of the table is then modelled by a binomial distribution (see Section 2.13).

```
$CALC  FCIG  =  1+%GT(CIG,0)+%GT(CIG,1)+%GT(CIG,3)+%GT(CIG,5)+
                    %GT(CIG,8):
       FPOL  =  1+%GT(POLL,55)+%GT(POLL,57.5)+%GT(POLL,60)+
                    %GT(POLL,62.5)+%GT(POLL,65)  $
```

Alternatively, FCIG and FPOL can be constructed using the $ASSIGN and $GROUP directives.

The variables FCIG and FPOL take the values 1 to 6: we have defined six categories for each of the two continuous variables CIG and POLL. Non-smokers are the first category of FCIG. The category boundaries are chosen by eye inspection to give roughly equal numbers in each marginal category; alternatively, $HIST may be used to estimate approximate percentiles.

We form two new variates TR and TN to contain the grouped numbers of successes and sample sizes for the 36 cells.

```
$TABULATE THE R TOTAL FOR FCIG ; FPOL INTO TR
  : THE N TOTAL FOR FCIG ; FPOL INTO TN $
$PRINT TR : TN $
```

```
0.        1.000   2.000   2.000  0.       2.000  0.      0.      0.      0.
0.        0.      0.      0.     0.       0.     0.      0.      0.
 2.000    2.000   1.000   2.000  0.       1.000   1.000   3.000  0.
 2.000    2.000   3.000   5.000   5.000   5.000   3.000   2.000
 6.000   11.00   14.00   11.00    7.000   5.000  17.00   11.00    8.000
 3.000    3.000   1.000   9.000   8.000   8.000   4.000   4.000   2.000
 6.000    6.000   4.000   2.000   3.000  0.       3.000   3.000   3.000
 2.000    5.000   4.000   6.000  11.00    7.000   9.000   4.000   2.000
```

The tabulation directive uses the convention that the last factor named varies most rapidly and the first least. Factor levels for cigarettes and pollution for the table variates TR and TN can now be defined:

```
$VAR 36 TCIG TPOL $
$CALC TCIG = %GL(6,6) : TPOL = %GL(6,1) $
```

The variates TCIG and TPOL each take values 1 to 6, but are of length 36. Cell sample sizes are spread fairly evenly from 2 to 11, with two larger values (14 and 17). One cell has no observations.

```
$UNIT 36
$FACTOR TCIG 6 TPOL 6 $
$YVAR TR $ERROR B TN $FIT TCIG + TPOL $
```

An error message results from the $FIT directive because of the cell with no observations. The binomial denominator must be non-zero. We solve this problem by defining a variate which is 1 except for cells with no observations, when it is zero. This variate is used both to adjust the zero binomial denominators and as a weight variate to remove these observations from the fit.

```
$CALC W = %NE(TN,0) : TN = TN+1-W
$WEIGHT W $FIT TCIG + TPOL $D E R $
```

The deviance of 14.88 on 24 df shows a good fit, and the largest residual is 1.486 for the 27th cell. Both main effects are necessary: the deviance for TCIG only is 24.97 on 29 df, a change of 10.09 on 5 df, and for TPOL only is 87.13 on 29 df, a change of 72.25 on 5 df. Cigarette consumption is much more important than smoke level.

Inspection of the parameter estimates for the main effects model shows that the estimates for TPOL generally increase with level, though the value for level 3 is considerably higher than for levels 2 or 4. This suggests a linear trend in logit with increasing TPOL. We construct a new variate LINP which takes the values of the factor TPOL.

```
$CALC LINP = TPOL $
$FIT -TPOL + LINP $D E R $
```

The deviance change is 4.59 on 4 df, and the largest residual is now 2.180 for the 27th cell. Inspection of the parameter estimates for TCIG shows a very surprising pattern: for levels 2 and 3 the estimates are -10.6, corresponding to a fitted probability of almost zero, and we see from the values of TR that there are no bronchitis sufferers at any air pollution level among smokers of not more than 3 cigarettes per day, though there are sufferers among non-smokers! It is for this reason that the response surface model was so complex: it was trying to reproduce both a zero proportion of bronchitis sufferers amongst light smokers and a non-zero proportion amongst non-smokers, as well as a high proportion amongst heavy smokers.

Further inspection of the parameter estimates suggests that the classification of FCIG is unnecessarily detailed: the fourth and fifth levels, with estimates 1.516 and 1.625, can be collapsed, as can the second and third levels.

```
$CALC MCIG = %EQ(TCIG,1) + 2*(%EQ(TCIG,2)+%EQ(TCIG,3)) +
             3*(%EQ(TCIG,4)+%EQ(TCIG,5)) + 4*%EQ(TCIG,6) $
$FACTOR MCIG 4 $
$FIT - TCIG + MCIG $D E $
```

```
scaled deviance = 19.498 (change =  +0.025) at cycle 10
           d.f. = 30     (change =  +2    ) from 35 observations

        estimate        s.e.       parameter
    1    -3.069        0.6620      1
    2    0.3265        0.1361      LINP
    3    -10.64         40.87      MCIG(2)
    4     1.574        0.5325      MCIG(3)
    5     2.473        0.5444      MCIG(4)
    scale parameter taken as  1.000
```

The deviance increases by 0.025 on 2 df. The interpretation of the model is that each step up the air pollution scale increases the logit by 0.327, i.e. multiplies the odds on chronic bronchitis by 1.39. Categories 3 and 4 of MCIG, smokers of 3–8 and more than 8 cigarettes a day have their odds on bronchitis multiplied by 4.83 and 11.86 respectively, compared with non-smokers. Light smokers (1–3 a day) have *no* observed cases of bronchitis.

We finally return to the original data and refit the model.

```
$UNIT 212
$CALC C =1 + %GT(CIG,0) + %GT(CIG,3) + %GT(CIG,8) $
$FACTOR C 4 $
$YVAR R $ERROR B N $LINK G $FIT C + POLL $D E $
```

Note that MCIG cannot be used in the $FIT directive since it is of length 36. The parameter estimate for POLL is 0.1241 with standard error 0.0537. The equal intervals for the LINP scale correspond to 2.5 units of POLL: multiplying the POLL coefficient and standard error by 2.5 gives 0.3103 and 0.1341, which agree very closely with the LINP values of 0.3265 and 0.1361. It can be verified that the addition of a quadratic term in POLL adds nothing to the model, and that using log POLL gives almost the same deviance as POLL. Further, the probit, complementary log–log and log–log links all give almost the same deviances. The final fitted model for $\theta = \text{logit}\, p$ is therefore taken as

$$\hat{\theta} = \underset{(3.282)}{-9.367} + \underset{(0.0537)}{0.1241}\ \text{POLL} - 8.380\ \text{C}(2) + \underset{(0.530)}{1.542}\ \text{C}(3) + \underset{(0.542)}{2.466}\ \text{C}(4)$$

The estimate of −8.380 for C(2) tries to reproduce on the logit scale the observed proportion of zero bronchitis sufferers. The large standard error for this parameter estimate occurs because the likelihood in this parameter is nearly flat at the estimated value and hence the second derivative of the log likelihood with respect to this parameter evaluated at the estimate is nearly zero.

Five of the previous eleven observations, (48, 59, 87, 122, 142) now have residuals larger than 2 in absolute value: these all correspond to non-smokers with bronchitis.

The paradoxical results for non-smokers and light smokers require explanation. Do non-smokers also include those who have given up smoking? Are the non-smokers much older than the smokers? Without more details of the original survey, these questions remain open.

This example shows the value of categorizing continuous explanatory variables when the response variable is binary. In many cases, as in the last two examples in this chapter, such categorization is routinely carried out before the data are analysed. Even without categorization, however, binary responses can become binomial if the number of explanatory variable categories is limited. The next example illustrates both this feature and the use of binary models for prediction.

```
$END
```

4.5 The prediction of binary outcomes

The data in subfile GHQ were published by Silvapulle (1981), and come from a psychiatric study of the relation between psychiatric diagnosis (as case or non-case) and the value of the score on a 12-item General Health Questionnaire (GHQ), for 120 patients attending a general practitioner's surgery. Each patient was administered the GHQ, resulting in a score between 0 and 12, and was subsequently given a full psychiatric examination by a psychiatrist who did not know the patient's GHQ score. The patient was classified by the psychiatrist as either a "case", requiring psychiatric treatment, or a "non-case". The question of interest was whether the GHQ score, which could be obtained from the patient without the need for trained psychiatric staff, could indicate the need for psychiatric treatment. Specifically, given the value of GHQ score for a patient, what can be said about the probability that the patient is a psychiatric case? Sex of the patient is an additional variable; both men and women patients are heavily concentrated at the low end of the GHQ scale, where the overwhelming majority are classified as non-cases. The small number of cases are spread over medium and high values of GHQ.

```
$INPUT 1 GHQ $
```

The input file gives the number of cases C and non-cases NC at each GHQ score, and the total number N = C + NC, classified by the factor SEX. We first plot the proportion of cases against GHQ for each sex separately:

```
$CALC P = C/N : %RE = %EQ(SEX,1) $
$PLOT P GHQ  '+' $
```

```
 1.1400 |
 1.0800 |
 1.0200 |                  +   +       +             +
 0.9600 |
 0.9000 |
 0.8400 |
 0.7800 |
 0.7200 |
 0.6600 |
 0.6000 |
 0.5400 |
 0.4800 |         +
 0.4200 |
 0.3600 |
 0.3000 |
 0.2400 |
 0.1800 |
 0.1200 |
 0.0600 |
 0.0000 +    +
----------:---------:---------:---------:---------:---------:---------:
        0.00      2.00      4.00      6.00      8.00     10.00     12.00
```

Males show a rapid change from non-cases to cases beween GHQ scores of 1 and 4.

```
$CALC %RE = %EQ(SEX,2) $
$PLOT P GHQ '+' $
```

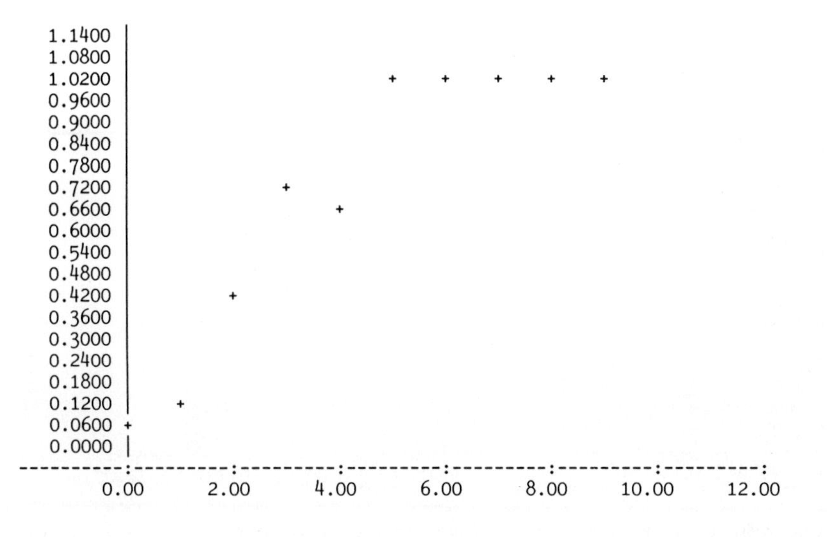

Females also show a rapid change, though there is a small proportion of cases even for GHQ zero.

Thus GHQ score is a good indicator of being a psychiatric case. To make this statement more precise we fit a model:

```
$YVAR C $ERROR B N $LINK G $FIT SEX+GHQ $D E $
```

The standard error of the sex parameter is large compared with the parameter estimate, suggesting that the sex effect is not needed in the model. This is confirmed by a formal test based on the deviance difference:

```
$FIT  -SEX $D E $
```

The change in deviance, from 4.942 to 5.744 is only 0.802, which is quite small compared to χ_1^2. Thus as far as the relation between "caseness" and GHQ is concerned, we can ignore the sex classification and use a single model for both sexes.

Our interest is in the probability of being a case, not in the logit of this probability, so we transform the fitted values back to the probability scale:

```
$CALC FP = %FV/N $
```

(note that the fitted values are for C, the number of cases), and finally plot observed and fitted proportions, pooled over sex:

```
$DEL %RE $
$PLOT P FP GHQ 'o+' $
```

```
1.1400 |
1.0800 |
1.0200 |                 o    2    2    4    2    2    2
0.9600 |                      2
0.9000 |                 2
0.8400 |
0.7800 |            o
0.7200 |            +
0.6600 |                 o
0.6000 |
0.5400 |
0.4800 |       o
0.4200 |       o
0.3600 |       2
0.3000 |
0.2400 |
0.1800 |
0.1200 |  3
0.0600 3
0.0000 o  o
----------:---------:---------:---------:---------:---------:---------:
        0.00      2.00      4.00      6.00      8.00     10.00     12.00
```

The analysis looks complete at this point, but if we examine the data more carefully a strange feature appears. Suppose we check the assumption of no interaction between SEX and GHQ by including their interaction in the model:

```
$FIT SEX+GHQ + SEX.GHQ $D E $
```

Very large positive and negative parameter estimates now appear for the GHQ score and its interaction with SEX, with even larger standard errors. Why is this happening?

The fitting of the full interaction model is equivalent to fitting separate GHQ models for each sex, with the slopes and intercepts different. This can also be achieved by fitting each sex separately using a weight variable.

```
$CALC MALE = %EQ(SEX,1) : FEMALE = %EQ(SEX,2) $
$WEIGHT FEMALE  $FIT GHQ  $D E $
$WEIGHT MALE   $FIT GHQ  $D E $
```

For males, convergence does not occur in 10 cycles. We increase the number of cycles to 30 and refit the model.

```
$CYCLE 30  $FIT .  $D E $
```

At convergence after 26 cycles, the deviance is zero, and the model fits exactly. The intercept and slope values, of -39.73 and 19.86, are such that at GHQ$=$2, the fitted logit from the model is zero, corresponding to the observed proportion of 0.5. *Any* large values of β_0 and β_1 with $\beta_1 > 0$ such that

$\beta_0 + 2\beta_1 = 0$ will give an exact fit to the data, in the sense that the fitted values for GHQ=0 or 1 and for GHQ>2 can be made arbitrarily close to the observed values. Thus the GLIM estimates in this example are determined solely by the convergence criterion in GLIM: if this is made stricter then $\hat{\beta}_0$ and $\hat{\beta}_1$ will be larger, but still with $\hat{\beta}_0 + 2\hat{\beta}_1 = 0$. The likelihood function approaches its maximum of 1 as $\beta_0 \to -\infty$, $\beta_1 \to +\infty$ with $\beta_0 + 2\beta_1 = 0$. Thus the maximum likelihood estimates of β_0 and β_1 do not exist, that is they are not finite. This causes GLIM no difficulty, however, since the convergence criterion is satisfied for finite values of these parameters.

Since the evidence for interaction is not very strong—the deviance reduction on adding the interaction is 3.40 on 1 df—the simpler main effects model can be retained, and the sex effect can be omitted from this model, as we have already seen.

The psychiatric value of the model involving only GHQ is in prediction: for a new patient (assumed similar to those in the study) with a value of GHQ of 2 (say), what can we say about the probability that this patient is a psychiatric case?

The fitted logit from the model at GHQ=2 is

$$-3.454 + 2\,(1.440) = -0.574$$

and the corresponding fitted probability is 0.360. For prediction purposes, however, this value is misleadingly precise. As in Section 2.15 confidence intervals for the true model proportion can be based on the transformed interval for the true logit, though these will not be accurate if the likelihood function in the parameters is not normal.

For the model

$$\text{logit } p = \beta_0 + \beta_1 \text{ GHQ}$$

the variance of a fitted logit at GHQ=X,

$$\widehat{\text{logit } p|\text{X}} = \hat{\beta}_0 + \hat{\beta}_1 \text{X}$$

can be obtained directly from the "variance of the linear predictor" %VL, as in Section 3.4, if X is a value in the data (if not, X can be introduced as an extra observation with weight zero, as in Section 3.4).

```
$WEIGHT  $FIT GHQ  $D  E  $
$EXTRACT %VL  $PRINT %LP : %VL $
```

We set out below the fitted value of the linear predictor, its standard error (the square root of %VL), the 95% confidence interval based on $\%\text{LP} \pm 1.96\sqrt{\%\text{VL}}$, and the transformed interval on the probability scale, for each value of X from 0 to 5.

X	%LP	s.e.	C.I.	Fitted p	C.I. for p
0	-3.454	0.562	-4.555	0.031	0.010
			-2.352		0.087
1	-2.013	0.377	-2.753	0.118	0.060
			-1.274		0.219
2	-0.573	0.382	-1.321	0.360	0.211
			0.175		0.544
3	0.867	0.570	-0.250	0.704	0.438
			1.984		0.879
4	2.307	0.825	0.691	0.909	0.666
			3.924		0.981
5	3.747	1.101	1.590	0.977	0.831
			5.905		0.997

Thus at X=2, the 95% confidence interval for p is (0.211,0.544).

A better approach, following that described in Section 2.15, is to base intervals for p on the *profile likelihood* function. This approach was used for the transformation parameter λ in the Box–Cox transformation family in Section 3.1

For a fixed X, define the parameter

$$\theta_X = \beta_0 + \beta_1 X.$$

Then

$$\beta_0 = \theta_X - \beta_1 X$$

and on substituting for β_0 in the model

$$\text{logit}\, p = \beta_0 + \beta_1\,\text{GHQ},$$

we obtain

$$\text{logit}\, p = \theta_X + \beta_1(\text{GHQ} - X).$$

We have reparametrized the model so that the model parameters are now θ_X and β_1. To construct the profile likelihood in θ_X, we maximize the likelihood in β_1 for each fixed value of θ_X. This is easily achieved using a sequence of $FIT directives in which the value of θ_X is taken as an offset, the value of X is subtracted from GHQ, and the single variable (GHQ − X) is fitted without an intercept. For example:

```
$CALC X = 2 : THETAX  = -2 : Z = GHQ-X $
$OFFSET THETAX $FIT Z-1 $
```

It is convenient to use the deviance directly, rather than to convert it back to a likelihood—using %EXP(−%DV/2). By evaluating the deviance over a grid of values of θ, we generate the "profile deviance" $-2p\ell(\theta)$, in the notation of Section 3.1. The profile deviance for p can then be obtained by transforming the scale from θ to p.

The subfile BINOMIAL contains a macro RPROB which computes and plots the profile deviance for the response probability from a model with a *single* explanatory variable. The macro requires as argument the name of the explanatory variable, and the specified value %X of the variable, the minimum value %A and maximum value %Z of p, and the grid increment %I all have to be explicitly declared. The Y-variable and error directives are assumed to have been previously specified.

```
$INPUT 1 BINOMIAL $
$CALC %X=2 : %A=0.2 : %Z=0.6 : %I=0.01 $
$USE RPROB GHQ $
```

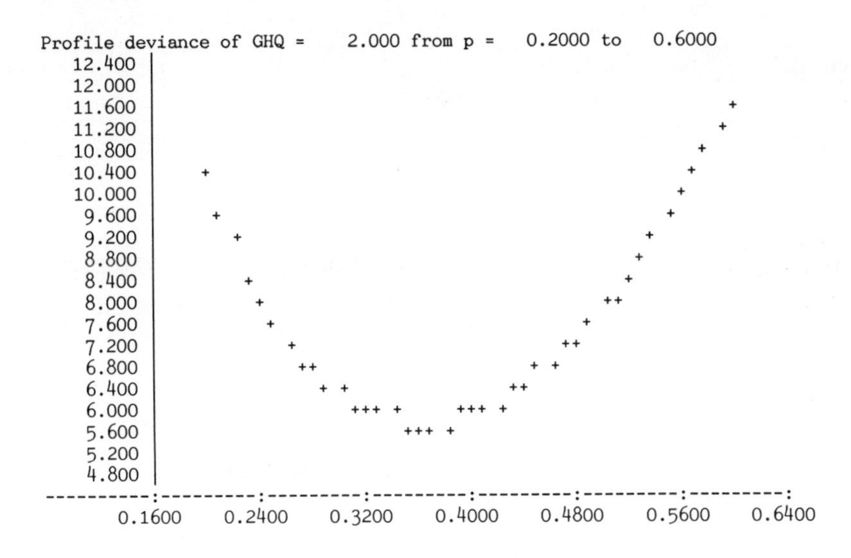

The endpoints of the 95% confidence interval for p based on the profile likelihood correspond to values of the deviance of 3.84+5.74 (the minimum value), i.e. 9.58. These can be determined by interpolation, or by tabulation over a finer grid interval, to be 0.212 and 0.552, in good agreement with the interval based on the linear predictor, because for this value of X the profile likelihood is closely normal on the logit scale. For extreme values of X, this is not the case: for $X=0$, the profile likelihood interval is (0.0087, 0.077) which is not in such good agreement with the previous interval (0.010, 0.087). However the *qualitative* conclusions are unaffected: a GHQ score of zero means that the patient is very unlikely to be a case (only 2 out of 60 patients with this score were cases) and a score of 5 or more means that the patient is very likely to be a case (none of the 15 patients with a score of 5 or more was not a case). For values from 1 to 4, however, the GHQ score is not a definite indicator of "caseness". Note that for a score of 4, the fitted probability is 0.909, strongly suggesting that such a patient would be a case: however the confidence interval includes values of p below 0.7, and we note that 1 out of 4 patients with this score was *not* a case.

In bioassay, a question which is often of interest is estimating the value of the dose variable at which the response probablity is 0.5. This is called the LD_{50} (other probability values like 0.9 may be of interest as well). Our context is somewhat different, but we illustrate the procedure for the GHQ example. Suppose we specify a probability p_0. What value X_0 of GHQ corresponds to a response probability of p_0?

Write

$$\theta_0 = \text{logit}\, p_0 = \beta_0 + \beta_1 X_0.$$

Then

$$X_0 = (\theta_0 - \beta_0)/\beta_1$$

is a nonlinear function of the parameters. The maximum likelihood estimate of X_0 is $\hat{X}_0 = (\theta_0 - \hat{\beta}_0)/\hat{\beta}_1$. Asymptotic standard errors for non-linear functions of the parameter estimates based on the information matrix can be seriously misleading in small samples, but the profile likelihood for X_0 is easily constructed using a macro very similar to RPROB. As before, we parametrize the model to θ_0 and β_1 by substituting for β_0 in terms of X_0.
Then

$$\text{logit}\, p = \theta_0 + \beta_1(\text{GHQ} - X_0)$$

Thus the specified value of θ_0 is taken as a fixed offset, and a grid of values of X_0 is used to construct the profile deviance. The macro LD in the subfile BINOMIAL computes and plots the profile deviance for the LD value corresponding to a specified p_0. This macro requires as argument the name of the explanatory variable, while the minimum value %A and maximum value

%Z of the value of X_0, the grid increment %I, and the value %P of p have to be explicitly declared. The Y-variable and error directives are assumed to have been previously specified. We illustrate first with the LD_{50}.

```
$CALC %A=0 : %Z=6 : %I=1 : %P=0.5 $
$USE LD GHQ $
```

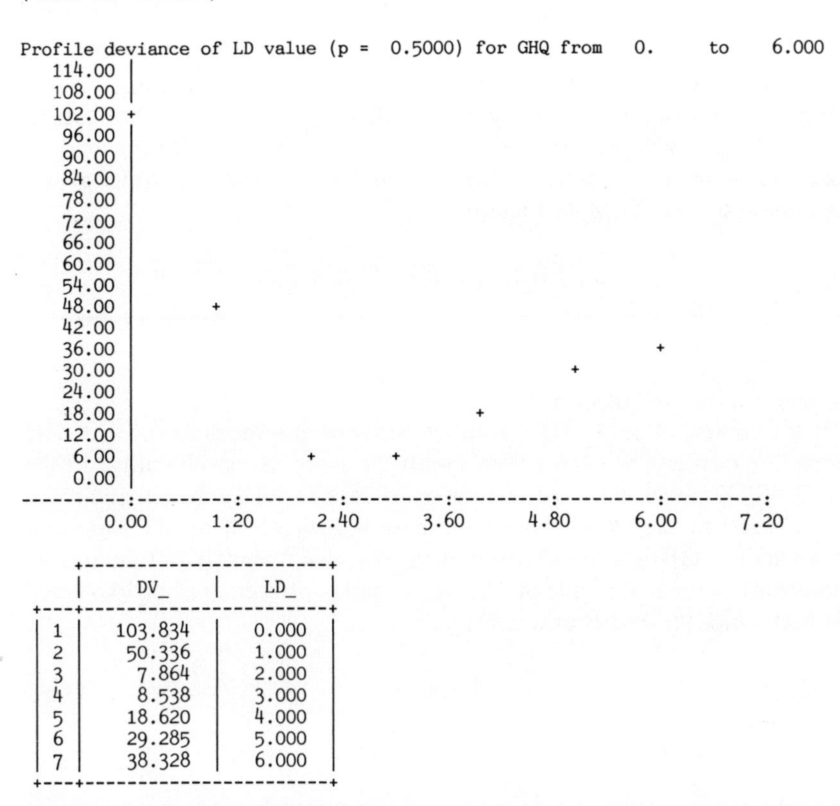

	DV_	LD_
1	103.834	0.000
2	50.336	1.000
3	7.864	2.000
4	8.538	3.000
5	18.620	4.000
6	29.285	5.000
7	38.328	6.000

The grid is the set of possible integer values of GHQ. The deviance is 7.86 at $X_0=2$ and 8.54 at $X_0=3$ (the minimum is 5.744 at $X_0=-\hat{\beta}_0/\hat{\beta}_1=3.454/1.44=2.40$). The LD_{50} at GHQ$=2$ is slightly better supported than at GHQ$=3$ (note that 0.5 is included in the 95% confidence intervals for p for both GHQ$=2$ and 3).

Similarly for the LD_{90}.

```
$CALC %A=2 : %Z=10 : %I=1 : %P=0.9 $USE LD $
```

Note that GLIM remembers that the argument to LD has been set to GHQ—it need not be respecified.

The deviance is 9.99 at $X_0=3$, 5.76 at $X_0=4$ and 8.26 at $X_0=5$. (In practical applications X_0 would be a continuous variable and the grids would be much finer.)

We conclude this discussion with an example of a macro for relative potency. In the comparison of potencies of two drugs, the model

$$\text{logit } p=\beta_0+\beta_1 x+\beta_2 z$$

is frequently used, where x is the logdose of drug, and z a (0,1) dummy variable identifying the drug. The *relative potency* is defined as the ratio of the doses of the two drugs which produce the same response probability. On the logdose scale, the parameter of interest is the difference in x for the two groups giving the same response. Thus we require

$$\beta_0+\beta_1 x_A=\beta_0+\beta_1 x_B+\beta_2$$

or

$$x_A-x_B=\beta_2/\beta_1,$$

another non-linear function.

In the context of our GHQ example, we ask: what is the difference in GHQ scores for men and women patients with the same case probability? In the GHQ+SEX model, we noted that the sex effect was not significant and could be omitted from the model; the ML estimate of β_2/β_1 is $0.7938/1.433=0.55$, or about half a GHQ unit. If $\beta_2=0$ then $x_A-x_B=0$, but if our interest is specifically in the magnitude of x_A-x_B, we can construct a profile likelihood for it. As with the earlier examples, let

$$\theta=\beta_2/\beta_1$$

and reparametrize the model by eliminating β_2:

$$\text{logit } p=\beta_0+\beta_1(x+\theta z)$$

Thus for a fixed value of θ, we form the composite variable $x+\theta z$ and fit a linear regression using this variable. The macro RELPOT requires as arguments the name of the variable x and the name of the group identifier z (which must be defined as a two-level factor). The minimum value %A and maximum value %Z of θ, and the grid increment %I, have to be explicitly declared.

```
$CALC %A = -1 : %Z = 2 : %I = 0.1 $USE RELPOT GHQ SEX $
$END
```

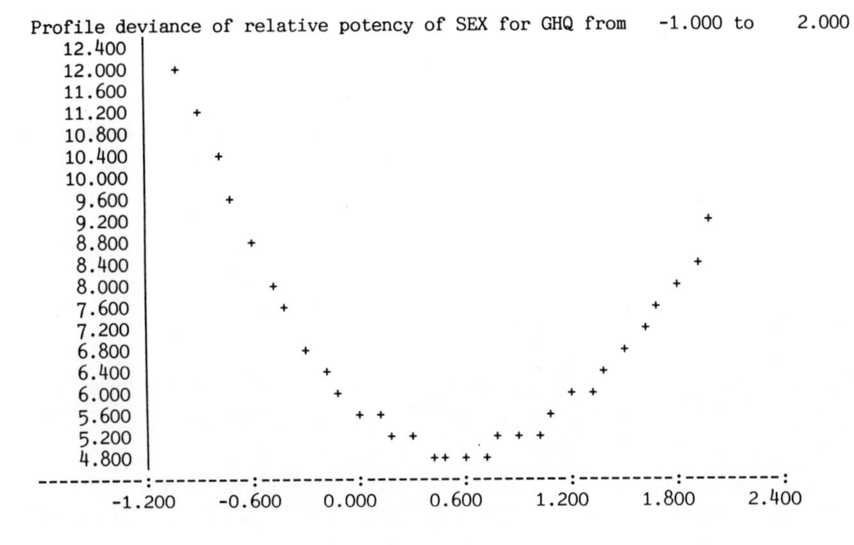

The approximate 95% confidence interval for θ is $(-0.61, 1.94)$, which is not symmetric about the MLE of 0.55.

4.6 Profile and conditional likelihoods in 2×2 tables

The 2×2 contingency table is an important special case, and its analysis is still a matter of some controversy (see for example Yates, 1984). Throughout the book we have used the likelihood ratio test, or equivalently the profile likelihood, to draw conclusions about a model parameter. The likelihood is maximized over the values of the other (nuisance) parameters in the model, for a fixed value of the parameter of interest. In small samples, both the justification for following this procedure and the exact coverage of the confidence interval are not altogether satisfactory, and other methods of eliminating nuisance parameters are of value.

Consider the 2×2 table in which a binary response is classified by a two-level group factor, with r_i successes in n_i trials with success probability p_i $(i=1,2)$.

The model

$$\text{logit } p_i = \beta_0 + \beta_1 z_i$$

where z_i is a (0,1) dummy variable taking the value 1 for the *first* level of the factor, is a saturated model since there are two parameters for the two response probabilities. The parameter of interest is the ratio of the odds on success $p_i/(1-p_i)$ in the two groups, which is called the *odds-ratio* $\theta = p_1(1-p_2)/p_2(1-p_1)$. In the above model, $\beta_1 = \log \theta$, the *log–odds ratio*, and β_0 is a nuisance parameter. We have already discussed in Section 4.5 profile likelihoods for several functions of model parameters: the same approach can be used to construct the profile likelihood for β_1.

An alternative approach introduced by Yates (1934) and Fisher is based on a *conditional likelihood.* Consider the conditional distribution of the number of successes r_1 in the first group given the total number of successes $r=r_1+r_2$. We have

$$P(r_1|r,n_1,n_2,\theta) = \binom{n_1}{r_1}\binom{n_2}{r-r_1}\theta^{r_1}/D(\theta)$$

where

$$D(\theta) = \sum_{s=s_1}^{s_2} \binom{n_1}{s}\binom{n_2}{r-s}\theta^s, \qquad \begin{aligned} s_1 &= \max(0,r-n_2), \\ s_2 &= \min(n_1,r). \end{aligned}$$

This is a non-central hypergeometric distribution which depends only on θ and hence β_1; the dependence on β_0 has been eliminated by conditioning on the total number of successes.

When r_1 and r_2 are observed, the *conditional likelihood* of θ is

$$CL(\theta) = P(r_1|r,n_1,n_2,\theta)$$

which is easily calculated provided n_1 and r are not too large, since $D(\theta)$ is a polynomial in θ of degree $s_2=\min(n_1,r)$. The likelihood in β_1 can then be obtained by log transforming the scale of θ.

The usual application of the conditional likelihood is to Fisher's "exact" test of the hypothesis $\theta=1$, which is equivalent to $p_1=p_2$. The value CL(1) is the (Fisher) "exact probability" of the observed table, that is, the hypergeometric probability of the observed table given fixed margins and $\theta=1$. In the usual application of the "exact" test this probability is assessed by comparison with the probabilities of "more extreme" tables, and a p-value assigned to the observed table which is the cumulative probability of the observed table and all more extreme ones. Opinions differ over whether this probability should be doubled (corresponding to a two-tailed test) or the probabilities of extreme tables in the other direction should be cumulated as well (Yates, 1984 gives a discussion).

These procedures do not lend themselves easily to interval construction, but the conditional likelihood can be used directly for this purpose.

We illustrate with a simple numerical example. In samples of $n_1=6$ and $n_2=7$, $r_1=4$ and $r_2=1$ successes are observed. The conditional likelihood is

$$CL(\theta) = \binom{6}{4}\binom{7}{1}\theta^4/D(\theta)$$

where

$$D(\theta) = \binom{6}{0}\binom{7}{5} + \binom{6}{1}\binom{7}{4}\theta + \binom{6}{2}\binom{7}{3}\theta^2 + \binom{6}{3}\binom{7}{2}\theta^3 + \binom{6}{4}\binom{7}{1}\theta^4 + \binom{6}{5}\binom{7}{0}\theta^5$$

Expanding out the binomial coefficients, we have

$$CL(\theta) = 105\theta^4/\mathrm{D}(\theta),$$
$$D(\theta) = 21 + 210\theta + 525\theta^2 + 420\theta^3 + 105\theta^4 + 6\theta^5$$

and a simple GLIM macro can be used to evaluate CL(θ). This macro is *not* listed in the Appendix as it lacks generality: the form of $D(\theta)$ depends on the given table.

```
$MACRO CONLIK
$DEL LTH TH DTH CL DV$
$CA %G=1+(%Z-%A)/%I$
$VAR %G LTH TH DTH CL DV$
$CA LTH = %A+%CU(%I)-%I
  : TH  = %EXP(LTH)
  : DTH = 21+210*TH+525*TH**2+420*TH**3+105*TH**4+6*TH**5
  : CL  = (105*TH**4)/DTH
  : DV  = -2*%LOG(CL) $
$PLOT DV LTH$
$LOOK LTH TH DV$
$DEL LTH TH DTH CL DV$
$ENDMAC
```

The macro plots and prints the deviance, equal to -2 times the log conditional likelihood, as a function of the log–odds ratio. To use the macro, we have to specify the grid increment %I and the smallest value %A and largest value %Z of the log–odds ratio β_1.

```
$CALC %I = 0.1 : %A = -1 : %Z= 2  $USE CONLIK $
```

The deviance continues to decrease at $\beta_1 = 2$, so we redefine the grid interval:

```
$CALC %A = -1 : %Z = 6  $USE CONLIK $
```

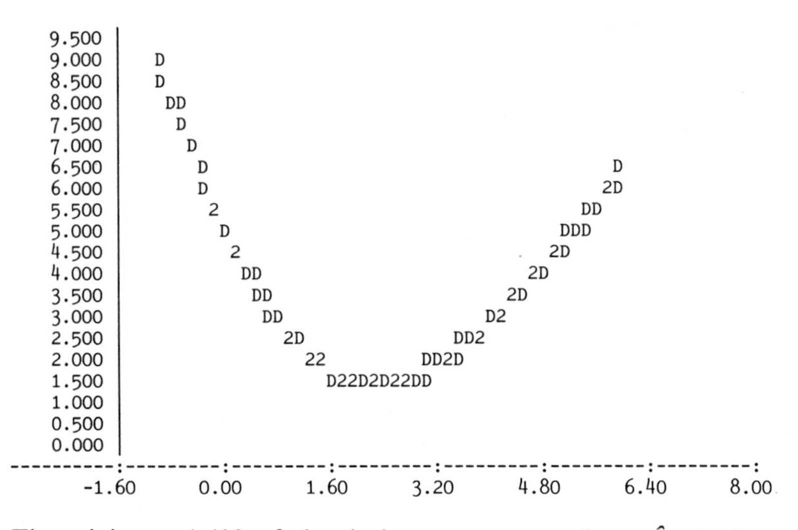

The minimum 1.408 of the deviance occurs at about $\hat{\beta}_1 = 2.23$, and an approximate 95% confidence interval, based on $1.41 + 3.84$, is $(-0.07, 5.41)$. The conditional likelihood interval (just) includes the value 0, corresponding to $p_1 = p_2$. Thus the conventional interpretation of the likelihood interval, relying on the validity of the approximate confidence coefficient, is that the hypothesis $p_1 = p_2$ would not be rejected at the conventional 5% level.

The conclusion from Fisher's "exact" test is expressed differently. The hypergeometric probabilities for $\theta = 1$ for the possible tables are given in Lindley and Scott (1984 Table 26). For $n_1 = 6$, $n_2 = 7$ and $r_1 + r_1 = 5$, the hypergeometric probabilities of $n_1 = 4$ and 5 are 0.082 and 0.005. Thus the probability of the observed value r_1, or the more extreme value 5, is 0.087, which would not lead to rejection of the hypothesis. (The probability of the other extreme event $r_1 = 0$ is 0.016.) The exact test assesses the probability of the observed event under the null hypothesis relative to those of other more extreme events under the same hypothesis. This is philosophically different from the conditional likelihood ratio test which assessed the probability of the observed event under the null hypothesis relative to that of the same event under other hypotheses.

We now compare the conditional likelihood with the profile likelihood. For the parameter β_1 in the logistic model, the profile likelihood can be obtained using the macro LOGODDS in Appendix 4. The same scalar values %I, %A and %Z are specified as for the conditional likelihood, and the name of a group dummy variable (coded 1 for group 1 and 0 for group 2) is specified as an argument to the macro.

```
$UNIT 2
$DATA R N
```

```
$READ
4 6 1 7
$ASS G = 1, 0 $
$INPUT 1 BINOMIAL $
$YVAR R $ERROR B N $LINK G
$CALC %I=0.1 : %A=-1 : %Z=6 %USE LOGODDS G $
```

```
Profile deviance of logodds for G from    -1.000 to     6.000
    9.500 |
    9.000 |
    8.500 |
    8.000 |    +
    7.500 |    +
    7.000 |     +
    6.500 |      +
    6.000 |      ++
    5.500 |        +
    5.000 |        +
    4.500 |         2
    4.000 |          +                                   +2
    3.500 |           2                                 +++
    3.000 |            +                              +++
    2.500 |             2                            2+
    2.000 |              2                          2+
    1.500 |               +2                      +2+
    1.000 |                 +2                  2++
    0.500 |                   2+2           22+
    0.000 |                      2+2+22+2+
----------:---------:---------:---------:---------:---------:---------:
      -1.60      0.00      1.60      3.20      4.80      6.40      8.00
```

The deviance here has a minimum of zero at log(12) = 2.485, the MLE of β_1.

The two deviance functions are plotted in Fig. 4.1, with that for the conditional likelihood corrected to a minimum of zero by subtracting 1.408. The profile likelihood is shifted towards larger values of β_1, and gives stronger evidence against $\beta_1 = 0$: the approximate 95% confidence interval is (0.03, 5.81).

In this example with quite small sample sizes, the conclusions about the log–odds ratio from the two analyses are not equivalent. In larger samples the conditional and profile likelihoods become equivalent, and so the computational difficulties of the conditional likelihood (the evaluation of high degree polynomials in θ) can be avoided in larger samples by computing the profile likelihood.

In small samples the use of the conditional likelihood (or Fisher's "exact" test) loses some information about the log–odds ratio. This is because the marginal distribution of the total number of successes, which is ignored in the conditional likelihood, depends on this parameter (Plackett, 1977). Against this has to be balanced the unknown confidence coverage of intervals based on the profile likelihood. In small samples we lack a unified (non-Bayesian)

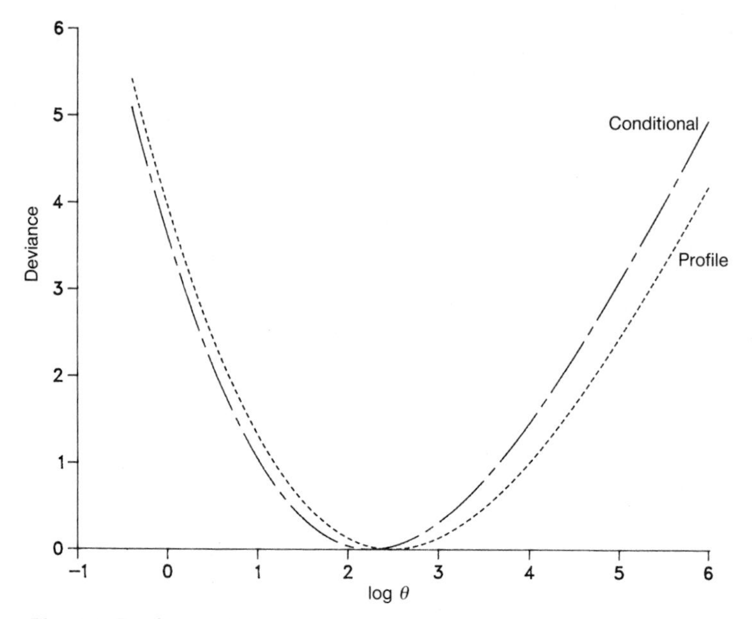

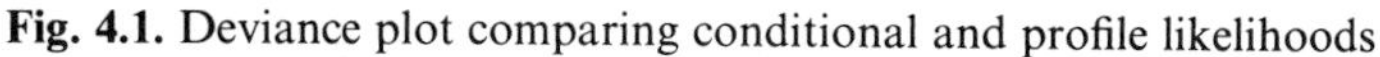

Fig. 4.1. Deviance plot comparing conditional and profile likelihoods

theory of inference in the presence of nuisance parameters. See Hinde and Aitkin (1987) for further discussion and a new approach.

```
$END
```

4.7 Multidimensional contingency tables with a binary response

We now consider two examples of contingency tables with a binary response. The first example in this section has a simple two-explanatory-factor structure, the second in the next section has a complex five-factor structure.

The subfile KULLBACK gives the number R of men diagnosed as having coronary heart disease (CHD) in an American study of 1329 men (the data are presented and analysed in Ku and Kullback, 1974). The serum cholesterol level and blood pressure were recorded for each man, and are reported in one of four categories, giving a 4×4 cross classified table in each cell of which the number R of men with CHD and the total number N of men examined are given. The table is reproduced below.

		Blood pressure in mm mercury (BP)			
		<127	127-146	147-166	>166
Serum	<200	2/119	3/124	3/50	4/26
Cholesterol	200-219	3/88	2/100	0/43	3/23
in mg/100cc	200-259	8/127	11/220	6/74	6/49
(CHOL)	>259	7/74	12/111	11/57	11/44

The explanatory variables were originally continuous, but have been categorized into ordered categories. We treat them first as unordered factors. BP and CHOL are defined on the subfile.

How is the proportion of men suffering from CHD related to blood pressure and serum cholesterol levels? Before fitting any models, we find the proportion suffering from CHD, and plot it against each of the factors:

```
$INPUT 1 KULLBACK $
$CALC P = R/N $
$PLOT P CHOL '+' $
$PLOT P BP '+' $
```

These plots show an overall increase of P with increasing level of each factor, but since the other factor is varying as well, the picture is not clear. We can plot P against BP separately for each value of CHOL by using different plotting characters:

```
$PLOT P BP 'A B C D' CHOL $
$PLOT P CHOL 'A B C D' BP $
```

The pattern of increase is fairly consistent at each level of the other factor.

Now we try fitting models on the logit scale:

```
$YVAR R $ERROR B N $LINK G
$FIT : BP : CHOL : CHOL+BP $
```

The two analysis of deviance tables are:

Source	dev	df	Source	dev	df
BP	23.56	3	CHOL	31.92	3
CHOL	27.09	3	BP	18.73	3
Interaction	8.08	9	Interaction	8.08	9

CHOL is the more important, but both main effects are necessary in the model. The interaction is not explicitly fitted, but we know that the interaction model is saturated and has a deviance of zero. The value of 8.076, near the mean (or median) of χ^2_9, shows that the main effects model with 7 parameters provides a good fit to the data.

To check that the model is fitting well at each sample point, we examine the (Pearson) residuals using $D R $. These are direct generalizations of the binary residuals of Section 4.3:

$$e_i = (r_i - n\hat{p}_i)/\{n_i\hat{p}_i(1-\hat{p}_i)\}^{\frac{1}{2}}$$

For the main effects model the largest residual is -1.35, showing a good fit.

Now we examine the parameter estimates to interpret the model:

```
$D E $
```

−3.482	0.349	1
−0.208	0.465	CHOL(2)
0.562	0.351	CHOL(3)
1.344	0.343	CHOL(4)
−0.041	0.304	BP(2)
0.532	0.332	BP(3)
1.200	0.327	BP(4)

The parameter estimates for both factors show a consistent increase as the factor level increases from 1 to 4, apart from level 2 which for each factor is very little different from level 1. This suggests that we try smoothing the proportions further by using the factor level as though it were a continuous variable, scoring the categories of each factor 1 to 4. We fit a regression model using one parameter for each score by redefining the factors to be variates:

```
$VAR CHOL BP $
$FIT CHOL + BP $D E R $
```

The deviance is increased to 14.85 on 13 df. This still looks a good fit, but the four parameters omitted have a deviance of 14.85 − 8.08, i.e. 6.77. Although this value is not extreme for χ^2_4, it is large enough for us to look for a possible poor fit at some points. The residual at unit 4 is 2.12, and that at unit 7 is −1.69. At unit 4 we have underestimated the observed proportion by a substantial amount: our model has smoothed the observed proportions too much.

We return to the two-factor model. Levels 1 and 2 of each factor differ very little. We try equating them by defining new factors with 3 levels, the first level being the same for levels 1 *and* 2 of the old factors:

```
$CALC CH = CHOL−%GE(CHOL,2) : B = BP−%GE(BP,2) $
$FACTOR CH 3 B 3 $
```

We can check by using $PRINT CH $ that this has had the desired effect.

```
$FIT CH + B $D E R $
```

The deviance increases from 8.08 on 9 df to 8.30 on 11 df, and the parameter estimates show nearly equal steps. Now we redefine CH and B as variables, and refit:

```
$VAR CH B $
$FIT CH + B $D E R  $
```

The deviance is 8.42 on 13 df and there are no large residuals. The parameter estimates of 0.61 for B and 0.72 for CH (with standard errors about 0.14) suggest that these could be equated: if $\beta_1 = \beta_2 = \beta$, then the model

$$\theta = \beta_0 + \beta_1 x_1 + \beta_2 x_2$$

reduces to

$$\theta = \beta_0 + \beta(x_1 + x_2),$$

a simple linear regression on the *total score* $x_1 + x_2$.

```
$CALC SCORE = B + CH $
$FIT SCORE $D E R $
```

The deviance of 8.74 on 14 df has increased very little, and the model fits well at all points, with a largest residual of -1.51 for unit 7, where the observed proportion is zero. The regression coefficient of 0.66 is halfway between those for B and CH.

How do we interpret the model? We have effectively a five-point scale for SCORE (2–6), and an increase of 1 point on this scale gives a predicted increase of 0.66 in the log-odds of having coronary heart disease, i.e. it *multiplies* the odds in favour of heart disease by $e^{0.66} = 1.93$, so the odds approximately double for each point up the scale. If the regression coefficient were $\log 2 = 0.6931$, the odds would *exactly* double. We try smoothing further by fixing the slope β at 0.6931, and estimating only the intercept β_0. This is achieved by defining an OFFSET of log 2 * SCORE:

```
$CALC OFS = %LOG(2) * SCORE $
$OFFSET OFS
$FIT  $D E $
```

Now we are fitting just the intercept as the SCORE slope is specified. The deviance is 8.86 on 15 df and the intercept is -5.047. The fitted logits, odds and probabilities at each point on the scale are:

Score	2	3	4	5	6
Logit	-3.660	-2.967	-2.276	-1.581	-0.888
Odds	0.0257	0.0515	0.1029	0.2058	0.4116
Probability	0.0251	0.0489	0.0933	0.1707	0.2916

The odds values are very nearly 1/40, 1/20, 1/10, 1/5, 2/5. Can we smooth the model any further? If the intercept had been -5.075 instead of -5.047, we would have had exactly the rounded-off odds values above. Since the standard error of $\hat{\beta}_0$ is 0.11, we might as well do the extra smoothing to present easily interpreted values. We use the method in Section 4.3 to fit a model with no estimated parameters.

```
$CALC OFS2 = -5.075 + OFS : ZERO = 0 $
$OFFSET OFS2
$FIT ZERO-1 $
```

The variable ZERO takes only zero values, and no intercept is fitted, so GLIM has fitted a completely specified model with no unknown parameters.

The deviance is 8.922. In presenting the data finally we can simplify by defining the score on the range 1–5 instead of 2–6, by subtracting 1. The observed and (fitted) probabilities are shown below using the original categorization, with the SCORE categories.

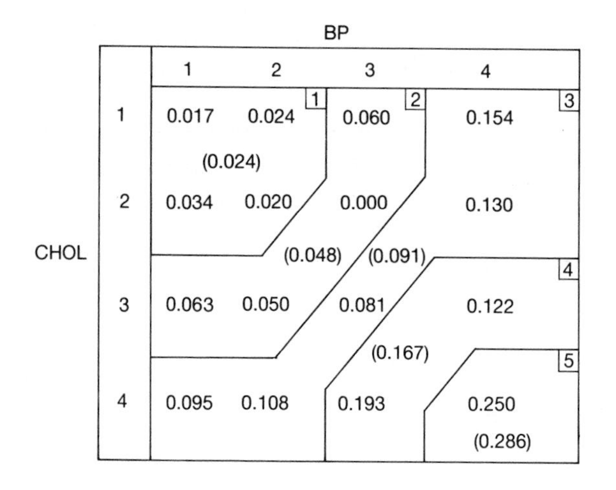

We do not quote standard errors for these fitted proportions (or fitted logits) because the purpose of presenting the smoothed fitted values is to give a simple interpretation to the data. Each step up the score scale doubles the odds on CHD, starting from odds of 1/40 at score 1. This gives a very simple and understandable relation between risk and the explanatory variables. A more precise statement might have been possible if the original values of blood pressure and serum cholesterol had been retained: the first example in this chapter shown how the model would then have been used. The categorization of the explanatory variables leads to a *simpler* interpretation, though we note that as far as the relation between CHD and the categorized variables is concerned, the lowest two categories for each variable do not need to be distinguished.

```
$END
```

4.8 The analysis of complex contingency tables with a binary response

The subfile BYSSINOSIS gives the number of workers in a survey of the US cotton industry suffering (YES) and not suffering (NO) from the lung disease byssinosis, together with the values of five cross-classifying categorical explanatory variables: the RACE, SEX and SMOKing habit of the worker, the length of EMPloyment in three categories, and the DUSTiness of the workplace in three categories. The data are presented and discussed in Higgins and Koch (1977).

The question of primary scientific interest was the relationship between the incidence of byssinosis and the dustiness of the workplace, but smoking habit and length of employment, and hence exposure to dust, are obviously relevant and need to be allowed for.

We first display the completely cross-classified table of proportions suffering from byssinosis.

```
$INPUT 1 BYSSINOSIS $
$CALC  N = YES+NO : P = YES/N $
```

A diagnostic message is printed here: the value of N in some cells is zero, and division by zero results in zero.

```
$ACC 2
$TPRINT P SMOK ; SEX ; EMP ; RACE ; DUST $
```

The table has a large number of zero values, but there is a noticeably high byssinosis rate at dust level 1. The accuracy setting gives two significant digits in the smallest value of p. Before modelling the table we examine marginal tabulations:

```
$ACC 3
$TAB THE P MEAN FOR DUST WITH N $
```

The first dust level has a 15.7% byssinosis rate; the other two are between 1% and $1\frac{1}{2}$%.

```
$TAB THE P MEAN FOR RACE WITH N $
```

The rate for whites is 2.6%, for non-whites 3.8%.

```
$TAB THE P MEAN FOR SEX WITH N $
```

The male rate of 4.4% is much higher than the female rate of 1.5%.

```
$TAB THE P MEAN FOR SMOK WITH N $
```

Smokers have a much higher rate, 3.9%, than non-smokers, 1.8%.

```
$TAB THE P MEAN FOR EMP WITH N $
```

Long and medium term employees have higher rates, 3.8% and 3.7%, than short-term employees, 2.3%.

```
$TAB THE P MEAN WITH N $
```

The overall byssinosis rate for all workers is 3.0%.

In Chapter 2 we observed the hazards of collapsing contingency tables over important explanatory variables: marginal tabulations may be misleading if explanatory variables are strongly associated. Such associations may be expected in survey data, and occur in this example, as is easily seen if we cross-tabulate by dust and race:

```
$TAB THE P MEAN FOR RACE ; DUST WITH N $
```

Now we see that the higher marginal byssinosis rate for non-whites is spurious: at each dust level the sample rate is in fact *lower* for non-whites than for whites! The higher marginal rate results from the much higher proportion of non-whites than of whites working in the high dust condition, as can be verified by a further tabulation:

```
$TAB THE N TOTAL FOR RACE ; DUST $
```

We now proceed to model the full table.

Seven cells in the five-way table have zero cases of both YES and NO, so have to be weighted out in the fit. Each empty cell corresponds to a non-estimable interaction parameter. We proceed to fit a sequence of logit models, with the main effects and interactions added in an arbitrary (but hierarchical) order.

```
$ACC
$CALC W=%NE(N,0) : N1=N+1-W $
$YVAR YES $ERROR B N1 $LINK G $WEIGHT W$ CYCLE 20$
$YVAR YES $ERROR B N1 $LINK G $WEIGHT W$ CYCLE 20$
%FIT : +DUST : +EMP : +SMOK : +SEX : +RACE :
      +DUST.EMP : + DUST.SMOK : DUST.SEX :
      +DUST.RACE : +EMP.SMOK : +EMP.SEX :
      + EMP.RACE : +SMOK.SEX : +SMOK.RACE :
      +SEX.RACE : +DUST.EMP.SMOK :
      +DUST.EMP.SEX : +DUST.EMP.RACE :
      +DUST.SMOK.SEX : +DUST.SMOK.RACE :
```

```
+DUST.SEX.RACE : +EMP.SMOK.SEX :
+EMP.SMOK.RACE : +EMP.SEX.RACE :
+SMOK.SEX.RACE $
```

The long sequences of models, computing time can be substantially reduced by using the directive $RECYCLE. This begins iterations of the IRLS algorithm for subsequent models with the iterative weights and adjusted dependent variate from the last FIT. This typically halves the number of cycles required for each model. For highly parametrized models with many zero observed proportions, however, repeated recycling can cause convergence difficulties.

The analysis of deviance table is shown below.

Source	Deviance	df
DUST	252.11	2
EMP	15.10	2
SMOK	11.43	1
SEX	0.30	1
RACE	0.32	1
DUST.EMP	4.51	4
DUST.SMOK	4.75	2
DUST.SEX	4.83	2
DUST.RACE	0.25	2
EMP.SMOK	1.94	2
EMP.SEX	2.04	2
EMP.RACE	7.33	2
SMOK.SEX	0.26	1
SMOK.RACE	1.26	1
SEX.RACE	0.45	1
DUST.EMP.SMOK	0.91	4
DUST.EMP.SEX	5.32	3*
DUST.EMP.RACE	1.52	4
DUST.SMOK.SEX	1.47	2
DUST.SMOK.RACE	0.07	2
DUST.SEX.RACE	1.31	2
EMP.SMOK.SEX	0.37	2
EMP.SMOK.RACE	1.22	2
EMP.SEX.RACE	0.00	2
SMOK.SEX.RACE	0.77	1
Estimable four- and five-way interactions	2.71	14

The immediately striking feature of this table is the very large deviance for DUST. However this is fitted first in the model, and it may be substantially less important when fitted after the other main effects. Interactions are much smaller, though some may be important. Note that DUST.EMP.SEX has only 3 df because one interaction parameter is not estimable.

The residual deviance for the estimable four- and five-way interactions is very small: the observed value of 2.71 is at the lower 0.06 percentage point of χ^2_{14}. This illustrates the failure of the asymptotic χ^2 distribution for the residual deviance when the table becomes very sparse, as the model more nearly reproduces the observed data. The failure of the residual deviance to follow the χ^2 distribution invalidates the simultaneous test procedure for model selection based on the likelihood ratio test, since this compares the size of the residual deviance from the current model to the χ^2 percentage point.

However, the same principles of model selection can be applied as in Section 3.12: we proceed by backward elimination from the saturated or "full" model, examining carefully the individual components of terms with multiple degrees of freedom, until a point is reached where the removal of any remaining term in the model produces a large change in deviance, so that no further simplification is possible without major distortion.

Inspection of the analysis of deviance table shows that all the three-way interactions are small and can be omitted, with the possible exception of DUST.EMP.SEX. We omit all the three-way interactions except this one, and examine its contribution. Typing of large numbers of interactions in specifying the model can be avoided by using the dot product of the main effects model with itself; GLIM interprets terms like DUST.DUST as main effects.

```
$FIT (DUST+EMP+SMOK+SEX+RACE).(DUST+EMP+SMOK+SEX+RACE)+
     DUST.EMP.SEX $
```

The deviance is 10.69, compared with 15.67 for the model with all two-way interactions. The deviance for DUST.EMP.SEX interaction is thus 4.98 when it is fitted as the first of the three-way interactions, and this interaction can be omitted as well.

```
$FIT - DUST.EMP.SEX $
```

Inspection of the analysis of deviance table shows that SMOK.SEX, SMOK.RACE and SEX.RACE can all be omitted, but EMP.RACE is large. We refit the model omitting the first three interactions and examine the parameter estimates to see which other terms can be omitted.

```
$FIT - SMOK.(SEX+RACE) - SEX.RACE  $D E $
```

The deviance increases by 1.96 on 3 df to 17.63. The parameter estimates for EMP.SMOK are both small, and this term can be omitted.

```
-0.0997   0.6260   EMP(2).SMOK(2)
-0.4368   0.4296   EMP(3).SMOK(2)
```

```
$FIT - EMP.SMOK $D E $
```

The deviance increases by 1.07 on 2 df to 18.70. The EMP.SEX parameter estimates are the least important, and we omit them:

```
-0.3540   0.9242   EMP(2).SEX(2)
-0.9906   0.6140   EMP(3).SEX(2)
```

```
$FIT - EMP.SEX $D E $
```

The deviance increases by 2.82 on 2 df to 21.52. The DUST.RACE terms are now the least important, and we omit them:

```
-0.8428   0.9740   DUST(2).RACE(2)
-1.151    0.7277   DUST(3).RACE(2)
```

```
$FIT - DUST.RACE $D E $
```

The deviance increases by 2.99 on 2 df to 24.51. The DUST.SMOK terms are now the least important and we omit them:

```
0.9957   0.5659   DUST(2).SMOK(2)
0.3782   0.4460   DUST(3).SMOK(2)
```

```
$FIT - DUST.SMOK $D E $
```

The deviance increases by 3.21 on 2 df to 27.72. Inspection of the parameter estimates suggests that DUST.EMP can be omitted, except perhaps for DUST(3).EMP(2):

```
-0.3618   0.7875   DUST(2).EMP(2)
-0.5624   0.5847   DUST(2).EMP(3)
-1.385    0.6786   DUST(3).EMP(2)
-0.5636   0.4234   DUST(3).EMP(3)
```

```
$FIT - DUST.EMP $D E $
```

The deviance increases by 5.52 on 4 df to 33.24. The EMP . RACE interactions are now quite small and can be omitted:

−0.7427	0.5575	EMP(2) . RACE(2)
−0.5947	0.4738	EMP(3) . RACE(2)

```
$FIT -EMP.RACE $D E $
```

The deviance increases by 2.48 on 2 df to 35.72. The DUST . SEX interaction is large and is retained, as are the marginal main effects of DUST and SEX. The RACE parameter estimate is 0.1127 with standard error 0.2065 and can clearly be omitted:

```
$FIT -RACE $D E $
```

The deviance increases by only 0.30 to 36.02. The SMOK and EMP main effects are large and must be retained. The fitted model at this point is given below (to three decimal places):

−1.753	0.156	1
−3.277	0.509	DUST(2)
−2.878	0.242	DUST(3)
0.464	0.251	EMP(2)
0.637	0.184	EMP(3)
−0.658	0.195	SMOK(2)
−0.999	0.611	SEX(2)
2.005	0.830	DUST(2) . SEX(2)
1.258	0.682	DUST(3) . SEX(2)

The two interaction terms differ by 0.75, or about one standard error, which suggests that they can be equated; the same is true for the main effect parameters. Equating DUST(2) and DUST(3) is equivalent to collapsing these two categories into one.

```
$FACTOR D2 2 $
$CALC D2 = 2 - %EQ(DUST,1) $
$FIT - DUST.SEX + D2 + D2.SEX $D E $
```

The aliased factor D2 is included in the model to allow a direct comparison between the parameter estimates for D2 . SEX and DUST . SEX.

The deviance increases by 1.43 on 1 df to 37.44, and the parameter estimates for DUST(2) and DUST(3) are now almost equal. We remove the three-level DUST factor and retain the two-level D2 factor.

```
$FIT - DUST $D E $
```

The deviance increases by 0.12 to 37.56. A similar recoding of EMP can be used to equate levels 2 and 3:

```
$FACTOR E2 2 $
$CALC E2 = 2 - %EQ(EMP,1) $
$FIT -EMP + E2  $D E R $
```

The deviance increases by 0.54 to 38.11. Only one residual exceeds 2.0 in magnitude: 2.27 at the 38th cell, where an observed proportion of $1/17=0.059$ with byssinosis is modelled by a fitted proportion of $0.144/17=0.008$. Thus the fitted model seems satisfactory. Parameter estimates are shown below, to three decimal places.

```
-1.758   0.157   1
-0.649   0.194   SMOK(2)
-0.993   0.611   SEX(2)
-2.949   0.227   D2(2)
 1.453   0.661   D2(2).SEX(2)
 0.592   0.172   E2(2)
```

Before proceeding to interpret this model, a word of warning is in order. We equated levels 2 and 3 of EMP because the difference between their parameter estimates was not large compared with the individual standard errors. Close examination of these parameter estimates suggests two other possible interpretations, however:

(1) since EMP(2) does not differ significantly from zero, levels 1 and 2 of EMP could instead be equated;

(2) since the values 0, 0.462 and 0.641 are roughly linear, we could replace the factor EMP by a variate taking the same values. We illustrate the latter model:

```
$CALC LINE = EMP $
$FIT -E2 + LINE  $D E $
```

The deviance is 37.93, slightly smaller than the value 38.11 when levels 2 and 3 of EMP are equated. We cannot choose between these models: the data do not provide enough evidence to discriminate between the model in which risk is the same for the two longer-term exposure categories and that in which risk increases steadily with exposure. To make such a discrimination the actual number of years worked in the industry for each individual would be necessary, or at least a finer classification of EMP. The parameter estimates for variables other than EMP are very little affected by this ambiguity: we base our interpretation on the model below:

-2.047	0.218	1
-0.665	0.194	SMOK(2)
-1.021	0.609	SEX(2)
-2.950	0.227	D2(2)
1.478	0.661	D2(2).SEX(2)
0.319	0.091	LINE

The odds on having byssinosis for smokers are greater than for non-smokers by a factor of $e^{0.665} = 1.94$, or approximately 2, and increase steadily (in this model) with increasing duration of employment in the industry. Interpretation of other features of the model is simplified if we tabulate the fitted proportions by the factors SMOK, EMP, SEX and D2. RACE can be omitted.

```
$CALC FP = %FV/N
$TAB THE FP MEAN FOR SMOK;EMP;D2;SEX WITH N $
```

The sample sizes can also be tabulated in the same way.

```
$TAB THE N TOTAL FOR SMOK;EMP;D2;SEX $
```

The fitted proportions and (sample sizes) are shown below so that the fitted proportion with byssinosis increases across and down the table. Note that two cells have no observations.

		Dust 2/3		Dust 1	
		Women	Men	Women	Men
Non-	EMP 1	0.0075(676)	0.0048(340)	0.0318(29)	0.0836(97)
smokers	EMP 2	0.0103(129)	0.0065 (82)	— (0)	0.112 (20)
	EMP 3	0.0141(537)	0.0090(248)	0.0586 (2)	0.147 (70)
Smokers	EMP 1	0.0145(687)	0.0092(667)	0.0601(29)	0.151(204)
	EMP 2	0.0198(137)	0.0126(277)	— (0)	0.196 (67)
	EMP 3	0.0270(275)	0.0173(695)	0.108 (2)	0.251(149)

It is now easily seen that there is a striking increase in the incidence of byssinosis for men in the dustiest working conditions. Another interesting feature is that smokers with less than 10 years' employment in the industry have the same incidence rates as non-smokers with more than 20 years' employment. The incidence of byssinosis is very high—one in four—amongst smoking men with more than 20 years' employment who are working in the dustiest conditions.

4.9 Overdispersion, omitted variables and model mis-specification

A phenomenon which occurs occasionally with binomial and Poisson data is *overdispersion*. When a contingency table is analysed, it may be found that high-order interactions are inexplicably significant, and sometimes even that no model other than the saturated model can provide a satisfactory representation of the data. This phenomenon frequently occurs with very large samples where the contingency table is not classified by some factors relevant to the response: the number of "successes" in each cell of the table then has a *mixed* binomial (or binomial mixture) distribution. That is, within a given cell the probability p of a success is not constant, but varies systematically with other factors which have not been recorded or included in the model. Since these factors are not identified, the success probability p behaves like a random variable with a probability distribution.

If Y given p has the binomial distribution $b(r|n,p)$, and p has a distribution with mean μ and variance σ^2, then

$$\begin{aligned} E(Y) = E[E(Y|p)] &= nE(p) = n\mu, \\ \text{var } Y &= \text{var}[\text{E}(Y|p)] + E[\text{var}(Y|p)] \\ &= n^2\sigma^2 + nE[p(1-p)] \\ &= n^2\sigma^2 + n[\mu-(\mu^2+\sigma^2)] \\ &= n\mu(1-\mu) + n(n-1)\sigma^2 \\ &> n\mu(1-\mu) \end{aligned}$$

except when $n=1$. Thus the effect of randomly varying p is to increase the variance of Y above that for the binomial distribution, leading to large residual deviances for models which would fit well if the random variation were correctly specified.

The same phenomenon occurs when the observations are not independent because the sample design is *clustered* (see Section 2.9). The omission of the cluster identification, or the failure to model the correlation between responses in the same cluster using a variance component model, will produce similar mixed binomial distributions within the cells of the table.

The difficulty in such cases is specifying and fitting a model for the distribution of p. The *beta-binomial* distribution has been frequently used (for example by Crowder, 1978 and 1979; Griffiths, 1973; Chatfield and Goodhardt, 1970 and Williams, 1975), in which p has a beta distribution. Williams (1982) considers a model in which the variance of the distribution of p is a specified function of the mean, $\sigma^2 = \phi\mu(1-\mu)$, so that

$$\mathrm{var}(Y) = n\mu(1-\mu)\{1+(n-1)\phi\}.$$

Williams (1982) gives a GLIM listing for fitting this model using quasi-likelihood (see McCullagh and Nelder, 1983).

A simpler procedure is to assume that $\mathrm{var}(Y) = n\mu(1-\mu)\psi$ where ψ is an unknown scale parameter, and to treat the residual deviance divided by its degrees of freedom, or the Pearson chi-square %X2 divided by %DF, as an estimate of ψ: $CALC %S=%DV/%DF. This is analogous to the residual mean square estimate of σ^2 in a normal model. Setting the scale parameter equal to %S using $SCALE %S $, and comparing the resulting scaled deviances for terms in the model with percentage points of the F-distribution, gives a procedure directly analogous to standard analysis of variance and regression procedures. McCullagh and Nelder (1983) give a discussion of this approach (see their entries under "Overdispersion").

This procedure is probably satisfactory for small amounts of overdispersion, but it is not a substitute for correct model specification, and it is impossible to test the goodness of fit of the model. Further, overdispersion will not necessarily be visible in a table even when it has been collapsed over important variables. We illustrate with the previous byssinosis example.

Suppose that DUST is not included as an explanatory variable in the model. For the five-way table, the sequence of models omitting DUST is now fitted:

```
$FIT : +EMP : +SMOK : +SEX : +RACE : +EMP.SMOK
  : +EMP.SEX : +EMP.RACE : +SMOK.SEX : +SMOK.RACE
  : +SEX.RACE : EMP.SMOK.SEX : +EMP.SMOK.RACE
  : + EMP.SEX.RACE : + SMOK.SEX.RACE
  : +EMP.SMOK.SEX.RACE $
```

The analysis of deviance table is shown below.

Source	dev	df
EMP	10.24	2
SMOK	21.33	1
SEX	25.73	1
RACE	23.30	1
EMP.SMOK	0.93	2
EMP.SEX	0.74	2
EMP.RACE	0.55	2
SMOK.SEX	0.26	1
SMOK.RACE	0.03	1
SEX.RACE	6.78	1
EMP.SMOK.SEX	1.85	2
EMP.SMOK.RACE	0.75	2
EMP.SEX.RACE	0.29	2
SMOK.SEX.RACE	1.59	1
EMP.SMOK.SEX.RACE	0.00	2
Residual	228.2	41

The residual deviance consists of the main effect of DUST and all its interactions with the other factors. The residual is very large, and it is clear that DUST is a very important variable and should be included in the model. Without it, large sex and race main effects appear because of their correlation with DUST: men and non-white people have higher byssinosis rates than women and white people, because the former are working more frequently in the higher dust conditions. From the full model including DUST we know that RACE is in fact irrelevant: there are no differences in incidence between races when we allow for dust condition of the workplace.

If, however, dust level had never been measured, the contingency table would have had only four dimensions. It can be verified, using the $TABULATE directive to collapse the five-way table over dust level, that the same deviances are obtained for the same terms in the four-way table analysis, except that of course there is no "residual" term: the model with all interactions up to the four-way interaction fits the four-way table exactly.

Thus collapsing of the five-way table over the most important explanatory variable leads to a mis-specified model for the four-way table. As in the simple 2×2 table example in Section 2.11, inappropriate collapsing of the table over an important variable leads to distorted conclusions about the importance of other explanatory variables.

This is, of course, a familiar problem in normal regression analysis. It illustrates the need, in contingency table analysis of observational data, to measure all relevant variables (so far as this can be known or determined). The resulting high dimension of the contingency table need not be a serious limitation on the analysis of binary responses: at the worst each cell of the table will contain either one observation which is a "success" of a "failure" or no observations, so the maximum length of the data vector is the number of binary responses. Cells with no responses do not need to be read as data, if each response is read with a level code for each explanatory variable. (For *multinomial* data, however, there *is* a serious limitation on the number of dimensions of the table: see Chapter 5.) Paradoxically, the higher the dimension of the table, the more likely we are to obtain a simple final model: "more is less".

5
Multinomial and Poisson response data

5.1 The Poisson distribution

Subfile CLAIMS gives the number of policyholders N of an insurance company who were "exposed to risk", and the number C of car insurance claims made in the third quarter of 1973 by these policyholders arranged as a contingency table, cross-classified by three four-level factors: DIST, the district in which the policyholder lived, CAR, the insurance group into which the car was placed, and AGE, the age of the policyholder. The definitions of the factors are given in Appendix 4. We wish to model the relation between the frequency of claims and the explanatory variables. The data come from Baxter *et al.* (1980).

At first sight the appropriate probability model appears to be binomial: we might reasonably suppose that each policyholder has a constant probability p of making a claim, so that the number C would be binomial $b(N,p)$, with p modelled as a logistic function of the explanatory variables. However, it is possible (though unlikely) for any policyholder to submit several claims during the quarter, so the binomial distribution may not be appropriate: we examine the use of the binomial distribution at the end of this section.

The *Poisson* distribution provides the standard model for count data. The number of claims R submitted by a policyholder during the quarter is modelled by

$$\Pr(R=r|\lambda)=e^{-\lambda}\,\lambda^r/r!, \qquad \lambda>0, \qquad r=0,1,2,\ldots$$

where λ is the mean number of claims, and is related to the linear predictor $\eta=\boldsymbol{\beta}'\mathbf{x}$ through

$$g(\lambda)=\eta$$

where the link function g is generally taken to be the log, guaranteeing positive fitted values:

$$\log\lambda=\eta=\boldsymbol{\beta}'\mathbf{x}$$

though the identity link is sometimes used. The fitting of the Poisson model in GLIM is achieved using $ERROR P and the appropriate link and model specifications.

In the i-th cell of the contingency table, there are N_i policyholders. The number of claims C_{ij} made by the j-th policyholder in the i-th cell is assumed to have a Poisson distribution with mean λ_i, and for each cell, the C_{ij} are assumed independent. The total number of claims

$$C_i = \sum_{j=1}^{N_i} C_{ij}$$

submitted by all the policyholders in the i-th cell is then the sum of N_i independent Poisson variables each with mean λ_i, and has a Poisson distribution with mean $\theta_i = N_i\lambda_i$.

If the C_{ij} were observed, the model $\log \lambda_i = \boldsymbol{\beta}'\mathbf{x}_i$ could be fitted directly. Since only C_i is observable, we can only fit a model to θ_i. We have

$$\begin{aligned}\log \theta_i &= \log N_i + \log \lambda_i \\ &= \log N_i + \boldsymbol{\beta}'\mathbf{x}_i\end{aligned}$$

Thus to specify the model correctly we must include the term $\log N_i$ as an explanatory variable with a coefficient of 1, that is, $\log N_i$ must be taken as an OFFSET for the model.

```
$INPUT 1 CLAIMS $
$YVAR C $ERROR P $LINK L
$CALC LN = %LOG(N)  $OFFSET LN
$FIT : + DIST : + CAR : +AGE : +CAR.AGE : + DIST.CAR :
    + DIST.AGE $
```

Here CAR.AGE is fitted before the other interactions as the one most likely to be important. The analysis of deviance table for the factors in the order fitted is:

Source	Deviance	df
DIST	12.73	3
CAR	87.24	3
AGE	84.87	3
CAR·AGE	10.51	9
DIST·CAR	7.38	9
DIST·AGE	6.24	9
DIST·CAR·AGE	27.29	27

The values of C are generally large, so the standard asymptotic χ^2 distribution theory for deviances should hold adequately. The three-way interaction term is not explicitly fitted as its deviance is just the residual deviance from the model with all two-way interactions. All the interactions are close to their degrees of freedom, suggesting that the main effects model with a residual deviance of 51.42 and 54 degrees of freedom provides a good fit. This model was used by Baxter *et al.* (1980). We check the fit at each sample point.

```
$FIT DIST + CAR + AGE  $D R $
```

The Pearson residuals for the Poisson distribution are

$$e_i = (y_i - \hat{\theta}_i)/\sqrt{\hat{\theta}_i}.$$

These are approximately standardized variables with mean 0 and variance approximately 1; the variance is approximate because of the estimation of β in the linear predictor.

There are two residuals exceeding 2 in absolute value: -2.279 at observation 9 and 2.101 at observation 17. Since the observed numbers of claims C are fairly large, the residuals e_i can be treated as approximately standard normal variables, and a normal quantile plot provides a check on their assumed distribution. We first need to give the GLIM expression for the residuals and store it in a macro before using QPLOT (see Section 2.10).

```
$INPUT %PLC QPLOT $
$MACRO PRS (%YV-%FV)/%SQRT(%FV) $ENDMAC
$USE QPLOT PRS $
```

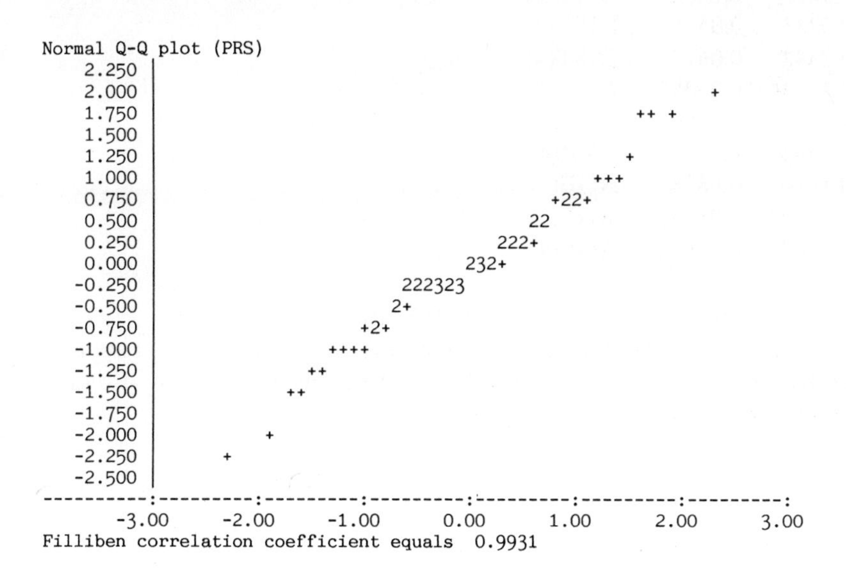

The fit to a straight line is very close, with a Filliben correlation of 0.993, so the assumption of a Poisson model is well supported.

Car type and age look very important but district less so. Can it be omitted from the model?

```
$FIT - DIST $
```

The deviance increases by 13.87 on 3 df, more than the reduction when DIST is fitted first. All three variables need to be retained.

In fitting the model, we have specified that LN has a fixed coefficient of 1. This is implicit in the structure of the model, but it may not be in accordance with the data. To check this, we fit LN explicitly, removing the offset.

```
$OFFSET
$FIT + DIST + LN $D E $
```

The deviance decreases by only 1.97 on 1 df, and the parameter estimate for LN is 1.202 with standard error 0.144. The fixed value of 1 *is* in accordance with the data. Now we examine the other parameter estimates.

```
$OFFSET LN
$FIT -LN  $D E $
```

−1.822	0.0768	1
0.0259	0.0430	DIST(2)
0.0385	0.0505	DIST(3)
0.2342	0.0617	DIST(4)
0.1613	0.0505	CAR(2)
0.3928	0.0550	CAR(3)
0.5634	0.0723	CAR(4)
−0.1910	0.0828	AGE(2)
−0.3450	0.0814	AGE(3)
−0.5367	0.0699	AGE(4)

The parameters for DIST(2) and DIST(3) are smaller than their standard errors, but that for DIST(4) is nearly four times its standard error. It appears that the first three districts can be amalgamated, only district four—London and other major cities—being kept separate.

```
$CALC D4 = %EQ(DIST,4)  $FIT -DIST +D4 $D E $
```

The deviance increases by 0.72 on 2 df to 52.14.

The CAR estimates and the AGE estimates show similar patterns of nearly linear increase or decrease with factor level, suggesting that each factor can be replaced by the corresponding variable.

```
$CALC LCAR = CAR  $FIT -CAR +LCAR $D E $
```

The deviance increases by 0.86 on 2 df to 52.99.

```
$CALC LAGE = AGE  $FIT -AGE +LAGE  $D E $
```

The deviance increases by 0.12 on 2 df to 53.11. The parameter estimates are shown below.

−1.853	0.080	1
0.219	0.059	D4
0.198	0.021	LCAR
−0.177	0.018	LAGE

The parameter estimates for LCAR and LAGE differ in magnitude by one standard error: it appears that we can equate their magnitudes.

```
$CALC SCORE = LCAR - LAGE  $FIT D4 + SCORE $D E $
```

The deviance increases by 0.58 on 1 df to 53.70.

−1.795	0.024	1
0.220	0.058	D4
0.186	0.014	SCORE

The coefficient of D4 is close to that for score: being in district 4 is nearly equivalent to one point on the SCORE scale. We try making them equivalent.

```
$CALC S2 = D4 + SCORE  $FIT S2  $D E $
```

The deviance increases by 0.32 on 1 df to 54.02 (with 62 df).

−1.789	0.022	1
0.188	0.014	S2

Thus each point up the SCORE scale adds 0.188 to the fitted mean log number of claims, so the fitted mean is multiplied by $\exp(0.188) = 1.207$. The SCORE scale runs from -3 to 3, with an extra point for district 4; the fitted mean number of claims for an individual with scale score -3 is $\exp(-1.789 - 3 \times 0.188) = 0.095$.

We present finally the observed mean numbers of claims per individual, C_i/N_i by CAR and AGE, for districts 1–3, and for district 4 separately, together with the fitted means. These are arranged in a table with the age classification reversed, so that SCORE increases from the top left-hand to the bottom right-hand corner. Within each cell of the table, the order of districts is

1 2,
3

with district 3 below district 1. We label the cells with the value of SCORE +4, which runs from 1 to 8.

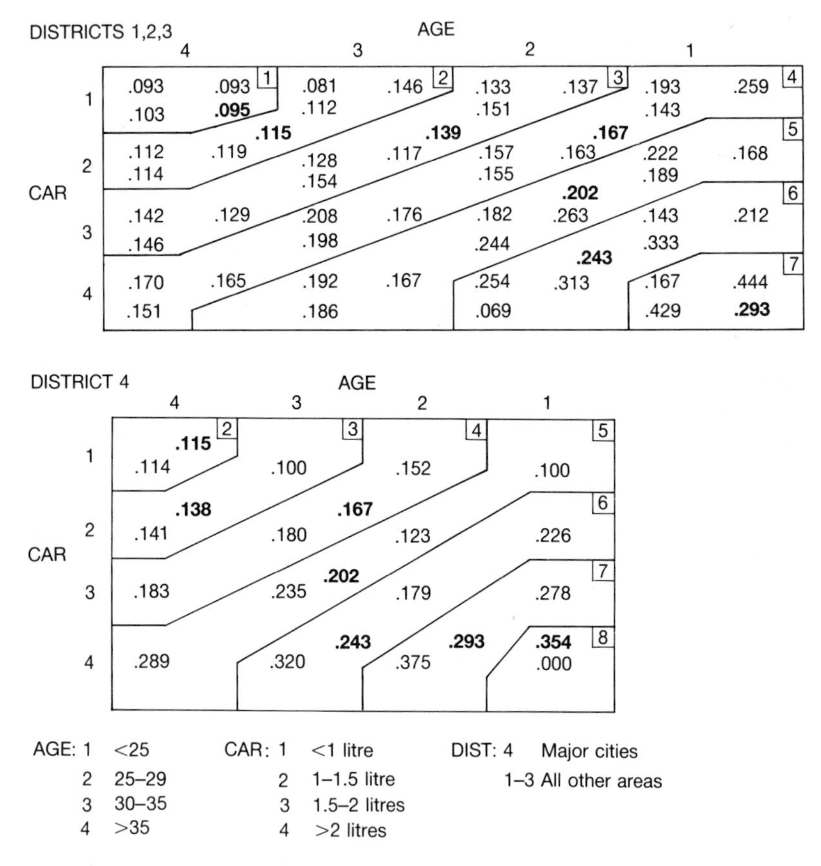

The fitted values accurately reproduce the observed means where the numbers exposed are large. The simplification of the table, as in the example in Chapter 4 on coronary heart disease, is very great: only two numbers are necessary to describe the whole 64-cell table.

It can be verified that a very similar main effects model results, with a deviance of 63.09 on 54 df, if C is assumed to be binomial $b(N,p)$ with a logit link for p (Baxter *et al.*, 1980). This is because the mean claim frequency is low, even in the highest score category, and so the probability of more than one claim in the quarter, which is $\{1-(\lambda+1)e^{-\lambda}\}$, is very small—0.049 for the largest mean value of 0.353.

```
$END
```

5.2 Overdispersion

In Section 4.9 we examined the effect of omitted variables on a binomial logit model, and discussed the problem of overdispersion. Similar problems arise with Poisson models for count data. An example is the data on absence from school, previously analysed using a lognormal model in Section 3.12. Days absent is recorded as a whole-number count which takes zero values for several children. These zero values had to be adjusted by a positive constant in the lognormal analysis before the log transformation could be used.

The Poisson distribution provides a more natural model for such data. We now reanalyse the data using this model. The log link is used to retain the multiplicative mean structure of the lognormal analysis.

```
$INPUT 1 QUINE $
$YVAR DAYS $ERROR P $LINK L
$FIT C*S*A*L $D E $
```

This large model will require substantial processing time on most computers.

The deviance is 1173.9 with 118 df for the saturated model. If the Poisson distribution holds and the regression model is correctly specified, the deviance should be of the same order as the df, or somewhat smaller for small cell sample sizes. Instead, it is ten times as large. Thus either the regression model is mis-specified, or the distribution of days absent within a cell is not Poisson.

The failure of the Poisson model is evident from inspection of the observations in individual cells. For example, the ten observations for the Aboriginal boy slow learners in the first form of high school are:

5 6 6 9 13 23 25 32 53 54

with a mean of 22.6. These observations could not possibly come from a single Poisson distribution: they are much too dispersed, with $\Sigma(y_i-\bar{y})^2/9=349.2$. A Poisson variable has variance equal to its mean: here the sample variance is 15.4 times as large as the mean.

One possible reason for overdispersion is that illnesses tend to produce clustering in absences: if a child is ill and absent one day, he or she is more likely to be away for a further day or days. Although we have excluded children with serious illnesses from the analysis, the pattern of absence from non-serious illnesses may still lead to overdispersion. A related possibility is that children vary in absence proneness and that an absence of (say) 54 days represents repeated absences for short periods.

If variation in absence-proneness, or clustering of absences, is related to the explanatory variables, then we run a serious risk of mis-specifying the model by any analysis: the only satisfactory course of action is to attempt to identify the missing variables which explain this heterogeneity.

If the variation is *not* related to the explanatory variables, then the simple analysis of Section 4.9 is available. Suppose that Y has a discrete distribution with $E(Y)=\mu$, var $(Y)=\sigma^2\mu$ and log $\mu=\boldsymbol{\beta}'\mathbf{x}$. Then the quasi-likelihood estimate of $\boldsymbol{\beta}$ (McCullagh and Nelder, 1983) is the same as for the Poisson distribution, and standard errors for $\hat{\boldsymbol{\beta}}$ may be obtained by multiplying those from the Poisson model by an estimate of σ^2, a simple scale factor obtained by dividing the residual deviance or the Pearson χ^2 by the residual degrees of freedom. This is very easily achieved.

```
$CALC %S = %DV/%DF $
$PRINT %S $
```

The scale factor is 9.948.

```
$SCALE %S $FIT . $D E $
```

The scale has to be specified as a number or a scalar.

The estimates remain the same, but the scaled deviance is now equal to the degrees of freedom, and the parameter standard errors have increased by a factor of 3.15 ($=\sqrt{9.948}$). Elimination of terms from the model can proceed using the scaled deviance increment as approximately $\chi^2{}_r$ (or $r\mathrm{F}_{r,\,118}\approx\chi^2_r$) just as for a Poisson (or normal) model.

We do not give the full model simplification here since the method parallels closely the lognormal analysis in Section 3.12. The C.A.L interaction is now larger, and 4.64 is concentrated in 1 df. The C.A interaction is much larger—9.29 on 3 df—so large that it should not be omitted. The presence of *two* large three-way interactions—C.A.L and C.S.L—considerably complicates the interpretation, since the cultural groups, or the learning groups, have to be modelled separately.

A difficulty with the quasi-likelihood approach is that it does not in general correspond to a probability model. If the Poisson rate (mean) λ is not constant within a cell but varies because of additional individual variation, this leads to a mean/variance relationship which is in general not that specified in the quasi-likelihood analysis. If, conditional on λ, Y is Poisson distributed with mean λ, and marginally λ has a distribution with mean μ and variance σ^2, then unconditionally Y has mean μ and variance $\mu+\sigma^2$, and this will not be proportional to μ unless σ^2 is proportional to μ. This is not the case, for example, when λ has a gamma distribution when Y is then negative binomial. We found in Section 3.12 that the log transformation (of DAYS + 1) stabilised variances, implying that standard deviation rather than variance is proportional to mean. The generality of the quasi-likelihood analysis is severely limited, and modelling of overdispersed data in general requires the explicit fitting of compound distributions. We do not pursue this point further, but note that GLIM macros for fitting the Poisson/normal compound distribution have been developed by Hinde (1982).

```
$END
```

5.3 Multicategory responses

Subfile MINERS gives the numbers of coalminers classified by radiological examination into one of three categories of pneumoconiosis (N—normal, M—mild pneumoconiosis, S—severe pneumoconiosis), and by period P spent working at the coalface. P is declared as a factor with eight levels, the levels corresponding to midpoints of class intervals: 5.8, 15.0, 21.5, 27.5, 33.5, 39.5, 46.0, and 51.5 YEARS worked at the coalface. The data are discussed in Ashford (1959) and in McCullagh and Nelder (1983, p. 113).

We first plot the proportions of miners in each category against years worked.

```
$INPUT 1 MINERS $
$CALC  T = N+M+S : NP = N/T : MP = M/T : SP = S/T $
$PLOT NP MP SP YEARS $
```

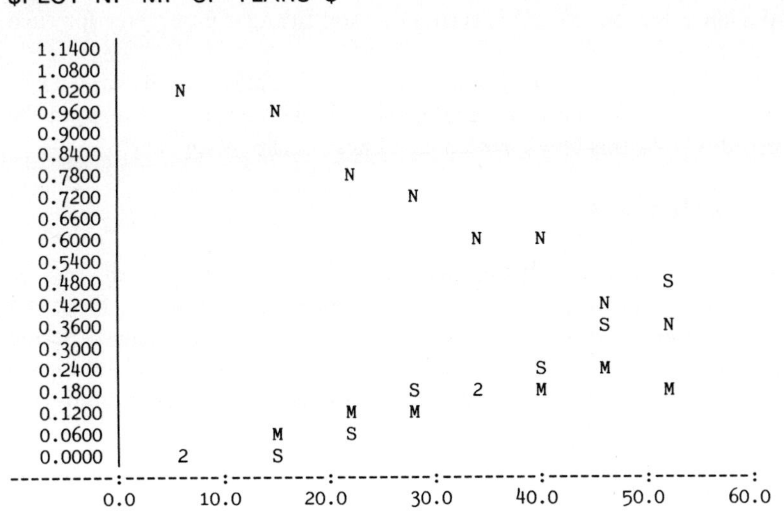

The proportions free of pneumoconiosis decline steadily with years, and those for the other two categories increase; there is a noticeable increase in the proportion with severe pneumoconiosis in the 7th and 8th periods.

How should multiple response proportions be modelled? One simple way would be to convert the three categories to two by collapsing two adjacent categories into one. For example, the second and third categories could be collapsed, and we could model the logit of the probability of no pneumoconiosis.

```
$YVAR N $ERROR B T $LINK G
$FIT : + P $D E $
```

```
scaled deviance =  97.56 at cycle  4
           d.f. =   7

scaled deviance = 0.0003 (change =  -97.56) at cycle 10
           d.f. = 0       (change =   -7    )
    (no convergence yet)

          estimate          s.e.      parameter
    1        13.30         47.35      1
    2       -10.47         47.35      P(2)
    3       -11.97         47.35      P(3)
    4       -12.31         47.35      P(4)
    5       -12.78         47.35      P(5)
    6       -12.87         47.35      P(6)
    7       -13.59         47.35      P(7)
    8       -13.86         47.35      P(8)
    scale parameter taken as   1.000
```

The GLIM message 'no convergence yet' appears when the second model is fitted but the deviance of 0.0003 indicates that convergence is very close. The exact deviance for this model is zero as we are fitting a parameter for each observation.

The fitted logits decline steadily with period, as we have already seen. Their large negative values reflect the zero proportion with pneumoconiosis at the first period. Could this be expressed as a linear decline with year?

```
$FIT YEAR $D E R $
```

The linear model does not fit well, with a deviance of 11.57 on 6 df, and the residuals are not satisfactory: although the largest is only 1.801, there is a systematic pattern of positive, negative and then positive residuals as period increases. This suggests a systematic failure of the model. We could attempt to improve the fit by using a higher-order polynomial, but we first examine the complementary log–log link.

```
$LINK C $FIT . $D E R $
```

The fit is considerably improved, with a deviance of 4.96, and the largest residual is only 1.00, but the same pattern of positive, negative and positive residuals appears. We try to remove the systematic sign changes in the residuals by a log transformation of years.

```
$CALC LY = %LOG(YEARS) $
$FIT LY  $D E R $
```

The fit is greatly improved again over the linear model, with a deviance of 1.41, and no systematic pattern now appears in the residuals. We calculate the fitted probabilities from the model for later plotting.

```
$CALC CNP = %FV/%BD $
```

Note the use of the system vector %BD, the binomial denominator. We now try the log transformation of years with the logit link.

```
$LINK G $FIT . $D E R $
```

The fit is not as good, with a deviance of 3.13, but the difference is not large. No residual pattern is evident. We plot the observed proportions and the two sets of fitted proportions against years.

```
$CALC LNP = %FV/%BD $
$PLOT CNP LNP NP YEARS $
$PRINT CNP :: LNP :: NP $
```

The fitted proportions agree closely, and an accurate plot (see Fig. 5.1) is necessary to see the difference between the fitted models. The complementary log–log model is flattening out at the 8th period, while the logit model continues to decline. However, the practical conclusions are the same.

While this analysis could be repeated for M and S, the resulting models for the proportions in the three categories would not in general give fitted probabilities summing to 1, and the models themselves might involve different functions of YEARS or different link functions. We need to model simultaneously the probabilities for all the categories. This is achieved in GLIM using the *multinomial logit model*.

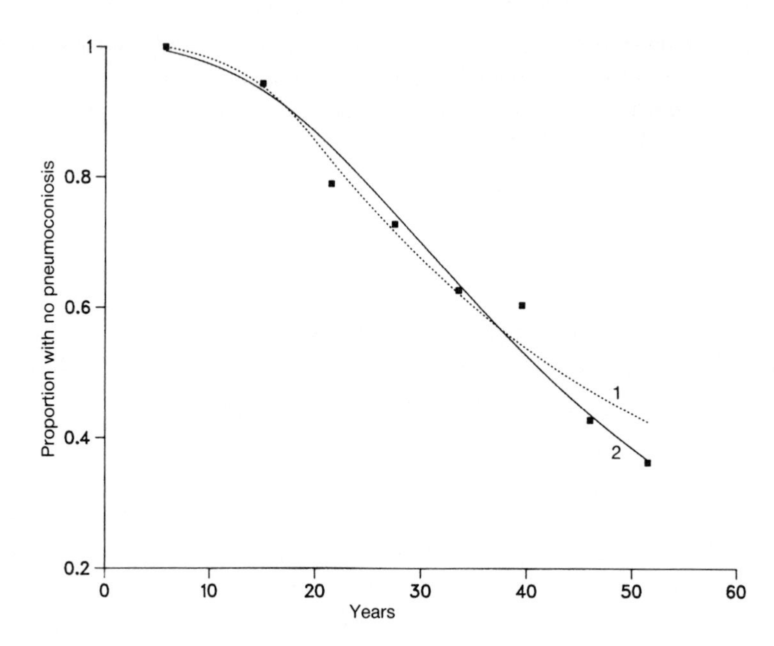

Fig. 5.1. Fitted models for probability of no pneumoconiosis in MINERS data (1) logit link (2) complementary log–log link

5.4 Multinomial logit model

Suppose the response variable has r categories, with category probabilities p_j, $j=1,2,\ldots, r$. For the moment we ignore any *ordering* which may exist in the response categories; models for ordered categories are considered in Section 5.7. The probability that in a sample of n observations we obtain n_j in the j-th category ($j=1,2,\ldots, r$) is

$$p(n_1,n_2,\ldots, n_r) = \frac{n!}{n_1!n_2!\ldots n_r!}\, p_1^{\,n_1}\, p_2^{\,n_2}\ldots p_r^{\,n_r}$$

with

$$\sum_{j=1}^{r} p_j = 1,$$

that is there are only $r-1$ distinct probabilities.

The multinomial logit transformation of $p_1,\ldots, p_r$ is the set of parameters $\theta_1, \theta_2,\ldots, \theta_r$, called multinomial logits, defined by

$$\theta_j=\log\,(p_j/p_1), \qquad j=1,\ldots, r$$

Note that $\theta_1 \equiv 0$, so there are only $r-1$ distinct logit parameters. This definition uses the convention that the first category is the *reference* category for the others: this is convenient in using GLIM because the dummy variable coding for factors in GLIM uses the same convention. However, any other category could be used as the reference category without changing the fit of the model: if we define

$$\theta_j{}^* = \log\,(p_j/p_r), \qquad j = 1, \ldots, r$$

for example, with $\theta_r{}^* = 0$, then the θ^*s are just a reparametrization of the θs, since

$$\theta_j{}^* = \theta_j + \log\,(p_1/p_r).$$

The multinomial logits play the same role in the analysis of multi-category response data as the binomial logit does in two-category response data. We could also define *multinomial probits* analogously to the binomial probits in Chapter 4, but we will not use these as they cannot be fitted in GLIM.

For the MINERS data, taking the normal category as the reference category, we can construct the *empirical logits* $\log\,(M_i/N_i)$ and $\log\,(S_i/N_i)$:

```
$CALC L2 = %LOG(M/N) : L3 = %LOG(S/N) $
$PRINT L2 : : L3 $
```

The warning message occurs because of the infinite values of L2 and L3 for the first observation. The logits are shown below.

L2:	$-\infty$	-3.24	-1.74	-1.95	-1.16	-1.19	-0.69	-0.69
L3:	$-\infty$	-3.93	-2.43	-1.48	-1.27	-1.06	-0.18	0.22

```
$PLOT L2 YEARS '+' : L3 YEARS '+' $
```

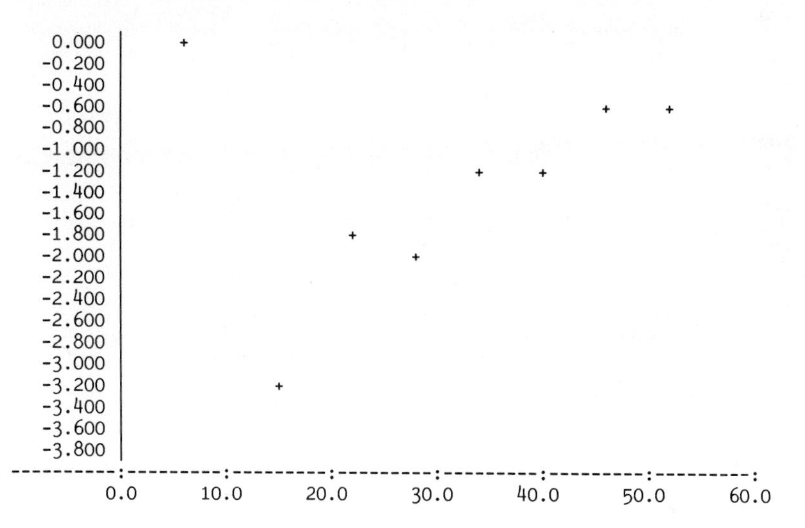

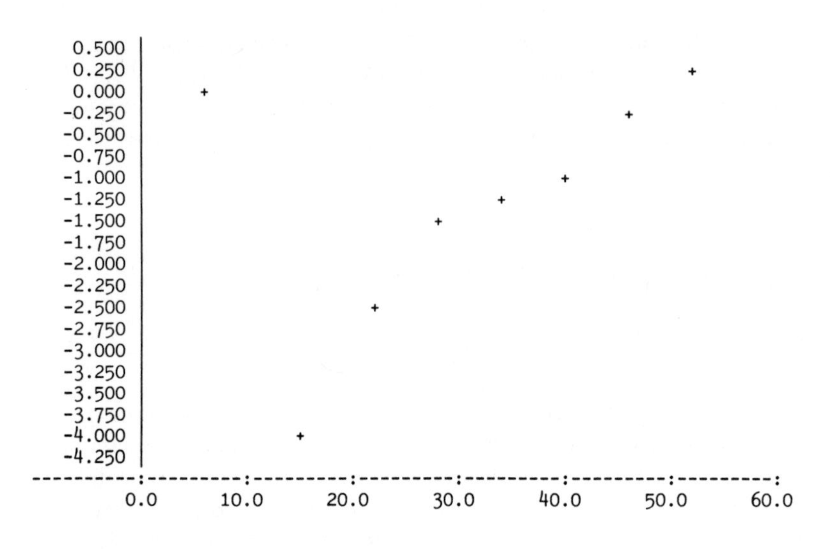

Ignoring the first point in each case, the logits show a rapid increase with YEARS. Curvature in the plots is very marked, clearly indicating the need for a log transformation of the YEARS scale.

```
$PLOT L2 LY '+' : L3 LY '+' $
```

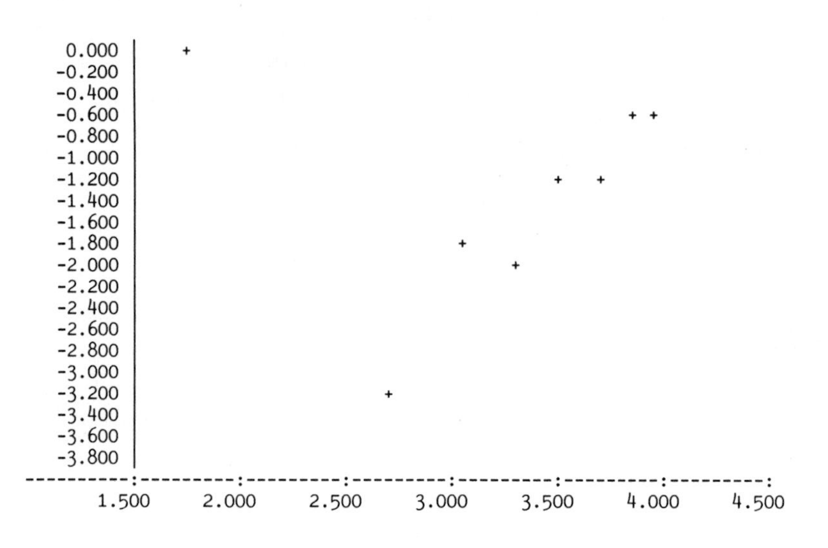

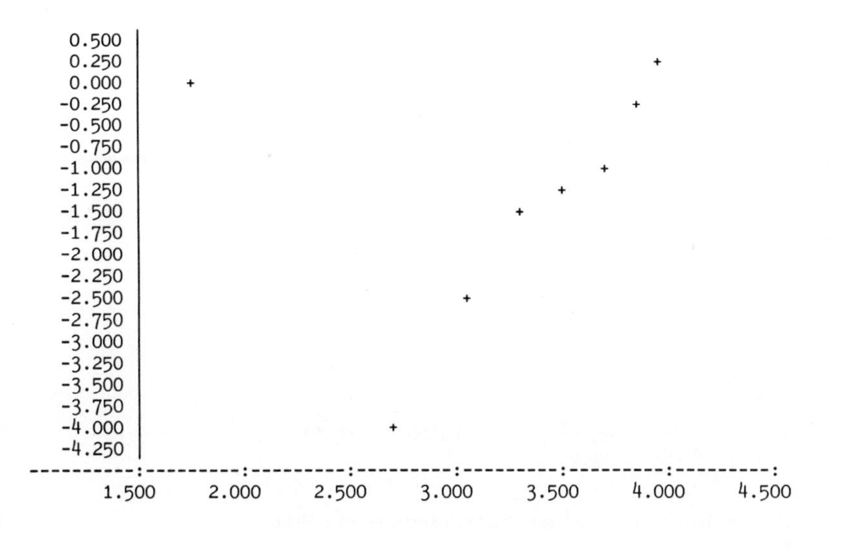

The trends are now nearly linear, with a steeper slope for the severe category.

We want then to fit simultaneously the multinomial logit models

$$\begin{aligned} \theta_{2i} &= \beta_{20} + \beta_{21} \log(\text{YEARS}_i) \\ \theta_{3i} &= \beta_{30} + \beta_{31} \log(\text{YEARS}_i) \end{aligned} \qquad i = 1, \ldots, 8,$$

assess the goodness of fit, and convert the fitted models back to fitted multinomial probabilities. This cannot be done directly, since GLIM does not include the multinomial distribution in its family of standard distributions. The model is fitted indirectly by using the relation between the multinomial and Poisson distributions, which we now describe.

```
$END
```

5.5 Relation between the multinomial and Poisson distributions

Suppose $N_1, N_2, \ldots, N_r$ are independent random Poisson variables with means $\lambda_1, \lambda_2, \ldots, \lambda_r$. Then the sum $N = N_1 + N_2 + \ldots + N_r$ also has a Poisson distribution with mean $\lambda = \lambda_1 + \lambda_2 + \ldots + \lambda_r$, and the *joint* distribution of the N_j, *conditional on* $N = n$, is the multinomial distribution

$$\Pr(N_1 = n_1, \ldots, N_r = n_r | N = n) = \frac{n!}{n_1! \ldots n_r!} p_1^{n_1} \ldots p_r^{n_r}$$

with

$$p_j = \lambda_j / \lambda.$$

This is easily shown analytically. We have

$$\Pr(n_1, \ldots, n_r) = \prod_{j=1}^{r} \exp(-\lambda_j)\lambda_j^{n_j}/n_j!$$

$$= \exp(-\lambda)\prod_j \lambda_j^{n_j}/n_j!$$

and

$$\Pr(n) = \exp(-\lambda)\lambda^n/n!$$

so that

$$\Pr(n_1, \ldots, n_r|n) = \Pr(n_1, \ldots, n_r)/\Pr(n)$$

$$= \prod_j (\lambda_j/\lambda)^{n_j} \, (n!/\prod_j n_j!)$$

as stated. The multinomial logit parameters θ_j are then

$$\theta_j = \log(p_j/p_1)$$
$$= \log(\lambda_j/\lambda_1)$$

We want to fit a regression model to each θ_j parameter (apart from θ_1):

$$\theta_{ji} = \boldsymbol{\beta}_j'\mathbf{x}_i$$

where i denotes the observation index, for the given set of explanatory variables $\mathbf{x}$. The regression coefficients $\boldsymbol{\beta}_j$ will in general be different for each j. The multinomial logit models are equivalent to

$$\log(\lambda_{ji}/\lambda_{1i}) = \boldsymbol{\beta}_j'\mathbf{x}_i$$

or

$$\phi_{ji} = \log \lambda_{ji} = \psi_i + \boldsymbol{\beta}_j'\mathbf{x}_i \qquad j = 1, \ldots, r; \quad i = 1, \ldots, n.$$

where $\boldsymbol{\beta}_1 = 0$, and $\phi_{1i} = \psi_i = \log \lambda_{1i}$. The ψ_i are a set of "nuisance parameters".

To fit the model in GLIM, we need to define a response factor with r levels corresponding to the multinomial response categories. The parameter $\boldsymbol{\beta}_j$ is then the coefficient of the *interaction* of the j-th level of the response factor with the explanatory variables $\mathbf{x}$. The fitted model must contain in addition a parameter for each observation; in GLIM terms, a "nuisance" factor with n levels has to be fitted as well. This becomes exceptionally demanding of computer time unless the explanatory variables form a contingency table of reasonably small dimension; in this case the nuisance factor can be written as the full cross-classification of the explanatory factors.

We illustrate the equivalence of multinomial and Poisson models with a simple two-category response for which the (binomial) logit model can be fitted explicitly. In Sections 2.11–2.14 we considered a 2×2 cross-classification of a binary survival response. We repeat the data in a slightly different form and the GLIM analysis below.

```
$UNIT 4
$DATA DEATH SURVIVE $
$READ
3  176  4  293  17  197  2  23
$FACTOR CLINIC 2 ATTEND 2 $
$CALC CLINIC = %GL(2,2) : ATTEND = %GL(2,1) :
  TOTAL = DEATH+SURVIVE $
$YVAR DEATH $ERROR B TOTAL $LINK G
```

The deviances are reproduced below for four of the models, and parameter estimates for two of these.

Model	Deviance	Estimates
1	17.828	-3.277 (0.200)
CLINIC	0.082	-4..205 + 1.756 CLIN(2) (0.381) (0.450)
ATTEND+CLINIC	0.043	
ATTEND*CLINIC	0.000	

To use the Poisson analysis, we have to define a response factor of length 8 and a COUNT vector which contains the SURVIVE and DEATH counts, and block the explanatory variables by duplicating their values in factors of length 8. This is achieved very easily using the $ASSIGN directive.

```
$VAR 8 COUNT RESP PCLIN PATT $
$ASSIGN PCLIN = CLIN,CLIN : PATT = ATTEND,ATTEND :
      COUNT = SURVIVE,DEATH  $
$FACTOR PCLIN 2  PATT 2  RESP 2 $
$CALC RESP = %GL(2,4) $
```

To fit the Poisson model we have to redefine the standard length of vectors ($UNIT) to be 8:

```
$UNIT 8
$YVAR COUNT  $ERROR P  $LINK L
$FIT  RESP + PCLIN*PATT $D E $
$FIT +RESP.PCLIN  $D E $
$FIT +RESP.PATT  $D E $
$FIT +RESP.PCLIN.PATT $D E $
```

It is easily verified that the deviances are identical to those above. For the first Poisson model the parameter estimates are given below.

$$\underset{(0.075)}{5.150} \underset{(0.200)}{-3.277}\,\mathrm{RESP}(2) \underset{(0.101)}{+0.179}\,\mathrm{PCLI}(2) \underset{(0.095)}{+0.506}\,\mathrm{PATT}(2)$$

$$\underset{(0.232)}{-2.653}\,\mathrm{PCLI}(2).\mathrm{PATT}(2)$$

The coefficient and standard error for RESP in the Poisson model are identical to the intercept estimate and standard error in the null logit model. The nuisance parameters for PCLIN and PATT serve only to reproduce the marginal totals over the RESP factor in each clinic/attendance cell. If these factors are omitted, the Poisson model does not correspond to the multinomial logit model because the marginal totals are being treated as random variables rather than fixed, as assumed in the logit model. It can be verified using $D R $ that the fitted counts added over the two response categories do reproduce the observed margins.

Similarly, for the second Poisson model above, the parameter estimates are

$$\underset{(0.075)}{5.173} \underset{(0.381)}{-4.205}\,\mathrm{RESP}(2) \underset{(0.103)}{+0.111}\,\mathrm{PCLI}(2) \underset{(0.095)}{+0.506}\,\mathrm{PATT}(2)$$

$$\underset{(0.450)}{+1.756}\,\mathrm{RESP}(2).\mathrm{PCLI}(2) \underset{(0.232)}{-2.653}\,\mathrm{PCLI}(2).\mathrm{PATT}(2)$$

Here the coefficients of RESP and RESP.PCLI are identical to those of the intercept and CLIN in the logit model.

If the vector of counts is constructed with DEATH above SURVIVE, the signs of the parameter estimates for terms involving RESP are reversed, since we are then modelling the logit of the survival instead of the death probability.

The need in GLIM to fit nuisance parameters for the complete cross-classification by the explanatory variables makes the computational burden of the Poisson model very considerable in large contingency tables with a multinomial response. In such cases alternative approaches which fit the multinomial model directly (e.g. Bock and Yates, 1973) or use iterative proportional fitting (Bishop, Fienberg and Holland, 1975, p. 83) are preferable.

```
$END
```

5.6 Fitting the multinomial logit model

We return now to the MINERS example, and construct the necessary vectors for the Poisson model.

```
$INPUT 1 MINERS $
$VAR 24 COUNT PYEAR PP RESP $
$ASSIGN COUNT = N,M,S : PYEAR = YEARS,YEARS,YEARS :
  PP = P,P,P $
$FACTOR PP 8 RESP 3 $
$CALC RESP = %GL(3,8) :PLY = %LOG(PYEAR) $
$UNIT 24
$YVAR COUNT $ERROR P $LINK L
$FIT RESP + PP  $D E $
```

This is equivalent to the null multinomial model.

```
$FIT + RESP.PYEAR   $D E $
```

This is equivalent to the multinomial model using PYEAR. The parameter estimates appear confusing because the coefficient of the *last* category of RESP has been set to zero instead of the first, as we might have expected. This is because the main effect of PYEAR has not been included in the model. The standard parametrization can be recovered by adding PYEAR to the model:

```
$FIT + RESP.PYEAR   $D E $
```

Note that PYEAR is *aliased*, because it can be expressed as a linear function of the dummy variables for the levels of PP. The fit of the model is consequently unchanged.

The deviance changes by 87.71 for the 2 df, and the residual deviance of 13.93 with 12 df looks reasonable. The largest residual is 1.658 at the 11th observation, but the residuals for the first and third categories show the same pattern of alternating signs as in the binomial logit analysis. The estimated slopes are 0.084 (s.e. 0.015) for the second category and 0.109 (s.e. 0.017) for the third (relative to the first). We try the log year model.

```
$FIT - PYEAR - RESP.PYEAR + PLY + RESP.PLY $D E R $
```

The deviance drops to 5.35, and the slopes are 2.165 (s.e. 0.457) for the second category and 3.067 (s.e. 0.564) for the third. There is no longer any obvious pattern in the residuals, and the model provides a good fit.

Since the estimated slopes differ by about twice either standard error, it seems unlikely that the true slopes could be equal. We can investigate this question using option S of $DISPLAY, which displays the standard errors of *differences* between parameter estimates.

```
$D S $
```

The standard error of the difference between the slopes is 0.668; the observed difference of 0.902 is 1.35 standard errors. The estimated slopes have a large *negative* covariance, so the variance of the difference is substantially larger than the individual variances.

We now fit the model using a single dummy variable to estimate the common slope.

```
$CALC R23 = %EQ(RESP,2) + %EQ(RESP,3) : RPLY = R23*PLY $
$FIT RESP + PP + RPLY  $D E R $
```

The deviance increases by 1.86 on 1 df to 7.21. The common slope is 2.576 (s.e. 0.386) and the largest residual is 1.678 at the 11th observation. We adopt this as the fitted model.

We finally construct and print the fitted proportions from the model. The fitted values from the Poisson model have to be regrouped into three vectors of length 8.

```
$VAR 8 NF MF SF T I $
$CALC I = %CU(1) : NF(I) = %FV(I) : MF = %FV(I+8) :
      SF = %FV(I+16) : T = N+M+S :
    NFP = NF/T : MFP = MF/T : SFP = SF/T $
$PLOT NFP MFP SFP YEARS $
$PRINT NFP : : MFP : : SFP $
```

The GLIM plot is not very informative: an accurate plot with observed and fitted values is shown in Fig. 5.2. In this case the fitted probabilities of the normal category are the same as those from the binomial logit model, but this happens only because the slopes for the mild and severe categories have been equated. If the RESP.PLY interaction is fitted instead of RPLY, the fitted probabilities from the binomial and multinomial models are different. The fitted probabilities for the mild and severe categories are very close.

5.7 Ordered response categories

The three-category response considered in Sections 5.3 and 5.6 had a natural *order* to the categories: normal, mild, severe. This is a common feature of categorical responses, and such responses often have many categories:

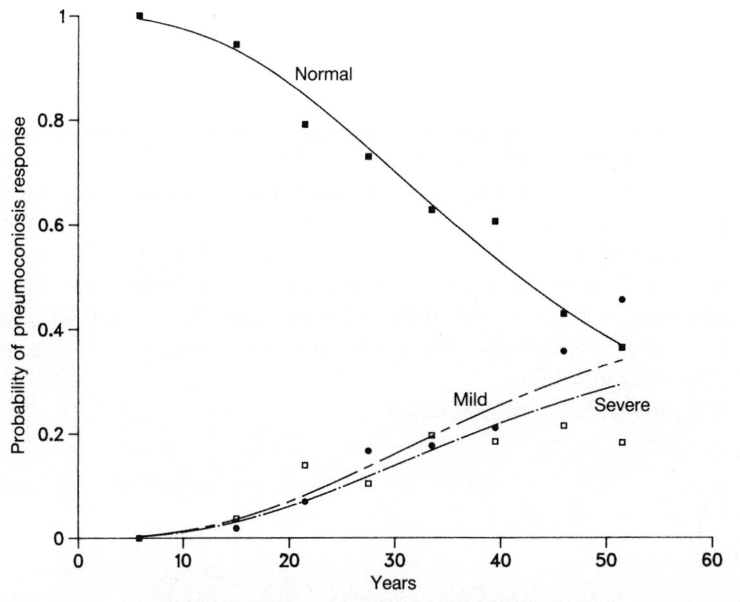

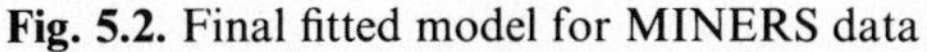

Fig. 5.2. Final fitted model for MINERS data

between five and nine is quite common. The number of parameters in the multinomial models becomes very large, and it is natural to look for simpler models with fewer parameters. Such simpler models make various assumptions about the relations among the category probabilities. We now consider a number of possibilities, in all of which only *one* regression coefficient vector $\boldsymbol{\beta}$ is used, rather than $(r-1)$ such vectors.

5.7.1 *Common slopes for the regressions*

The general model, in the notation of Section 5.5 is

$$\theta_1 = 0$$

$$\theta_j = \log (p_j/p_1) = \boldsymbol{\beta}_j'\mathbf{x} \qquad j = 2, \ldots, r.$$

In the MINERS example of Section 5.6 we found that the slope estimates $\hat{\beta}_{21}$ and $\hat{\beta}_{31}$ could be equated. If the slopes are equal across categories for *all* the explanatory variables, excluding the intercepts, we can write

$$\theta_1 = 0$$

$$q_j = \gamma_j + \boldsymbol{\beta}'\mathbf{x}, \qquad j = 2, \ldots, r.$$

Then

$$p_1 = 1/(1 + \delta e^{\boldsymbol{\beta}'\mathbf{x}})$$

$$p_j = \frac{e^{\gamma_j + \boldsymbol{\beta}'\mathbf{x}}}{1 + \delta e^{\boldsymbol{\beta}'\mathbf{x}}}, \qquad j = 2, \ldots, r.$$

Where

$$\delta = \sum_{j=2}^{r} \exp(\gamma_j).$$

Since δ can be expressed as e^{α}, it can be absorbed as an additional intercept term into $\boldsymbol{\beta}'\mathbf{x}$. The model for p_1 is then an ordinary binomial logit model for category 1 relative to the other pooled categories. This explains why the fitted logit model for the normal proportion in Section 5.3 gave the same fitted probabilities as those for the multinomial logit model in Section 5.6 in which the regression coefficients β_{21} and β_{31} were equated. The models for p_2 and p_3 are however not binomial logit models since the exponents in the numerator and denominator differ.

If the intercepts are also equated, then

$$\theta_1 = 0$$

$$\theta_j = \boldsymbol{\beta}'\mathbf{x}, \qquad j = 2, \ldots, r$$

and

$$p_1 = 1/\{1 + (r-1)e^{\boldsymbol{\beta}'\mathbf{x}}\}$$

$$p_j = e^{\boldsymbol{\beta}'\mathbf{x}}/\{1 + (r-1)e^{\boldsymbol{\beta}'\mathbf{x}}\}, \qquad j = 2, \ldots, r.$$

Identical models are now being fitted for all but the first category: the second and third categories are indistinguishable and are effectively being collapsed.

We continue the analysis of the MINERS example of Section 5.6, where we have already fitted a common slope for the response categories in the regression on PLY. Equal intercepts are fitted by replacing the RESP factor in this model by R23:

```
$FIT - RESP + R23 $D E $
```

The deviance increases by only 0.44 on 1 df to 7.65 with 14 df, not surprising since the estimates for RESP(2) and RESP(3) in the previous model were -10.38 and -10.23 with common standard error 1.34. Note that this deviance does not correspond to the deviance of 3.13 with 6 df for the logit model in Section 5.3 because the variation between the second and third categories is suppressed in that model. The difference of 4.52 with 8 df is the variation due to constraining these two categories to have the same response probabilities.

5.7.2 *Linear trend over response categories*

Suppose that the regression coefficients $\boldsymbol{\beta}_j$, rather than being equal, increase in equal steps with the category index (apart from the intercepts):

$$\theta_1 = 0$$

$$\theta_j = \gamma_j + (j-1)\boldsymbol{\beta}'\mathbf{x} \qquad j = 2, \ldots, r.$$

Thus the regression coefficients $\boldsymbol{\beta}_j$ have a *linear trend* over the given ordering of the response categories. In using the Poisson representation to fit this model, we replace the interaction between response factor and explanatory variables by its linear component.

We now define the linear component of RESP, and its interaction with PLY:

```
$CALC LRES = RESP-1 : INT = LRES*PLY $
```

If the slopes 2.165 and 3.067 were estimates of β and 2β, then β would be roughly $5.232/3 = 1.744$.

```
$FIT RESP + PP + INT  $D E $
```

The deviance increases by 1.70 on 1 df to 7.05 on 13 df, and the estimate of β is 1.726 (s.e. 0.258).

Thus the model with $\beta_{31} = 2\beta_{21}$ fits just as well as the previous model with $\beta_{31} = \beta_{21}$. As is often the case, different smoothed structures fit the data equally well. In the common slopes model the probabilities of mild and severe pneumoconiosis increase at the same rate relative to the normal group; in the linear trend model, the severe probability increases much more rapidly. We cannot distinguish between these two interpretations.

If the *intercepts* γ_j also increase in equal steps, (Haberman, 1974; Goodman, 1979), we can write

$$\theta_j = (j-1)\boldsymbol{\beta}'\mathbf{x} = b(j-1)$$

where

$$b = \boldsymbol{\beta}'\mathbf{x},$$

and

$$p_j = p_1 e^{b(j-1)} \qquad j = 2, \ldots, r$$

which is equivalent to

$$p_j = (1-\theta)\theta^{j-1}/(1-\theta^r) \qquad j = 1, \ldots, r$$

with

$$\theta = e^b.$$

Thus the category probabilities $p_1, \ldots, p_r$ follow a *truncated geometric distribution* with parameter $\theta = e^{\boldsymbol{\beta}'\mathbf{x}}$.

The *practical* interpretation of this distribution is less interesting in general than the model simplification resulting: the regression coefficients are simply proportional.

Is the linear trend in the intercept model reasonable for the MINERS data? We examine the intercepts:

−7.429	0.877	RESP(2)
−13.36	1.83	RESP(3)

The intercept for category 3 is very close to twice that for category 2: it looks as though the model will fit well.

```
$FIT - RESP + LRES $D E $
```

Surprisingly, the deviance increases by 17.09 on 1 df to 24.14 with 14 df. Why does this happen? The answer becomes clear if we examine the correlation between the RESP(2) and RESP(3) estimates.

```
$FIT + RESP - LRES $D C $
```

The correlation is 0.9785. Thus the (asymptotic) variance of $\hat{\gamma}_3 - 2\hat{\gamma}_2$ is $(1.83^2 - 4 \times 0.9785 \times 1.83 \times 0.877 + 4 \times 0.877^2)$, i.e. 0.1438, so the standard error is 0.379, and $(\hat{\gamma}_3 - 2\hat{\gamma}_2)/\text{s.e.}(\hat{\gamma}_3 - 2\hat{\gamma}_2) = 3.95$.

Thus the truncated geometric distribution is untenable for the proportions of normal, mild and severe pneumoconiosis, though the model with a linear trend in the slopes fits well.

5.7.3 *Proportional slopes*

A weaker constraint than the linear trend in Section 5.7.2 is proportional slopes. Suppose

$$\begin{aligned} \theta_1 &= 0 \\ \theta_j &= \gamma_j + \delta_j \boldsymbol{\beta}'\mathbf{x}, \qquad j = 2, \ldots, r; \end{aligned}$$

then the regression coefficients β are proportional across categories (except for the intercepts). This model is called the *stereotype model* by Anderson (1984); it is not easily fitted in GLIM. If there is only one explanatory variable, the model imposes no constraint on the full multinomial logit model.

5.7.4 *Other models*

A number of other useful models for ordered multinomial responses are discussed by McCullagh (1980), McCullagh and Nelder (1983), and Anderson (1984). We refer to them only briefly here because they cannot in general be fitted in GLIM.

Cumulative probabilities may be modelled directly using the same link functions as for the binomial. Define the cumulative probability P_j for the j-th category by

$$P_j = \sum_{s=1}^{j} p_s \quad j = 1, \ldots, r-1,$$

$$P_r = 1.$$

Then

$$g(P_j) = \boldsymbol{\beta}_j'\mathbf{x} \qquad j \neq r$$

is a model for the cumulative probability, with g the logit, probit or CLL link function. This model has as many parameters as the multinomial, so is of interest only if restricted. A useful restriction is

$$g(P_j) = \theta_j + \boldsymbol{\beta}'\mathbf{x}$$

where θ_j is an intercept parameter, since then $\boldsymbol{\beta}$ is invariant under the collapsing together of adjacent categories. If g is the logit link, this model is called the *proportional odds model.* If g is the CLL link, the model is called the *proportional hazards model.*

This class of models can be expressed in terms of a *latent variable representation.* If

$$g(P_j) = \theta_j + \boldsymbol{\beta}'\mathbf{x}$$

for some link function g, then

$$P_j = H(\theta_j + \boldsymbol{\beta}'\mathbf{x})$$

where H is the cumulative distribution function inverse to the link function g (see Section 4.2). Then the category probabilities p_j can be represented as

$$p_j = H(\theta_j + \boldsymbol{\beta}'\mathbf{x}) - H(\theta_{j-1} + \boldsymbol{\beta}'\mathbf{x}).$$

Suppose Z is an unobserved continuous random variable with a location/scale parameter distribution with cumulative distribution function $H\{(z-\mu)/\sigma\}$, and let $\phi_1, \ldots, \phi_{r-1}$ be fixed cut-point values of Z. We observe only the interval (ϕ_{j-1}, ϕ_j) in which Z lies, with probability

$$p_j = H((\phi_j - \mu)/\sigma) - H((\phi_{j-1} - \mu)/\sigma) \qquad j = 1, \ldots, r$$

with $\phi_0 = -\infty$, $\phi_r = +\infty$. This is identical to the above model if

$$\phi_j/\sigma = \theta_j$$

and

$$-\mu/\sigma = \boldsymbol{\beta}'\mathbf{x}.$$

Thus an ordered cumulative probability model always has an interpretation in terms of modelling an underlying latent variable, though no such interpretation is necessary to use the model.

These models are discussed by Bock (1975, Chapter 8), McCullagh (1980) and Anderson (1984). The cumulative distribution function H is usually taken to be normal or logistic. Programs for fitting some of these models include MULTIQUAL (Bock and Yates, 1973) and PLUM (McCullagh, 1980); it is possible to fit the grouped normal (latent variable) model in GLIM using a composite link function (Thompson and Baker, 1981).

5.8 An Example

A survey of student opinion on the Vietnam War was taken at the University of North Carolina in Chapel Hill in May 1967 and published in the student newspaper. Students were asked to fill in "ballot papers", available in the Student Council building, stating which policy out of A, B, C or D they supported. Responses were cross-classified by sex and by undergraduate year or graduate status:

Policy A: The US should defeat the power of North Vietnam by widespread bombing of its industries, ports and harbours and by land invasion.

B: The US should follow the present policy in Vietnam.

C: The US should de-escalate its military activity, stop bombing North Vietnam, and intensify its efforts to begin negotiation.

D: The US should withdraw its military forces from Vietnam immediately.

The data are stored in the subfile VIETNAM, and are reproduced below.

		Policy				
	Year	A	B	C	D	Total
Males	1	175	116	131	17	439
	2	160	126	135	21	442
	3	132	120	154	29	435
	4	145	95	185	44	469
	Grad	118	176	345	141	780
Females	1	13	19	40	5	77
	2	5	9	33	3	50
	3	22	29	110	6	167
	4	12	21	58	10	101
	Grad	19	27	128	13	187

What can we conclude about students' attitude to the war? A first question concerns the response rates: what proportion of students actually responded in the survey? The numbers of students enrolled at the University, and the numbers responding (summed over the four categories A, B, C and D) are shown below, with the response rates.

	Year					
	1st	2nd	3rd	4th	Graduate	Total
Males						
Responses	439	442	435	469	780	2565
Enrolled	1768	1792	1693	1522	3005	9780
Response rate	0.248	0.247	0.257	0.308	0.260	0.262
Females						
Responses	77	50	167	101	187	582
Enrolled	487	326	772	608	1221	3414
Response rate	0.158	0.153	0.216	0.166	0.153	0.170

Overall, only 26% of male and 17% of female students expressed an opinion by filling in a ballot paper. The response rate is somewhat higher than average for 4th-year males and 3rd-year females, and slightly lower than average for the other groups. We could model these response rates, treating the populations as samples, but the pattern of response is clear.

Given such low response rates, it seems difficult to draw any conclusion about the view of the student population. We have no way of knowing whether response is itself related to intensity of feeling, and therefore possibly to the policy chosen. We can, however, investigate how the policy chosen varies by sex and year for those who responded. This is a different question; whether the conclusions are applicable to the population as a whole is not just open, but unresolvable.

```
$INPUT  1 VIETNAM $
$CALC T = A+B+C+D : AP = A/T : BP = B/T :
  CP = C/T : DP = D/T $
$CALC %RE = %EQ(SEX,1) $
$PLOT AP BP CP DP YEAR $
$CALC %RE = %EQ(SEX,2) $
$PLOT AP BP CP DP YEAR $
$DEL  %RE $
```

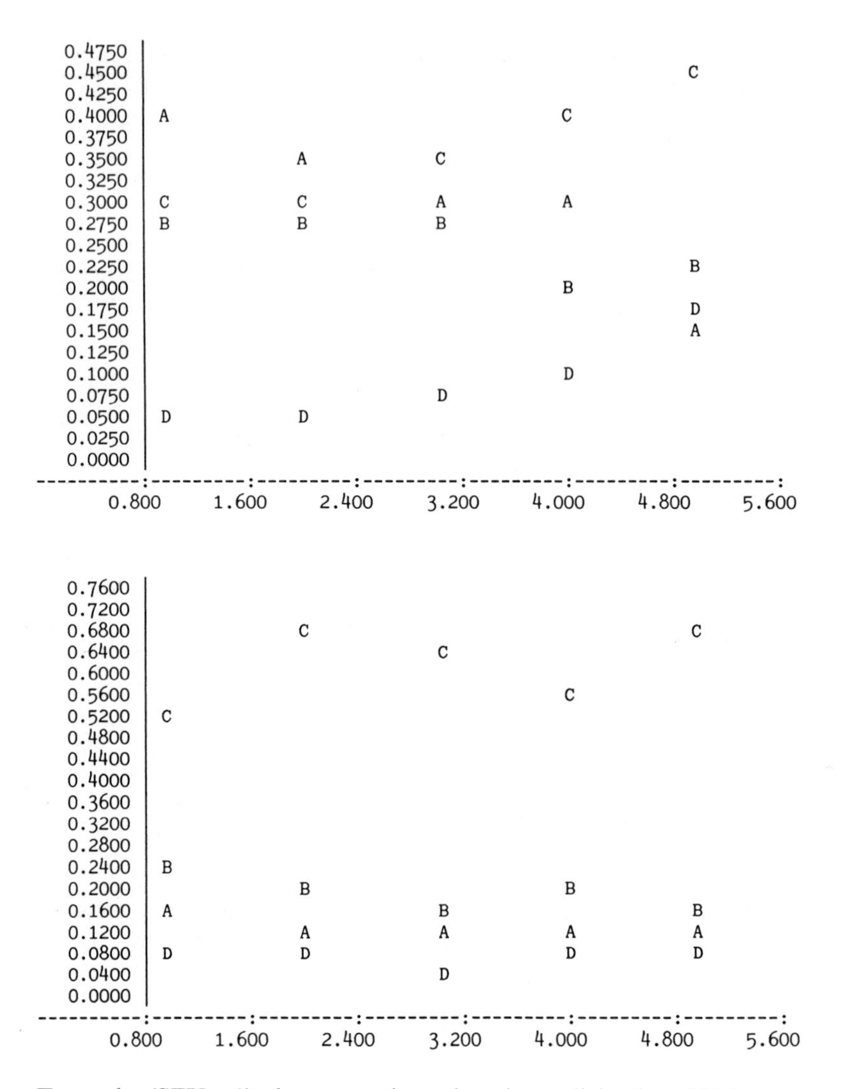

For males (SEX = 1), the proportions choosing policies A and B decrease, and those choosing C and D increase steadily with year. For females, policy C is consistently favoured but there is substantial variation with year.

We proceed to model the proportions as though the respondents were randomly sampled from the student population. The variable response rates noted above are irrelevant in the analysis since we are including sex and year in the model (see Section 2.9 for a discussion of unequal sampling fractions). We first block the factors and policy response into vectors of length 40.

```
$VAR 40 COUNT PSEX PYEAR POLICY $
$ASSIGN COUNT = A,B,C,D : PSEX = SEX,SEX,SEX,SEX :
                          PYEAR = YEAR,YEAR,YEAR,YEAR $
$FACTOR PSEX 2 PYEAR 5 POLICY 4 $
$CALC POLICY = %GL(4,10) $
$UNIT 40
$YVAR COUNT $ERROR .P $LINK L
$FIT POLICY + PYEAR*PSEX $
```

This is the null model on the multinomial logit scale: the POLICY and PYEAR*PSEX terms are included to reproduce the marginal totals. The fit is very poor: strong differences in response proportions are present. We now add the sex and year interactions with response:

```
$FIT +PSEX.POLICY : +PYEAR.POLICY -PSEX.POLICY :
 +PSEX.POLICY $
```

Large changes in deviance show that both interactions are necessary. The only term not yet fitted is the three-way interaction. Although the deviance of 19.19 on 12 df does not indicate a bad fit, it is possible that a large contribution to this deviance comes from only a few degrees of freedom of this interaction term. We have already seen a different response pattern in the plots and this should be investigated further. We will look at the responses for the two sexes separately.

```
$CALC MALE = 2 - PSEX $WEIGHT MALE
```

Only males are now being analysed, so no sex terms should be fitted (if they are, they will be aliased and ignored).

```
$FIT POLICY +PYEAR : +POLICY.PYEAR $D E $
```

The large deviance for the interaction term (203.1) shows that proportions adopting each policy differ significantly over years. How can we describe these differences simply from the parameter estimates (listed below with their standard errors)?

```
  0.172   0.169   POLI(2).PYEA(2)
  0.316   0.174   POLI(2).PYEA(3)
 -0.012   0.178   POLI(2).PYEA(4)
  0.811   0.169   POLI(2).PYEA(5)
  0.120   0.164   POLI(3).PYEA(2)
  0.444   0.166   POLI(3).PYEA(3)
```

```
0.533    0.160    POLI(3).PYEA(4)
1.362    0.157    POLI(3).PYEA(5)
0.301    0.344    POLI(4).PYEA(2)
0.816    0.327    POLI(4).PYEA(3)
1.139    0.307    POLI(4).PYEA(4)
2.510    0.283    POLI(4).PYEA(5)
```

We have two orderings: by year and by policy. Within each policy, the logit increases with year, but not smoothly: there is a large jump in the logit in each case from year 4 to year 5, and for policy 2 (B) the logit for year 4 is small and negative, whereas it is large and positive for the other years. Thus a linear trend over year will not fit all years, though it might fit years 1–4.

The year pattern of logits becomes stronger with increasing policy, suggesting that a linear trend over policy might fit, that is, the regression coefficients on year are proportional over policy (see Fig. 5.3). We try this model.

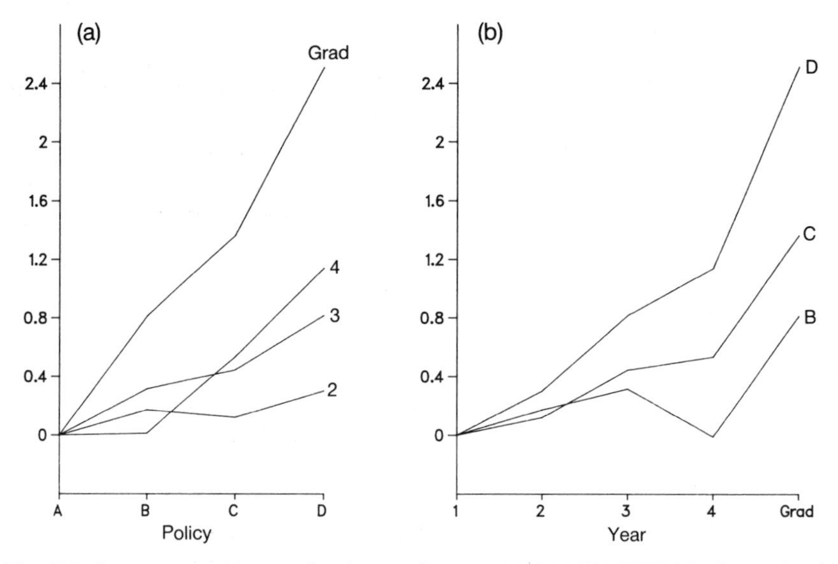

Fig. 5.3. Parameter estimates for interaction term POLICY.PYEAR for males in VIETNAM data; (a) ordered by POLICY (b) ordered by YEAR

```
$CALC LINP = POLICY $
$FIT -POLICY.PYEAR + LINP + PYEAR.LINP $D E $
```

The deviance is 13.91 on 8 df—a reasonable fit. We examine the residuals.

```
$D R $
```

There is a large residual of -2.29 at the 14th observation, where the observed number 95 of fourth-year students choosing policy B—the current policy—is not close to the model fitted value of 120.04. This feature is visible in Fig. 5.3. We refit the above model excluding the 14th observation.

```
$CALC W = 1 : W(14) = 0 : WT = MALE*W $
$WEIGHT WT
$FIT . $D E $
```

The deviance is now 4.66 on 7 df: the model fits very closely. Should we look for a further model which fits better, or is there a sensible interpretation of this model, *and* the outlying 14th observation?

A linear trend in policy means a linear increase in the log odds against policy A with each step along the policy scale: the policy options have been expressed in a smoothly ordered way. The outlying observation 14 does not fit this trend: fourth-year male students choose option B much less often than option A, instead of slightly less often as would be predicted from the linear increase in the odds against A.

We can give a plausible explanation for this. Fourth-year male students faced the draft and the possibility of serving in the war at the end of their academic year, unless they obtained deferment, for example by becoming graduate students. A policy of continuing the present conflict, rather than either bringing it to a presumably rapid end by invasion, or withdrawing to relatively safe positions, might have been viewed as the policy most likely to endanger their own lives. It is not surprising to find that this policy was much less favoured than A or C among fourth year students. Students in other undergraduate years did not face this immediate prospect.

The parameter estimates for the PYEAR.LINP interaction are listed below, with standard errors.

```
0.070    0.072    PYEA(2).LINP
0.232    0.072    PYEA(3).LINP
0.303    0.069    PYEA(4).LINP
0.750    0.066    PYEA(5).LINP
```

They show a strong increase, with a large jump for graduate students. For the undergraduate years, a linear increase with year looks a reasonable model.

```
$CALC UG = %NE(PYEAR,5) : W2 = WT*UG :
      LINY = PYEAR-1 : LYLP = LINY*LINP $
$WEIGHT W2
$FIT . : -  PYEAR.LINP - LINP + LYLP $D E $
```

The deviance for the first fit is 1.73 with 5 df: this is increased to 2.38 on 7 df on the replacement of the interaction by its linear trend for undergraduates (the aliased term LINP is dropped also). Graduates are excluded from this model: we bring them back in by adding an additional linear policy term for graduates.

```
$CALC GLP = (1-UG)*LINP $
$WEIGHT WT
$FIT  + GLP $D E $
```

The estimates of LYLP and GLP are shown below with standard errors.

Estimate	s.e.	Parameter
0.107	0.022	LYLP
0.331	0.071	GLP

The deviance increases by 2.93 on 2 df (for GLP and PYEA(5)) to 5.31 on 9 df. The 14th observation is still weighted out: if it is included in the model (using $WEIGHT MALE) the deviance increases to 14.45 on 10 df.

We now turn to the females.

```
$CALC FEMALE = 1 - MALE $
$WEIGHT FEMALE
$FIT POLICY + PYEAR $D E $
```

The null logit model gives a good fit, with a deviance of 13.26 on 12 df, and the largest residual is −1.42. There are no differences other than those attributable to sampling variation among the proportions choosing the four policies. The parameter estimates for the policy main effect are 0.391, 1.648 and −0.652 for B, C and D, relative to zero for A. The corresponding fitted proportions are obtained from

$$\hat{p}_j = e^{\hat{\theta}_j} / \sum_{j=1}^{r} e^{\hat{\theta}_j}, \qquad \hat{\theta}_1 = 0,$$

and are 0.122, 0.180, 0.634 and 0.064. These are simply the marginal proportions giving each response, collapsed over year.

The device of fitting separate models for each sex is useful as the model structures are very different. The same result can be achieved by fitting the following single model to the complete data.

```
$CALC V1 = MALE*LYLP : V2 = MALE*GLP $
$WEIGHT W
$FIT POLICY + PYEAR*PSEX + FEMALE + POLICY.FEMALE + V1 + V2
$D E $
```

The deviance of 18.57 on 21 df is the sum of the values 5.31 for males and 13.26 for females. The 14th observation is still excluded: the fitted value of 138.06 corresponds badly to the observed value of 95. The corresponding Pearson residual can be directly calculated: it is not listed by $D R $ since this observation is weighted out.

```
$CALC (%YV(14) - %FV(14))/%SQRT(%FV(14)) $
```

The residual is −3.67. Note that this is a "jack-knife" residual, in the sense defined in Section 2.10, since this point has not been used in fitting the model.

We now construct the fitted proportions from the model. For this purpose we need to include a term in the model which exactly reproduces the observed value for observation 14. This is easily achieved.

```
$CALC D14 = 0 : D14(14) = 1 $
$WEIGHT
$FIT + D14 $D E $
```

Equivalent models are being fitted, so the deviance is unchanged.

```
$VAR 10 I AF BF CF DF $
$CALC I = %CU(1) :
    AF = %FV(I) : BF = %FV(I+10) : CF = %FV(I+20) :
    DF = %FV(I+30) :
    AFP = AF/T : BFP = BF/T : CFP = CF/T :
    DFP = DF/T $
$CALC %RE = %EQ(SEX,1) $
$PLOT AFP AP BFP BP CFP CP DFP DP YEAR
  'a A b B c C d D' $
```

Note that MALE cannot be used in the calculation of %RE as it is of length 40, while the vectors being plotted are of length 10. The model fits the observed proportions closely. Repeating the plot with %RE set equal to %EQ(SEX,2) gives the constant probabilities over year from the null model for females.

Both fitted models are shown in a high quality plot in Fig. 5.4.

Thus 63% of female students responding choose policy C, while 18% choose B, 12% A and 6% D. These patterns are consistent over years. For undergraduate males, support for policy A declines steadily with year, while that for policy B increases slightly and then decreases. Support for C increases rapidly, but that for D only slowly. For male graduate students, there is a steep decline in support for A, and increases in B, C and D.

We note finally that graduate students tend to come from outside North

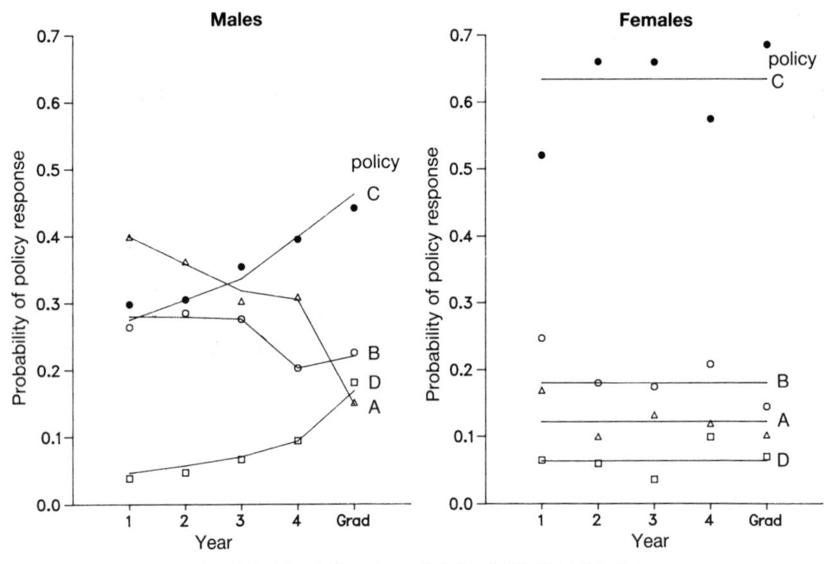

Fig. 5.4. Final fitted model for VIETNAM data

Carolina, while undergraduates are predominantly from within North Carolina. This does not affect responses for females, but it may be a partial explanation for the different responses for graduate males.

```
$END
```

5.9 Structured multinomial responses

In many studies with observational data, several categorical responses may be observed. The associations between these categorical response variables may be of interest, and particularly the effect of explanatory variables on the strength of these associations. The subfile TOXAEMIA contains the number of women giving birth to their first child who showed toxaemic signs (hypertension and/or proteinurea) during pregnancy. The data may be found in Brown *et al.* (1983) and were collected in Bradford between 1968 and 1977 on 13 384 United Kingdom women. The four possible response patterns are cross-classified by CLASS, giving five categories of social class (I–V), and SMOK, a three-level factor relating to the number of cigarettes smoked per day during pregnancy (0, 1–19, 20+). The counts in the four response categories are given by the variables HU, HN, NU and NN, representing those with hypertension and proteinurea, hypertension only, proteinurea only and neither symptom.

How do the prevalence and association of these symptoms vary with CLASS and SMOK? We first look at the marginal proportions with each combination of symptoms, and then with each single symptom.

```
$INPUT 1 TOXAEMIA $
$CALC  H = HU+HN : U = HU+NU : T = H+NU+NN : HUP = HU/T :
       HNP = HN/T : NUP = NU/T : NNP = NN/T $
$PLOT  HUP HNP NUP NNP CLASS 'A B C D' $
$PLOT  HUP HNP NUP NNP SMOK 'A B C D' $
```

The high proportions with neither symptom mask the other symptom combinations in these plots and apart from the different general levels of the four response categories little can be concluded. We replot these individually making use of the plotting character facility to simultaneously plot against CLASS and SMOK.

```
$PLOT HUP CLASS 'O L H' SMOK $
  :   HNP CLASS 'O L H' SMOK $
  :   NUP CLASS 'O L H' SMOK $
  :   NNP CLASS 'O L H' SMOK $
```

Again, no simple picture emerges but there are indications of CLASS and SMOK effects and perhaps some interaction between these and the presence of hypertension.

```
$CALC  HP = H/T : UP = U/T $
$PLOT  HP CLASS 'O L H' SMOK $
   :   UP CLASS 'O L H' SMOK $
```

Again we see similar relationships with the response variables.

We could perform separate analyses for H and U, but this would not tell us about the extent of association between the symptoms and how it varies with CLASS and SMOK. This is, however, easily examined if we construct the log–odds ratio for each of the implicit 2×2 tables for each class/smoking combination.

```
$CALC LODDS = %LOG(HU*NN/(HN*NU)) $
```

The "invalid function/operator argument(s)" message is due to the two zero observed counts. We plot the log–odds ratio against CLASS and SMOK excluding the undefined values.

```
$CALC %RE = %NE(LODDS,0)
$PLOT LODDS CLASS 'O L H' SMOK $
$DEL %RE $
```

The log–odds ratio appears fairly constant at about 1.2. Note that the undefined values are both associated with very small sample sizes.

To perform a full analysis of these data with the four response categories we need to use the Poisson model representation. This requires the blocking of the variables into vectors of length 60, as described in Section 5.5. However, we now need to take account of the structure of our responses; hypertension and proteinurea are both defined as two-level factors; we do not have a single four-level response factor.

```
$VAR 60 COUNT BCLASS BSMOK BH BU $
$ASSIGN COUNT = NN, HN, NU, HU :
        BCLASS = CLASS, CLASS, CLASS, CLASS :
         BSMOK = SMOK, SMOK, SMOK, SMOK $
$CALC       BH = %GL(2,15) :
            BU = %GL(2,30) $
$FACTOR BCLASS 5 BSMOK 3 BH 2 BU 2 $
$UNIT 60
$YVAR COUNT $ERROR P $LINK L $
$FIT BCLASS*BSMOK + BH*BU $
```

The BCLASS*BSMOK term corresponds to a set of nuisance parameters which serve to reproduce the marginal totals of the women in each class/smoking combination. The BH*BU term reproduces the overall pattern of symptom prevalence. Consequently this is equivalent to the null model on the multinomial logit scale: the marginal totals being fixed for each explanatory variable combination and a common response pattern being fitted. The deviance is 179.03 on 42 df: the null model is a poor fit and there are strong patterns of association to be modelled. It is convenient to carry out model selection using backward elimination from the full model as in Section 4.8, recalling that we must retain the terms BH*BU and BCLASS*BSMOK in any model.

```
$FIT BH*BU*BCLASS*BSMOK - BH.BU.BCLASS.BSMOK $D E $
```

The deviance of 12.68 on 8 df shows that we can exclude the four-way interaction from the model (over one third of this deviance is attributable to the two zero counts in the table). The three factor interaction terms involving BH.BU relate to the association between the log–odds ratio and the explanatory variables. Since in our plot of the log–odds ratio there was little evidence of any association with CLASS or SMOK we try removing these terms. Inspection of the parameter estimates shows that those for BH.BU.BCLASS are small. To shorten computing times for model fitting, we use the $RECYCLE directive.

```
$RECYCLE
$FIT - BH.BU.BCLASS $D E $
```

The deviance increases by 2.97 on 4 df to 15.65.

```
$FIT - BH.BU.BSMOK $D E $
```

The deviance increases by 6.64 on 2 df to 22.29 on 14 df, showing the data to be barely consistent with a model having constant log–odds ratio. We now try removing the other three-way interactions. The parameter estimates for BU.BCLASS.BSMOK are all small, but those for BH.BCLASS.BSMOK are not.

```
$FIT - BU.BCLASS.BSMOK : - BH.BCLASS.BSMOK $D R $
```

The deviance increases are 3.55 and 14.64, both on 8 df. The deviance is now 40.47 on 30 df, so although the model is still just adequate there is evidence that some component of the BH.BCLASS.BSMOK interaction is needed in the model. This is also indicated by the large standardized residual (2.108) for unit 20, and the presence of a number of other moderately large residuals. We re-introduce this interaction to examine the parameter estimates.

```
$FIT + BH.BCLASS.BSMOK $D E $
```

The parameter estimates for the three-way interaction are listed below, with standard errors.

−0.163	0.278	BH(2).BCL(2).BSMO(2)
0.201	0.661	BH(2).BCL(2).BSMO(3)
−0.566	0.251	BH(2).BCL(3).BSMO(2)
−0.314	0.593	BH(2).BCL(3).BSMO(3)
−0.565	0.269	BH(2).BCL(4).BSMO(2)
−0.220	0.609	BH(2).BCL(4).BSMO(3)
−0.365	0.295	BH(2).BCL(5).BSMO(2)
−0.031	0.645	BH(2).BCL(5).BSMO(3)

The only large coefficients are for BH(2).BCLA(3).BSMO(2) and BH(2).BCLA(4).BSMO(2)—medium smokers in social cases 3 and 4. The negative estimates for these two cases are almost identical. For heavy smokers the standard errors are large because few women smoked heavily during pregnancy. The estimate for medium smokers in social class 5 is also negative and fairly large. We pool the estimates for these three groups of medium smokers by defining a single dummy variable for them.

```
$CALC DH = %EQ(BH,2)*%GE(BCLASS,3)*%EQ(BSMOK,2) $
```

We equate the estimates for non-smokers and heavy smokers by omitting the full interaction and retaining only DH. The heavy smokers could also be pooled with the medium smokers instead, giving a different model.

```
$FIT - BH.BCLASS.BSMOK + DH $D E R $
```

The deviance increases by 4.353 on 7 df to 30.19 on 29 df and there are no large residuals. Inspection of the remaining parameter estimates suggests that both BU.BSMOK, and BH.BCLASS are unimportant

```
$FIT - BU.BSMOK : -BH.BCLASS $D E $
```

The deviance increases are all small, giving in total an increase of 3.82 on 6 df to a deviance of 34.01 on 35 df. The parameter estimates for the BU.BCLASS and BH.BSMOK interactions are listed below, with standard errors.

−0.456	0.167	BU(2).BCLA(2)
−0.177	0.142	BU(2).BCLA(3)
−0.013	0.155	BU(2).BCLA(4)
0.170	0.171	BU(2).BCLA(5)
−0.098	0.104	BH(2).BSMO(2)
−0.386	0.085	BH(2).BSMO(3)

The only large coefficients are for BU(2).BCLA(2) and BH(2).BSMO(3). We define dummies for these categories

```
$CALC U2C2 = %EQ(BU,2)*%EQ(BCLASS,2) :
      H2S3 = %EQ(BH,2)*%EQ(BSMOK,3) $
$FIT + H2S3 - BH.BSMOK $D E $
```

The deviance increases by 0.91 on 1 df.

```
$FIT + U2C2 - BU.BCLASS $D E $
```

Surprisingly, the deviance increases by 12.44 on 3 df. The estimates BU(2).BCLA(3) and BU(2).BCLA(5) are both only about one standard error away from zero (the reference category value), but they have opposite signs and cannot be equated. We retain the full BU.BCLASS interaction

```
$FIT + BU.BCLASS - U2C2 $D E $
```

No further simplification of the model is possible. On the multinomial logit scale we have the following terms: a class effect on the U logits, an effect for lower class medium smokers and for all heavy smokers on the H logits, and a constant interaction between the logits of U and H. These terms and their standard errors are given below.

```
 1.372   0.061   BH(2).BU(2)
-0.458   0.167   BU(2).BCLA(2)
-0.171   0.142   BU(2).BCLA(3)
-0.008   0.155   BU(2).BCLA(4)
 0.175   0.171   BU(2).BCLA(5)
-0.478   0.045   DH
-0.380   0.085   H2S3
```

There is only one large residual (−2.02) for the 37th point, all other residuals being small. To construct the fitted proportions for the four categories from the model, we have to regroup the fitted values from the Poisson model back into four vectors of length 15.

```
$VAR 15 HUFP NUFP HNFP NNFP I $
$CALC I = %CU(1) : NNFP = %FV(I)/T : HNFP = %FV(I+15)/T :
      NUFP = %FV(I+30)/T : HUFP = %FV(I+45)/T $
$PLOT HUFP HNFP NUFP NNFP CLASS 'A B C D' $
$PLOT HUFP HNFP NUFP NNFP SMOK 'A B C D' $
```

To interpret the final model, we see that the positive association between hypertension and proteinurea is constant over the other variables and is quite strong: the odds ratio of $e^{1.372}$ is nearly 4. Those women with one symptom present are four times as likely to have the second symptom present as those with the first symptom absent. The presence of proteinurea varies over social class, with lower rates in social classes 2 and 3, medium rates in classes 1 and 4, and a higher rate in class 5. Heavy smokers, and medium smokers in social classes 3 to 5, have lower rates of hypertension than non-smokers, or medium smokers in classes 1 and 2.

It will be found that the same models are obtained by separate binomial logit models for H and U. This occurs because the association between the symptoms is constant: if it had varied with the explanatory variables, different models would have been obtained.

6
Survival data

6.1 Introduction

Over the last 20 years there has been a rapid development of probability models and statistical analysis for technological and medical survival data. Many studies have been made of the length of life or of periods of remission of animal or human subjects being treated for serious diseases. These studies have used probability models for duration of life which originated in engineering reliability studies of the operating lifetimes of electrical and mechanical systems. In this chapter we give an account of the main probability models and their use in data analysis. The recognition of the special form of the likelihood in an important class of these distributions allows their simple fitting in GLIM. Many books are now available on survival analysis; a concise account can be found in Cox and Oakes (1984), and a detailed discussion in Kalbfleisch and Prentice (1980).

This chapter is restricted to a discussion of single events: we do not deal with models for recurrent events, like repeated spells of remission or of unemployment.

6.2 The exponential distribution

The exponential distribution plays a central role in survival analysis. Although few systems have exponentially distributed lifetimes, most of the useful survival distributions are directly related to the exponential distribution. We first consider its properties as a survival distribution. In this chapter we use the notation t for a non-negative response variable, which will frequently be a survival time.

The density function of the exponential random variable T for survival time t is

$$\begin{aligned} f(t) &= \Pr(t < T < t + \mathrm{d}t)/\mathrm{d}t \\ &= (1/\mu)\, e^{-t/\mu} \qquad \mu, t > 0 \end{aligned}$$

where μ is the mean of the distribution. The cumulative distribution function is

$$F(t) = 1 - e^{-t/\mu} \qquad \mu, t > 0.$$

A fundamental concept in survival analysis is that of the *hazard function*

$h(t)$, which is the conditional density function at time t given survival up to time t:

$$h(t)dt = \Pr(t < T \leq t + \mathrm{d}t \mid T > t)$$
$$= f(t)\mathrm{d}t/[1 - F(t)].$$

It is convenient to introduce a notation for the upper tail probability: we define the *survivor function*

$$S(t) = P(T > t) = 1 - F(t).$$

Then

$$h(t) = f(t)/S(t).$$

The hazard function can be interpreted as the instantaneous failure rate at time t. Any continuous probability distribution can be specified equivalently by its density, survivor or hazard function. In particular the density and survivor functions can be obtained from the hazard function by

$$S(t) = \exp(-H(t))$$
$$f(t) = h(t)\exp(-H(t))$$

where the function

$$H(t) = \int_0^t h(u)\mathrm{d}u$$

is called the *integrated hazard function*. The usefulness of the hazard function is clear in the servicing and repair of mechanical or electrical systems. When the system has been in operation for some time, the appropriate servicing or repair schedule depends on the probability of failure, conditional on the previous operating time. Historically, the effect of explanatory variables on survival time has been expressed by modelling the hazard function; as we shall see, such models are models for functions of the parameters of the probability distributions.

For the exponential distribution

$$h(t) = 1/\mu$$
$$H(t) = t/\mu.$$

Thus the exponential distribution has *constant hazard*: the probability of death at time t is not dependent on the length of previous lifetime. The exponential distribution represents the lifetime distribution of an item which does not age or wear: the instantaneous probability of failure is the same no matter how long the item has already survived. In most applications this strong property does not hold, and so the exponential distribution has limited application as a survival distribution

In a more general context, if "death" is replaced by an event which can occur repeatedly in time, then these events occur in a time-homogeneous Poisson process with rate $\lambda = 1/\mu$ if the times between events are independent exponential variables with the same mean μ. Thus the probability of r events in time t has the Poisson distribution

$$P(r|\lambda,t) = e^{-\lambda t}(\lambda t)^r/r! \qquad r = 0,1,2,\ldots$$

This relation between the exponential and Poisson distributions is used to advantage in Section 6.7 for maximum likelihood estimation with censored data.

6.3 Fitting the exponential distribution

The exponential distribution is available as a standard distribution in GLIM as it is a special case of the *gamma* distribution, considered in Section 6.9. The error distribution is specified as G, and the $SCALE parameter has to be fixed at 1.

We illustrate with a simple example from Feigl and Zelen (1965) already introduced in Chapter 1. Subfile FEIGL contains the survival TIMEs in weeks of 33 patients suffering from acute myelogeneous leukaemia, and the values of two explanatory variables, white blood cell count WBC in thousands and a positive or negative factor AG, defined as 1 for positive, 2 for negative.

We first plot TIME against WBC.

```
$INPUT 1 FEIGL $
$PLOT TIME WBC $
```

There is a heavy concentration of points along both the horizontal and vertical axes. We try the log scales, and use a different plotting symbol for the two levels of AG.

```
$CALC LT = %LOG(TIME) : LWBC = %LOG(WBC) $
$PLOT LT LWBC '+ -' AG $
```

There is a general decline in log survival time with increasing LWBC, and the negative AG group appears to have shorter survival times. We try fitting the exponential distribution using a log link function for the mean with LWBC as the explanatory variable interacting with the factor AG.

Thus the model is $\log \mu = -\log h(t) = \boldsymbol{\beta}'\mathbf{x}$, so a log–linear model for the mean is also a log–linear model for the hazard function.

```
$YVAR TIME $ERROR G $SCALE 1
$LINK L $FIT AG*LWBC $D E $
```

The deviance is 38.56 on 29 df.

It might be thought that the residual deviance from the model could be used for testing the goodness of fit, since the standard application of large sample theory would suggest that the deviance is distributed as chi-squared with $n-p$ df. However, as $n \to \infty$, the number of parameters in the saturated model also goes to infinity and this violates one of the regularity conditions for the validity of this asymptotic distribution.

The parameter estimates and standard errors are shown below to three decimal places.

5.150	0.501	1
−1.874	0.785	AG(2)
−0.482	0.174	LWBC
0.329	0.267	AG(2).LWBC

In this model the observed and estimated expected information matrices are different because the log link function is not the canonical link for the exponential distribution. See Section 2.16 and Appendix 1 for details, and also Section 6.7. The interaction appears non-significant and might be omitted from the model. We note also that the slope of the regression in the second AG group is $(-0.482+0.329)$ i.e. -0.153, so that an alternative simplification of the model might be achieved by setting this slope to zero, as in the SOLV example in Section 2.8.

```
$FIT - AG.LWBC $D E $
$CALC AG1 = %EQ(AG,1) : Z = AG1*LWBC $
$FIT AG + Z $D E $
```

The deviance is 40.32, an increase of 1.76 on 1 df, for the first model and 39.37, an increase of 0.81, for the second model. We cannot choose between these two models on the basis of the data, because the slope of the regression in the second AG group is poorly defined. The fitted models are

$$\widehat{\log \mu} = \underset{(0.412)}{4.730} - \underset{(0.349)}{1.018}\,\text{AG(2)} - \underset{(0.132)}{0.304}\,\text{LWBC}$$

and

$$\widehat{\log \mu} = \underset{(0.501)}{5.150} - \underset{(0.560)}{2.263}\,\text{AG(2)} - \underset{(0.174)}{0.482}\,\text{Z}$$

The log link, while a natural choice for positive survival times, is not the only possible choice. The mean survival time can also be related to the linear predictor using the *reciprocal link*, giving a *linear hazard model*

$$\mu^{-1} = \lambda = \boldsymbol{\beta}'\mathbf{x}.$$

Here λ is the mean rate of dying, or failure rate (per unit time). This choice of link function is sometimes thought natural because sufficient statistics exist for the model parameters on this scale. This property does not, however, assist the model fitting in any way, and it does not prevent the occurrence of negative fitted values from the linear model.

```
$LINK R  $FIT AG*LWBC $D E $
```

The deviance is 39.78, and the parameter estimates and standard errors are shown below.

0.00598	0.00359	1
0.02196	0.02638	AG(2)
0.00585	0.00238	LWBC
0.00581	0.01116	AG(2).LWBC

The interaction is small and can be omitted.

```
$FIT - AG.LWBC  $D E $
```

The deviance increases by 0.26 to 40.04 and the final model is

$$\widehat{\mu^{-1}} = \underset{(0.00340)}{0.00580} + \underset{(0.01461)}{0.03441}\,\mathrm{AG(2)} + \underset{(0.00232)}{0.00611}\,\mathrm{LWBC}$$

With the reciprocal link an equally good fit is obtained if WBC instead of LWBC is used as the explanatory variable.

Finally, we consider the identity link. The original analysis by Feigl and Zelen used this link, with

$$\mu = 1/\lambda = \boldsymbol{\beta}'\mathbf{x}.$$

```
$LINK I  $FIT  AG*LWBC $D E $
```

The deviance is 40.50, and the interaction, though large (estimate 15.69) does not appear significant (s.e. 9.31). Omitting the interaction gives an error message:

```
$FIT - AG.LWBC $D E $

** fitted mean out of range for error distribution, at [AG.LWBC $]

The fitted mean for unit 32 is less than or equal to zero and the error is G.
This indicates an inappropiate model.
```

The negative fitted value corresponds to observation 32 with the largest value of WBC in the AG negative group and a small value of survival time. This difficulty is avoided with the log link because with this link the fitted values can never be negative.

The three interaction models have very similar deviances, so it is clear that we cannot choose among them from statistical considerations. A formal comparison could be based on the likelihood ratio test, since all three models can be expressed in the form

$$\boldsymbol{\beta}'\mathbf{x} = \eta = (\mu^{\theta} - 1)/\theta, \qquad \mu = (1 + \theta\eta)^{1/\theta}$$

with $\theta = -1$, 0 and 1 for the reciprocal, log and identity links. The corresponding deviances are 39.78, 38.56 and 40.50: the log link has the smallest deviance, but all three fit the data equally well.

The choice of a final model is made difficult, as in the SOLV example of Section 2.8, by the small sample size. We choose the parallel slopes model AG+LWBC with the log link as a reasonable description.

We now consider model criticism for the exponential distribution.

6.4 Model criticism

How do we know that the exponential distribution is a reasonable choice? Probability plotting of approximately standardized exponential variables provides a guide, as for the normal distribution.

If T_i is exponential with mean $\mu_i = 1/\lambda_i$, then $T_i/\mu_i = \lambda_i T_i$ is standard exponential with mean 1.

If $\mu_i = g(\eta_i)$ with η_i the linear predictor, then

$$u_i = t_i/\hat{\mu}_i$$

with

$$\hat{\mu}_i = g(\hat{\boldsymbol{\beta}}'\mathbf{x}_i)$$

The u_i have approximately standard exponential distributions, though they are not independent, as with normal residuals.

The survivor function for the standard exponential is

$$S(u) = e^{-u}$$

so that

$$-\log S(u) = u$$

and

$$\log(-\log S(u)) = \log u.$$

Thus a probability plot of the u_i is obtained by plotting the empirical survivor function $\hat{S}(u)$ of the u_i against $S(u) = e^{-u}$. It will be convenient to plot on the log scale or iterated log scale (i.e. log (−log) scale) as this may indicate the form of departure from the exponential assumption if this is incorrect (see Section 6.14).

We now construct these plots, using the model adopted in Section 6.3 above.

```
$LINK L  $FIT AG+LWBC $
$CALC U = %YV/%FV $
```

The u_i are closely related to the GLIM residuals. These are

$$e_i = \frac{t_i - \hat{\mu}_i}{\hat{\mu}_i} = u_i - 1,$$

as can easily be verified:

```
$DISP R $
$PRINT U $
```

We first calculate the empirical survivor function.

```
$SORT OU U $
$CALC S = 1 - %CU(1)/
        (%NU+1) :
      LS = -%LOG(S) :
     LLS =  %LOG(LS) $
```

The survivor function estimate uses $(n+1)$ in the denominator to avoid a zero value for the last observation (see Section 2.10 for a discussion of plotting positions). We now plot $-\log \hat{S}(u)$ and u against u, followed by the equivalent plot on the iterated log scales.

```
$PLOT LS OU OU '+ *' $
$CALC LOU = %LOG(OU) $PLOT LLS LOU LOU '+ *' $
```

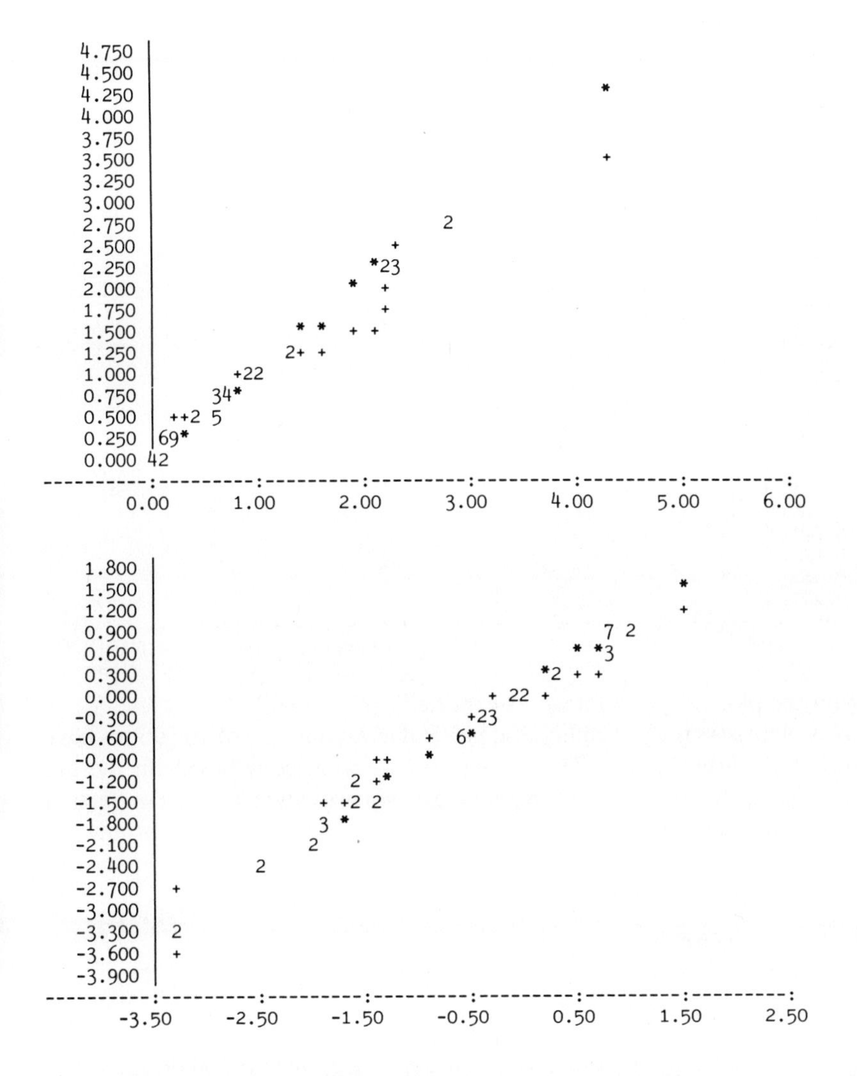

Both plots show a reasonable fit to the theoretical straight line (represented by asterisks). The exponential distribution assumption seems satisfactory.

A disadvantage of both the above plots is that for large u, when the number of observations on which $\hat{S}(u)$ is based becomes small, the variability of $\hat{S}(u)$ becomes very great and it is difficult to decide whether a real failure of the probability model is occurring. Similarly the variability of log $\hat{S}(u)$ becomes very great when u is small. Since $\hat{S}(u)$ is essentially an estimated binomial probability for each u, it can be "variance stabilized" using the arc sine transformation. Thus we plot $\sin^{-1}\sqrt{\{\hat{S}(u)\}}$ against $\sin^{-1}\sqrt{e^{-u}}$, in addition to the plots of $-\log \hat{S}(u)$ against u and $\log(-\log \hat{S}(u))$ against $-\log u$. The plot should be a straight line through the origin with slope 1.

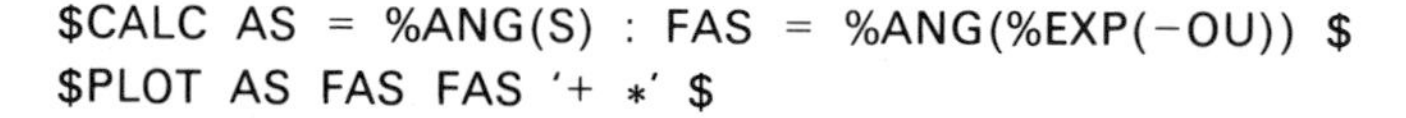

```
$CALC AS = %ANG(S) : FAS = %ANG(%EXP(-OU)) $
$PLOT AS FAS FAS '+ *' $
```

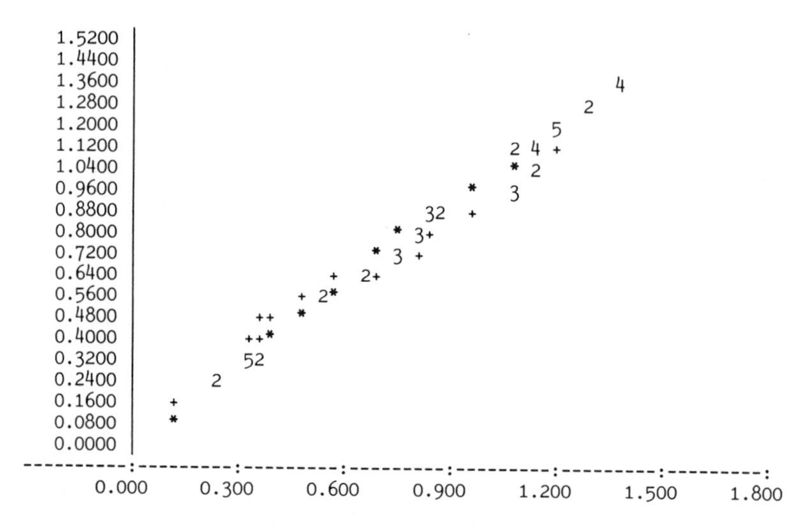

Again the plot is closely linear, as expected.

Now we assess the possibility that particularly influential observations have affected the fitted model. The assessment of influence is based on the hat matrix from the iteratively reweighted least squares algorithm, as in Section 4.3. The diagonal elements of this matrix are easily obtained as the product of the variance of the linear predictor and the iterative weight variate. First we fit the interaction slope model.

```
$FIT AG * LWBC $
$EXTRACT %VL $CALC H = %VL * %WT : I = %CU(1) $
$PRINT H $PLOT H I '+' $
```

Using the value $2(p+1)/n = 0.242$ as a rough guide, we find that observations 2 ($h_2 = 0.297$) and 21 ($h_{21} = 0.282$) are potentially influential. Both are on the edge of the variable space, with the two smallest values of WBC. Deleting these observations in turn makes no substantial difference to the fitted model, and we conclude that the data are represented adequately by the exponential distribution with parallel regressions on the log scale (although other regression models also fit the data adequately).

Figure 6.1 shows the complete data on the log–log scale and the parallel slopes models for survival time using the log and reciprocal links. The fitted models agree closely relative to the random variation in the data.

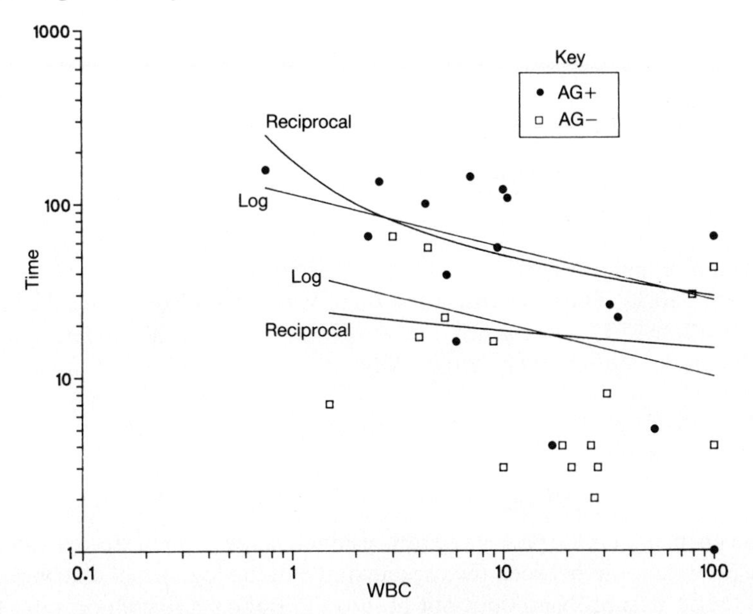

Fig. 6.1. Fitted parallel slopes models to FEIGL data, assuming exponential distribution with log and reciprocal link functions

One aspect of this Figure may seem strange. The fitted models for each group seem to lie above most of the points for that group. This is a consequence of the exponential assumption: on the log scale of time the distribution is extreme value (Section 6.12) with scale parameter 1. This distribution has negative skew: large negative, and small positive values are characteristic of it (see Fig. 6.3 in Section 6.12).

6.5 Comparison with the normal family

The Box–Cox normal transformation family discussed in Chapter 3 provides an alternative family of possible survival distributions. Does the exponential distribution provide a better representation? We first try fitting the interaction model without the log transformation of WBC:

```
$ACC 5
$INPUT %PLC BOXCOX $
$ERROR N
$MACRO YVAR TIME $ENDMAC
$MACRO MODEL AG*WBC $ENDMAC
$USE BOXCOX $
```

Specifying a grid for λ of $-2(0.5)2$, we find the minimum of $-2p\ell(\lambda)$ near $\lambda=0$. Replotting over a finer grid we find the minimum of $-2p\ell(\lambda)$ of 311.17 occurs at about $\lambda=0.10$, with an approximate 95% confidence interval of $(-0.15, 0.37)$. The value at $\lambda=0$ is 311.86. The log scale for time is strongly indicated.

Should WBC be log transformed as well?

```
$MACRO MODEL AG*LWBC $ENDMAC
$USE BOXCOX $
```

The minimum of $-2p\ell(\lambda)$ of 308.00 now occurs near $\lambda=0.20$, and the value at $\lambda=0$ is 310.11. The log transformation of WBC at $\lambda=0$ decreases the value of $-2p\ell(0)$ by 1.75: there is not much to choose between the models, but the LWBC model is slightly preferred. We now fit this model to log time.

```
$USE BOXFIT $
Value of lambda?
$DIN? 0
```

The normal quantile plot of the raw residuals shows a good straight-line fit. How do we choose between the exponential and the lognormal distributions for TIME? This problem does not fit into the standard likelihood ratio test theory because the two distributions are not in the same family and have different numbers of parameters. We discuss the problem further in Section 6.20, where the two distributions are embedded in a larger family; here we simply compare the maximized log likelihoods for the two models.

Further problems arise in this comparison, because in both cases there are constants omitted from the calculation of the likelihood which have to be included if comparisons are to be made between different probability models.

For the Box–Cox tranformation family of Section 3.1, the complete maximized log likelihood is (for fixed λ)

$$\ell_1(\hat{\boldsymbol{\beta}},\hat{\sigma},\lambda)=-\frac{n}{2}\log(2\pi)+n\log|\lambda|-n\log\hat{\sigma}+(\lambda-1)\Sigma\log t_i-\frac{n}{2}$$

and hence

$$-2\ell_1(\hat{\boldsymbol{\beta}},\hat{\sigma},\lambda)=n\{\log(2\pi)+\log(\mathrm{RSS}/n)+1-2\log|\lambda|\}-2(\lambda-1)\Sigma\log t_i.$$

The deviance printed out by the Box–Cox macro is

$$D_1=n\log(\mathrm{RSS}/\lambda^2)-2(\lambda-1)\Sigma\log t_i$$

so that

$$-2\ell_1(\hat{\boldsymbol{\beta}},\hat{\sigma},\lambda)=D_1+n\{\log(2\pi/n)+1\}.$$

For the exponential distribution, the GLIM deviance is

$$D_2 = -2 \log [L(\hat{\boldsymbol{\beta}})/L\{(\hat{\mu}_i)\}]$$

where

$$L\{(\hat{\mu}_i)\} = \prod_i \frac{1}{\hat{\mu}_i} e^{-t_i/\hat{\mu}_i}$$

and

$$\hat{\mu}_i = t_i$$

giving

$$L\{(\hat{\mu}_i)\} = e^{-n}/\prod_i t_i.$$

Thus

$$\begin{aligned} -2\ell_2(\hat{\boldsymbol{\beta}}) &= -2 \log L(\hat{\boldsymbol{\beta}}) \\ &= D_2 + 2(n + \Sigma \log t_i). \end{aligned}$$

To compare the actual maximized likelihoods we have to adjust the deviances in each case by the constant factors. For the lognormal model AG∗LWBC, $D_1 = 310.11$, and to calculate the true value of $-2 \log L_{\max}$, we use

```
$CALC 310.11 + %NU*(%LOG(2*%PI/%NU)+1) $
```

which gives

$$-2\ell_1(\hat{\boldsymbol{\beta}},\hat{\sigma},0) = 288.38.$$

For the exponential model with log link, $D_2 = 38.56$, and we calculate the true value of $-2 \log L_{\max}$ using

```
$CALC %C = %CU(%LOG(TIME)): 38.56+2*(%NU+%C) $
```

which gives

$$-2\ell_2(\hat{\boldsymbol{\beta}}) = 291.32.$$

Thus the exponential distribution provides a slightly worse fit, possibly because it has no free scale parameter. For the lognormal distribution the fitted model is

$$\underset{(0.603)}{5.425} - \underset{(0.945)}{2.525}\,\text{AG(2)} - \underset{(0.209)}{0.818}\,\text{LWBC} + \underset{(0.321)}{0.583}\,\text{AG(2).LWBC}$$

Again the interaction term can be omitted, giving a final model

$$\underset{(0.514)}{4.802} - \underset{(0.436)}{0.988}\,\text{AG(2)} - \underset{(0.165)}{0.571}\,\text{LWBC}$$

The slope estimate is greater in magnitude for the lognormal than for the exponential model (−0.304), but the AG difference is very similar. Survival appears to decline more rapidly with WBC under the lognormal distribution than under the exponential, but since these distributions fit equally well, we cannot be precise about the rate of decline of survival with WBC.

The fitted models for both distributions are shown in Fig. 6.2.

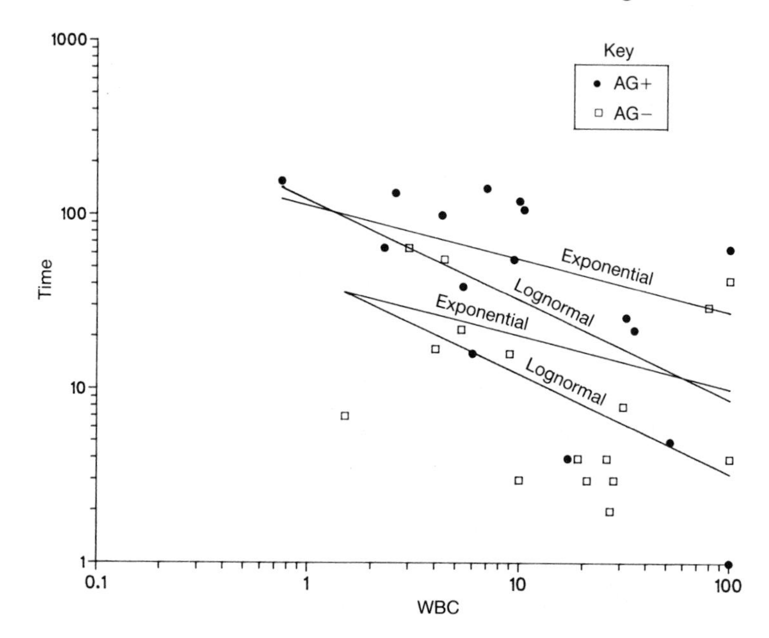

Fig. 6.2. Fitted parallel slopes models to FEIGL data, assuming (i) exponential distribution, log link, (ii) lognormal distribution, identity link

We consider this dataset further in Sections 6.9, 6.11 and 6.17, where we compare these models with those produced assuming gamma, Weibull and piecewise exponential distributions.

6.6 Censoring

A characteristic feature of survival data is the occurrence of *censored* observations, i.e. observations on which the complete lifetime or survival time is not observed. Censoring occurs for a number of different reasons. In industrial experiments on destructive testing of components, testing is usually terminated after a fixed time, or after the failure of some fixed proportion of components on test. The remaining components have not failed and their complete lifetimes are not observed. In medical studies censoring arises from the withdrawal of patients from the study because they have moved away or have stopped returning for follow-up, and from the need to analyse the data at various stages of the study before complete lifetimes are observed on all patients. Censoring is generally on the right; that is, the observed time is less than the actual survival time. Much less commonly it may be on the left; that is, the observed time is *greater* than the actual survival time. We will assume that censoring is on the right for most of this chapter; left-censoring is considered in Section 6.18.

Throughout this chapter we shall assume that the censoring is uninformative, i.e. that the censored lifetimes do not occur because of factors associated with the survival time which would have been observed. This assumption is similar to that of data missing at random in Section 3.13. Censoring would be informative if, for example, patients were withdrawn from a drug treatment study because they were suffering unexpected or severe effects of the treatment which might have reduced their survival times.

As with missing data, it is prudent to compare by logit regression the group of censored observations with the group of uncensored observations to see if censoring is systematically related to the explanatory variables. Systematic effects would be a warning of the possible failure of the assumption of uninformative censoring.

6.7 Likelihood function for censored observations

When some observations have censored lifetimes, the likelihood function is changed, and, in general, maximum likelihood estimation can no longer be carried out using the standard GLIM algorithm. Nevertheless, for survival distributions related to the exponential (and also for the logistic and log-logistic distributions in Section 6.18) it is still possible to use GLIM model fitting procedures. We illustrate with the exponential distribution.

For each individual in the sample we observe the vector of explanatory variables $\mathbf{x}_i$, and a pair of variables (t_i, w_i). The *censoring indicator* w_i takes the value 1 if the survival time t_i for the i-th observation is uncensored, and zero if it is censored. Thus when $w_i = 1$, the true survival time $T_i = t_i$, while when $w_i = 0$, it is known only that $T_i > t_i$. The contribution of a censored observation to the likelihood is thus the probability $S(t_i)$.

The likelihood function can then be expressed as

$$L(\boldsymbol{\beta}) = \prod_{i=1}^{n} \{f(t_i)\}^{w_i} \{S(t_i)\}^{1-w_i}$$

$$= \prod_{i=1}^{n} \{h(t_i)\}^{w_i} S(t_i).$$

For the exponential distribution with mean μ_i, we have

$$h(t_i) = 1/\mu_i = \lambda_i$$

$$S(t_i) = e^{-\lambda_i t_i}$$

so that

$$L(\boldsymbol{\beta}) = \prod_i \lambda_i^{w_i} e^{-\lambda_i t_i}$$

$$= \prod_i (\lambda_i t_i)^{w_i} e^{-\lambda_i t_i} / \prod_i t_i^{w_i}$$

The term in the denominator is not a function of the parameter vector $\boldsymbol{\beta}$ and can be omitted. The remaining term is identical to the likelihood function for a set of n observations w_i having independent Poisson distributions with mean $\lambda_i t_i$ (the term $1/w_i!$ is missing, but this is a constant and can be omitted). Since w_i is only either 0 or 1, this result looks strange. Where does the Poisson distribution come from?

The Poisson result is easily understood if we visualise a set of n independent time-homogeneous Poisson processes as in Section 6.2, the i-th having rate λ_i. The number of events for the i-th process in a time interval of length t then has a Poisson distribution with mean $\lambda_i t$. We observe the i-th process until either the first event occurs, or a fixed time has elapsed without the event occurring. In the first case, one event occurs at time t_i; in the second, no event has occurred by time t_i.

The Poisson likelihood for the censored and uncensored exponential survival times allows us to fit models to the hazard rate λ using the Poisson error model (Aitkin and Clayton, 1980). Write θ_i for the Poisson mean:

$$\theta_i = \lambda_i t_i.$$

If the *linear hazard model* is used, then

$$\lambda_i = \boldsymbol{\beta}' \mathbf{x}_i$$

and

$$\theta_i = \boldsymbol{\beta}'(t_i \mathbf{x}_i).$$

Thus if each explanatory variable $\mathbf{x}_i$ (including the unit vector $\mathbf{1}$) is multiplied by t_i, the model can be fitted by declaring w_i as the response variable with Poisson error and identity link function.

If the *log linear hazard model* is used, then

$$-\log \mu_i = \log \lambda_i = \boldsymbol{\beta}' \mathbf{x}_i$$

and

$$\log \theta_i = \log \lambda_i + \log t_i$$

$$= \boldsymbol{\beta}' \mathbf{x}_i + \log t_i.$$

This model is particularly simple to fit: the same error and response variable specifications are used, but now with log link function and an offset of log t_i.

This approach can also be used for uncensored observations, by defining the censoring indicator as a vector of 1s. We illustrate with the example of Section 6.3:

```
$INPUT 1 FEIGL $
$CALC LT = %LOG(TIME) : LWBC = %LOG(WBC) : W = 1 $
$YVAR W $ERROR P $LINK L $OFFSET LT
$FIT AG*LWBC  $D E $
```

We obtain the same parameter estimates as in Section 6.3 (to within the accuracy set by GLIM's convergence criterion), though with opposite signs for the parameters, but the standard errors are different (e.g. 0.732 for the s.e. of AG(2), rather than 0.785). This is because the expected information matrix for the Poisson model treats t_i as a constant explanatory variable, while in the exponential model t_i is treated as a random variable and is replaced in the expected information matrix by its expected value μ_i.

It may seem strange that we can fit a model to a constant response variable, but the offset introduces the variation to be modelled by the explanatory variables.

If the *reciprocal hazard model* is used, then

$$\lambda_i = 1/\boldsymbol{\beta}'\mathbf{x}_i$$

giving

$$\mu_i = \boldsymbol{\beta}'\mathbf{x}_i$$

and the model is a linear model for the *mean*. Then

$$\theta_i = \lambda_i t_i = t_i/\boldsymbol{\beta}'\mathbf{x}_i,$$

so that

$$\theta_i^{-1} = \boldsymbol{\beta}'(\mathbf{x}_i/t_i).$$

The model can be fitted by scaling each explanatory variable (including the unit vector **1**) by dividing by t_i, and using the reciprocal link function.

The Poisson representation of the likelihood applies more generally to *proportional hazard* models, as we shall see in Section 6.15.

6.8 Probability plotting with censored data: the Kaplan–Meier estimator

The presence of censoring complicates our assessment of the probability distribution assumption for the response variable. If T_i is exponential with mean μ_i, then T_i/μ_i is standard exponential, but if the true lifetime is censored at t_i, then t_i/μ_i is a censored value from the standard exponential distribution. Thus when a model has been fitted to mixed censored and uncensored observations, the $u_i = t_i/\hat{\mu}_i$ are themselves mixed censored and uncensored values from the (approximately) standard exponential distribution. The u_i are easily obtained from the Poisson fit as the GLIM fitted values: $u_i = \theta_i = \hat{\lambda}_i t_i = t_i/\hat{\mu}_i$. However it is not clear how to construct the empirical cumulative distribution function of the u_i to assess the correctness of the probability distribution. The *product-limit* or *Kaplan–Meier* (1958) estimator of the survivor function is used for this purpose.

We begin by considering the set of all discrete times $0 < a_1 < a_2 < \ldots < a_N$ at which deaths or censorings may occur. There is no real loss of generality in assuming that these times are discrete, because the finite precision of measurement means that the values of survival time actually recorded can take only a finite (though possibly large) number of values (see Appendix 1 for a discussion of measurement precision). Let A_j denote the time-interval $(a_{j-1}, a_j]$ with $a_0 = 0$. For the continuous survival distribution with density $f(t)$, survivor function $S(t)$ and hazard $h(t)$, we define the corresponding functions for the resulting grouped discrete distribution:

$$\begin{aligned} f_j &= P(T \in A_j) \\ &= S(a_{j-1}) - S(a_j) \\ &= s_j - s_{j+1}, \end{aligned}$$

where

$$\begin{aligned} s_j &= P(T > a_{j-1}) \\ &= f_j + f_{j+1} + \ldots + f_N, \end{aligned}$$

and

$$\begin{aligned} h_j &= P(T \in A_j | T > a_{j-1}) \\ &= f_j / s_j. \end{aligned}$$

Then

$$h_j = (s_j - s_{j+1})/s_j$$

so that

$$s_{j+1}/s_j = 1 - h_j$$

whence

$$s_{r+1} = \prod_{j=1}^{r} s_{j+1}/s_j = \prod_{j=1}^{r} (1 - h_j)$$

since $s_1 = 1$. The survivor function at time t can therefore be expressed as

$$S(t) = \prod_{j: a_j < t} (1 - h_j).$$

We estimate the survivor function by considering the passage of each individual through time, shown in the table below.

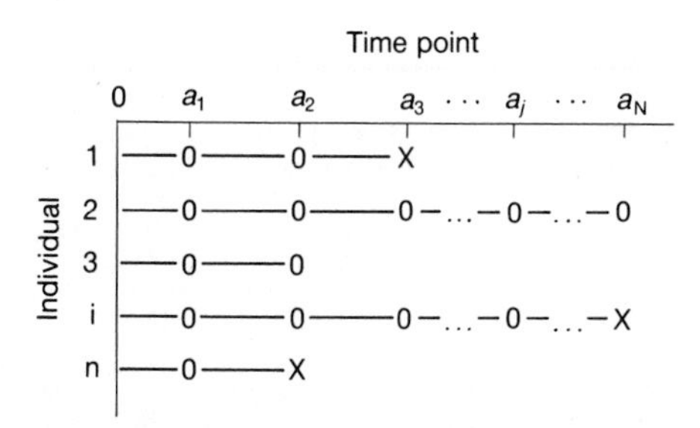

The circles represent censoring and the crosses death. The i-th individual survives to the beginning of the interval A_{j_i} and is either finally censored or dies at the end of the interval. Let w_{ij} be the censoring indicator for the i-th individual at a_j; the hazard in A_j is h_j, and the survivor function at a_{j-1} is s_j, and at a_j is s_{j+1}.

The contribution L_i to the likelihood for the i-th individual is f_{j_i} if the individual dies in A_{j_i}, and s_{j_i+1} if he survives beyond the end a_j of the interval. Thus

$$
\begin{aligned}
L_i &= f_{j_i}^{w_{i,j_i}} s_{j_i+1}^{1-w_{i,j_i}} \\
&= (h_{j_i} s_{j_i})^{w_{i,j_i}} \{s_{j_i}(1-h_{j_i})\}^{1-w_{i,j_i}} \\
&= s_{j_i} h_{j_i}^{w_{i,j_i}} (1-h_{j_i})^{1-w_{i,j_i}} \\
&= h_{j_i}^{w_{i,j_i}} (1-h_{j_i})^{1-w_{i,j_i}} \prod_{j=1}^{j_i-1} (1-h_j) \\
&= \prod_{j=1}^{j_i} h_j^{w_{ij}} (1-h_j)^{1-w_{ij}}
\end{aligned}
$$

since all the w_{ij} for $j < j_i$ are zero. The likelihood function over all individuals is then

$$
L = \prod_{i=1}^{n} L_i = \prod_{j=1}^{N} \prod_{i \in R_j} h_j^{w_{ij}} (1-h_j)^{1-w_{ij}}
$$

where R_j is the *risk set* of individuals in the j-th interval, that is the set of individuals not already dead or withdrawn (censored).

The likelihood function is the product of a set of N binomial likelihoods, one for each interval. In the j-th interval, the number of individuals at risk is

$$
\sum_{i \in R_i} 1 = r_j,
$$

and the number dying is

$$\sum_{i \in R_j} w_{ij} = d_j.$$

The constant probability of death for each individual is h_j, so we have immediately that the MLE of h_j is

$$\hat{h}_j = d_j/r_j$$

and therefore that of $S(t)$ is

$$\hat{S}(t) = \prod_{j:a_j < t} (1 - d_j/r_j).$$

This is the *product-limit* or *Kaplan–Meier* estimator of the survivor function. Note that when $d_j = 0$, $\hat{h}_j$ is zero: a change in the estimated survivor function occurs only in intervals with observed deaths. Thus in computing the Kaplan–Meier estimator we need to consider only the times at which deaths occur: censoring times at which no deaths occur can be omitted from the set of a_j. When there is no censoring, the Kaplan–Meier estimator reduces to the usual empirical survivor function. Thus the usual empirical cumulative distribution function is in fact the nonparametric maximum likelihood estimator of the cumulative distribution function $F(t)$ when no assumption is made about the form of F. This provides a justification for the use of plotting positions i/n in empirical cumulative distribution function plotting (see Section 2.10), though the last point usually has to be suppressed in the plot.

We illustrate with a simple example. Subfile GEHAN contains 42 observations of remission times in weeks of leukaemia patients. A randomized treatment group was treated with 6-mercaptopurine, the other group was a control. The data are presented and analysed in Gehan (1965). We first examine the treatment group, in which censoring is heavy. The observations t for this group are given below, in increasing order, with the censoring indicator w.

t	6	6	6	6	7	9	10	10	11	13	16	17	19	20	22	23	25	32	32	34	35
w	1	1	1	0	1	0	1	0	0	1	1	0	0	0	1	1	0	0	0	0	0

The first censored observation at 6 weeks we take to just *exceed* 6 weeks, and similarly for that at 10 weeks. The survivor function changes value only at the observed remission times $t_j = 6, 7, 10, 13, 16, 22$ and 23 weeks. We construct a table of t_j, r_j, d_j, h_j and $\hat{S}(t)$.

j	1	2	3	4	5	6	7
t_j	6	7	10	13	16	22	23
r_j	21	17	15	12	11	7	6
d_j	3	1	1	1	1	1	1
h_j	0.1429	0.0588	0.0667	0.0833	0.0909	0.1429	0.1667
$1-h_j$	0.8571	0.9412	0.9333	0.9167	0.9091	0.8571	0.8333
$\Pi(1-h_j)$	0.8571	0.8067	0.7529	0.6902	0.6275	0.5378	0.4482

The Kaplan–Meier estimator can be computed using the macro KAPLAN, stored in subfile KAPLAN. The macro requires as arguments the name of the variable containing the censored and uncensored survival times, and the censoring indicator which must be coded 1 for uncensored and 0 for censored observations. It produces a plot of the Kaplan–Meier estimator against the survival variable named as the macro argument. The survivor function estimate is held in a variate called ZA_, and the *uncensored* values of the survival variable in ZC_.

We illustrate with the full set of data from GEHAN in which the response variable is T, the censoring indicator is W and the treatment group factor is G. We fit an exponential distribution with a two-group model. The fitted values are then the standardized exponential values u_i.

If the largest observation is uncensored, the survivor function estimate at this point will be zero. When the log of the survivor function is calculated, this will give a value of zero which will appear out of place in the plot. The system vector %RE can be used to suppress the plotting of this point, using for example

```
$CALC %RE = %NE(ZA_,0) $
$INPUT 1 GEHAN KAPLAN $
$YVAR W $ERROR P $LINK L
$CALC OFS = %LOG(T) $OFFSET OFS
$FIT G $D E $USE KAPLAN %FV W $
```

The hazard in the second group is significantly lower than in the first, with an estimated group difference of 1.526 with standard error 0.396. The survivor function appears to decrease exponentially. We plot $-\log \hat{S}(u)$ against u, $\log\{-\log \hat{S}(u)\}$ against $\log u$, and $\sin^{-1}\sqrt{\{\hat{S}(u)\}}$ against $\sin^{-1}\sqrt{e^{-u}}$.

```
$CALC LS = -%LOG(ZA_) : LLS = %LOG(LS) :
      AS =  %ANG(ZA_) : AU = %ANG(%EXP(-ZC_)) :
      LU =  %LOG(ZC_) $
$CALC %RE = %NE(ZA_,0) $
$PLOT LS  ZC_ ZC_ '+ *' :
     LLS  LU LU '+ *' :
      AS  AU AU '+ *' $
$DEL %RE $
```

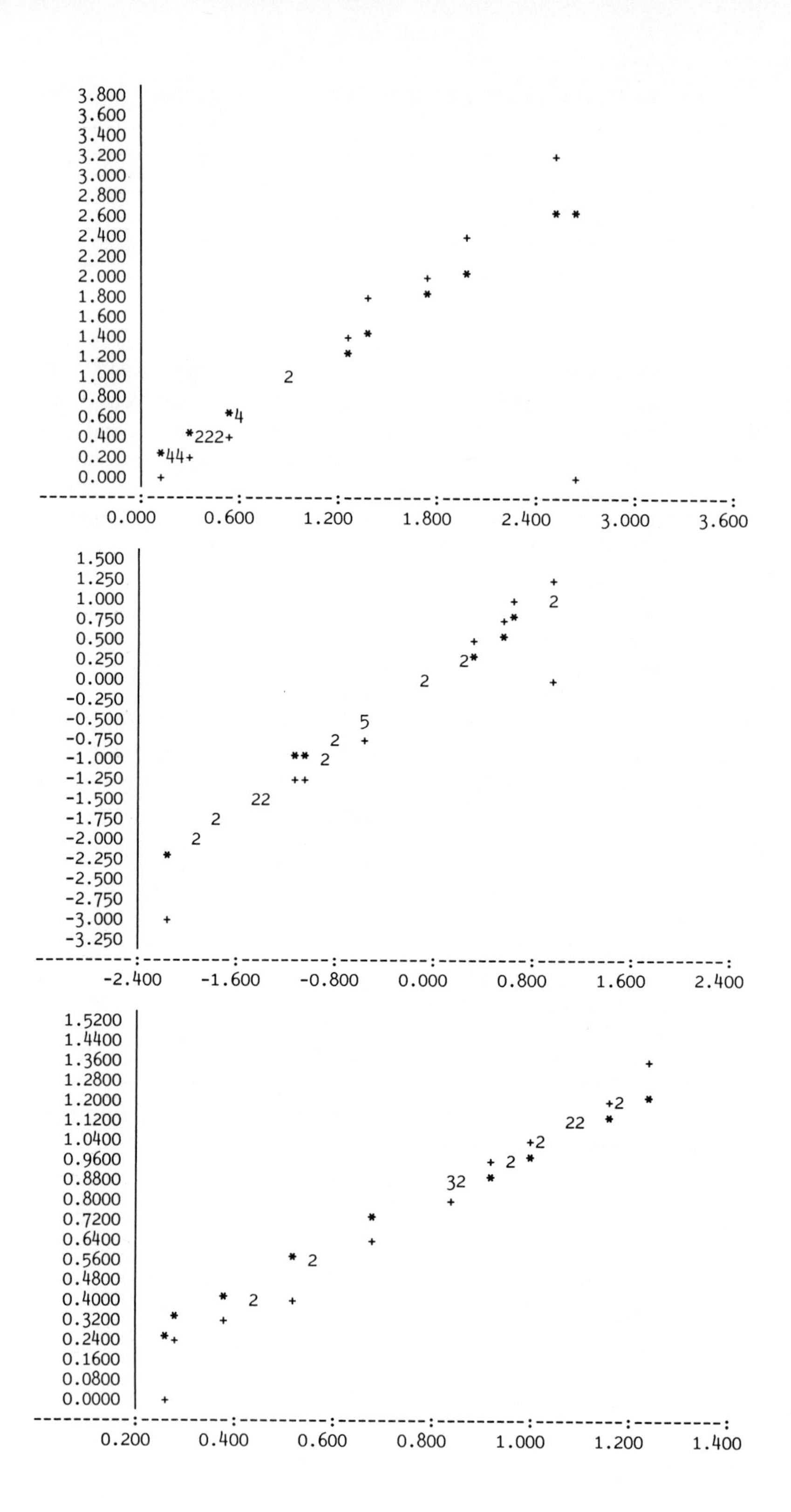
3.800
3.600
3.400
3.200
3.000
2.800
2.600
2.400
2.200
2.000
1.800
1.600
1.400
1.200
1.000
0.800
0.600
0.400
0.200
0.000
0.000 0.600 1.200 1.800 2.400 3.000 3.600
1.500
1.250
1.000
0.750
0.500
0.250
0.000
-0.250
-0.500
-0.750
-1.000
-1.250
-1.500
-1.750
-2.000
-2.250
-2.500
-2.750
-3.000
-3.250
-2.400 -1.600 -0.800 0.000 0.800 1.600 2.400
1.5200
1.4400
1.3600
1.2800
1.2000
1.1200
1.0400
0.9600
0.8800
0.8000
0.7200
0.6400
0.5600
0.4800
0.4000
0.3200
0.2400
0.1600
0.0800
0.0000
0.200 0.400 0.600 0.800 1.000 1.200 1.400

There is some evidence of curvature in the first and third plots, and the second appears linear but with slope slightly greater than 1. The exponential distribution does not appear completely satisfactory. In Section 6.11 we consider the Weibull distribution as an alternative model for survival time.

We now consider survival distributions related to the exponential.

6.9 The gamma distribution

Suppose that an observed lifetime is a *sum* of times for r separate stages of life, each of which is exponentially distributed with the same mean θ. Then the total lifetime has a gamma distribution with parameters r and θ:

$$f(t) = \frac{1}{\Gamma(r)\theta^r} e^{-t/\theta} t^{r-1} \qquad t, r, \theta > 0$$

The value of r can more generally be any positive real number.

The mean of the distribution is $r\theta$, and we adopt the GLIM parametrization using $\mu = r\theta$, writing

$$f(t) = \frac{r^r}{\Gamma(r)\mu^r} e^{-rt/\mu} t^{r-1} \qquad t, r, \mu > 0.$$

The variance is μ^2/r.

Generalized linear models can be fitted in GLIM using the gamma distribution, but full maximum likelihood is available only for r *known*. A further restriction is that only uncensored observations can be handled; these two restrictions greatly limit the usefulness of the distribution in GLIM for survival analysis. The case $r=1$ gives the exponential distribution already discussed.

We illustrate the case of general r with a log-linear model $\log \mu_i = \boldsymbol{\beta}'\mathbf{x}_i$. The log likelihood is

$$\log L = nr \log r - n \log \Gamma(r) - r\Sigma \log \mu_i - r\Sigma(t_i/\mu_i) + (r-1)\Sigma \log t_i,$$

and

$$\frac{\partial \log L}{\partial \boldsymbol{\beta}} = r\Sigma\{(t_i/\mu_i) - 1\}\mathbf{x}_i$$

$$\frac{\partial^2 \log L}{\partial \boldsymbol{\beta}\, \partial \boldsymbol{\beta}'} = -r\Sigma(t_i \mathbf{x}_i \mathbf{x}_i'/\mu_i).$$

Thus the MLE of $\boldsymbol{\beta}$ does not depend on r, since r cancels from $\partial \log L/\partial \boldsymbol{\beta} = 0$, but the observed and expected information matrices contain r as a multiplicative constant. Thus the MLE of $\boldsymbol{\beta}$ for the gamma distribution will be the same as that for the exponential distribution, and the standard errors will be $1/\sqrt{r}$ times those for the exponential. The MLE $\hat{r}$ of r is the solution of

$$n(1+\log r)-n\psi(r)-\Sigma\log\hat{\mu}_i-\Sigma\,(t_i/\hat{\mu}_i)+\Sigma\,\log\,t_i=0$$

where $\psi(r)$ is the digamma function, which is not available in GLIM.

For the saturated model with $\hat{\mu}_i=t_i$, the log likelihood is

$$\log\,L_{\text{sat}}=nr\,\log\,r-n\,\log\,\Gamma(r)-r\Sigma\,\log\,t_i-nr+(r-1)\Sigma\,\log\,t_i$$

and the goodness-of-fit likelihood ratio test statistic for a given model is therefore

$$\begin{aligned}\text{dev} &= -2(\log\,L_{\text{model}}-\log\,L_{\text{sat}})\\ &= 2r\{\Sigma(t_i/\hat{\mu}_i)-n-\Sigma\,\log\,(t_i/\hat{\mu}_i)\}\end{aligned}$$

This test statistic, like the information matrix, depends on r as a scale factor and therefore cannot be used for model simplification without knowing the value of r. When no scale parameter has been explicitly set by the user, GLIM calculates the deviance using the above expression with $r=1$ (the value corresponding to the exponential distribution) and uses the mean deviance as an estimate of $1/r$ for the purpose of calculating standard errors of the parameters. It is possible to set explicitly the scale parameter to a known value by calculating the value of $1/r$ and using this as an argument for the $SCALE directive. The deviance and standard errors of the parameters will then be calculated using this specified value or r. The mean deviance estimate is not unbiased since

$$E[2\{t_i/\mu_i-1-\log\,(t_i/\mu_i)\}]=2\{\log\,r-\psi(r)\}$$

which approaches $1/r$ for large r, but it may be useful as a starting value in the numerical solution of the likelihood equation for r. A full maximum likelihood analysis requires the ML estimation of r for each fitted model and the evaluation of the maximized likelihood at this value of r. Alternatively a profile likelihood for r can be constructed by evaluating the true log-likelihood at $\hat{\boldsymbol{\beta}}$ for each r. Since r is not usually itself of scientific importance, this procedure is generally of little interest.

We illustrate with the FEIGL data of Section 6.3.

```
$INPUT 1 FEIGL $
$CALC LWBC = %LOG(WBC)
$YVAR TIME  $ERROR G  $LINK L
$FIT AG*LWBC  $D E $
```

No argument is specified for $SCALE, so GLIM estimates $1/r$ from the mean deviance as 1.329. The parameter estimates are the same as those from the exponential distribution, but the standard errors are different by a factor of $\sqrt{1.329}$ because of the different scale parameter value. Finding the MLE of r (outside GLIM) is straightforward: we first calculate $\Sigma \log (t_i/\hat{\mu}_i)$ and $\Sigma t_i/\hat{\mu}_i$:

```
$CALC %A = %CU(%LOG(%YV/%FV)): %B = %CU(%YV/%FV) $
$PRINT %A : %B $
```

The values are -19.277 and 33.000 (to 3 d.p.). Then $\hat{r}$ is the solution of

$$1+\log r-\psi(r)=1.5842.$$

Tabulation over a suitable grid of values of r gives $\hat{r}=0.99$, which is very close to $r=1$ (the value for the exponential distribution). Within the class of gamma distributions the exponential is almost the best fit (see Section 6.11 for a similar result for the Weibull distribution, and Section 6.17 for the piecewise exponential distribution for these data). The likelihood ratio test statistic for the hypothesis $r=1$ can be easily evaluated: we have

$$\log L(\hat{r}) = n\hat{r}\log \hat{r}-n\log \Gamma(\hat{r})-\hat{r}\Sigma \log \hat{\mu}_i-\hat{r}\Sigma(t_i/\hat{\mu}_i)+(\hat{r}-1)\Sigma \log t_i$$

$$\log L(1) = \qquad -\Sigma \log \hat{\mu}_i-\Sigma(t_i/\hat{\mu}_i)$$

so that

$$-2\{\log L(1)-\log L(\hat{r})\}=$$

$$2[n\hat{r}\log \hat{r}-n\log \Gamma(\hat{r})-(\hat{r}-1)\{\Sigma \log (t_i/\hat{\mu}_i)-\Sigma(t_i/\hat{\mu}_i)\}]$$

which in this example is 0.776, obviously non-significant.

We do not proceed further with this example, since it can be treated as exponential. Note that the GLIM mean deviance estimate $(1.329)^{-1}$, of r differs considerably from the MLE.

6.10 The Weibull distribution

The Weibull distribution was first used for the strength of materials corresponding to a form of weakest link model: specifically, if $Z_1, \ldots, Z_r$ are independently distributed on $(0, \infty)$ and $V = \min(Z_j)$ then the distribution of V, suitably standardized, approaches the Weibull distribution as r increases.

The Weibull distribution can be expressed as a power transform of the exponential. Suppose U has an exponential distribution with mean $\mu = \lambda^{-1}$, and $T = U^{1/\alpha}$ where $\alpha > 0$. Then T has a Weibull distribution with density, survivor and hazard functions

$$f(t) = \alpha\lambda t^{a-1} e^{-\lambda t^\alpha}, \qquad t > 0$$

$$S(t) = e^{-\lambda t^\alpha}$$

$$h(t) = \alpha\lambda t^{\alpha-1}.$$

The mean of the distribution is $\Gamma(1+\alpha^{-1})\mu$ and the median is $(\mu \log 2)^{1/\alpha}$. If $\alpha = 1$, then the hazard function is constant and T has an exponential distribution; for $\alpha > 1$ the hazard function is monotone increasing with t, while for $\alpha < 1$ it is monotone decreasing. The form of the density varies considerably with the *shape parameter* α.

If the explanatory variables $\mathbf{x}$ do not affect the shape parameter, then the Weibull distribution has the important property of a *proportional hazard function*. The effect of the explanatory variables in any regression model for λ is simply to multiply the hazard by some constant: the functional form in t remains the same.

The Weibull distribution has the further property of being an *accelerated failure* (or life) *time distribution*. If T_1 and T_2 have Weibull distributions with the same shape parameter α but different parameters λ_1 and λ_2, then

$$S_1(t) = e^{-\lambda_1 t^\alpha}$$

$$S_2(t) = e^{-\lambda_2 t^\alpha}$$

$$= S_1(\phi t)$$

where $\phi = (\lambda_2/\lambda_1)^{1/\alpha}$. Thus the survivor function for T_2 is the same as that for T_1 if time for T_1 is scaled by the factor ϕ: death or failure is *accelerated* for T_2 relative to T_1 if $\phi > 1$.

Other distributions with this property include the lognormal and log-logistic (see Sections 6.18, 6.19); a characteristic property of random variables T with accelerated failure time distributions is that $\log T$ has a location and scale parameter distribution with constant variance (Cox and Oakes, 1984 p. 65).

6.11 Maximum likelihood fitting of the Weibull distribution

The Poisson likelihood representation for the exponential distribution can be easily generalized to the Weibull distribution (Aitkin and Clayton, 1980). Given observations (t_i, w_i, x_i) with w_i the censoring indicator, the likelihood function is, from Section 6.7,

$$L(\boldsymbol{\beta},\alpha) = \prod_{i=1}^{n} \{h(t_i)\}^{w_i} S(t_i).$$

Assuming a constant shape parameter, we have for the Weibull distribution,

$$h(t_i) = \alpha\lambda_i t_i^{\alpha-1}$$
$$S(t_i) = e^{-\lambda_i t_i^{\alpha}},$$

and we write

$$\theta_i = \lambda_i t_i^{\alpha} = H(t_i)$$

where H is the integrated hazard function.

Then

$$L(\boldsymbol{\beta},\alpha) = \prod_i (\alpha\theta_i/t_i)^{w_i} e^{-\theta_i}$$

$$= \alpha^{\Sigma w_i} \prod_i \theta_i^{w_i} e^{-\theta_i} \Big/ \prod_i t_i^{w_i}.$$

Again the term

$$\prod_i t_i^{w_i}$$

can be ignored, and the term

$$\prod_i \theta_i^{w_i} e^{-\theta_i}$$

is a Poisson likelihood, but there now is an additional term $\alpha^{\Sigma w_i}$ in the full likelihood. The log of the Weibull hazard function is

$$\log h(t_i) = \log \alpha + (\alpha - 1) \log t_i + \log \lambda_i,$$

and if λ_i itself has a log-linear model:

$$\log \lambda_i = \boldsymbol{\beta}'\mathbf{x}_i$$

then the hazard is log–linear in the explanatory variables:

$$\log h(t_i) = \log \alpha + (\alpha - 1) \log t_i + \boldsymbol{\beta}'\mathbf{x}_i.$$

Since

$$\log \theta_i = \alpha \log t_i + \boldsymbol{\beta}'\mathbf{x}_i,$$

if α were known the log-linear hazard model could again be fitted using the log–linear Poisson model for w with a known offset $\alpha \log t$. On the other hand if $\boldsymbol{\beta}$ were known, α could be estimated from the likelihood equation

$$\frac{\partial \ell}{\partial \alpha} = 0.$$

Writing $n_1 = \Sigma w_i$, the number of uncensored observations, we have

$$\frac{\partial \ell}{\partial \alpha} = n_1/\alpha + \Sigma(w_i - \theta_i) \log t_i$$

and hence

$$\hat{\alpha}^{-1} = \Sigma(\hat{\theta}_i - w_i)\log t_i/n_1.$$

Joint maximum likelihood estimation of $\boldsymbol{\beta}$ and α can thus be carried out by the method of successive relaxation. First α is fixed at $\alpha_0 = 1$, giving an exponential distribution, and $\boldsymbol{\beta}$ is estimated from the Poisson model taking $\log t_i$ as an offset. Then α_0' is estimated from the above likelihood equation using the fitted values $\hat{\theta}_i$. A new estimate $\alpha_1 = (\alpha_0 + \alpha_0')/2$ is then used to define the offset $\alpha_1 \log t_i$ for a new fit of the Poisson model, and this process is continued until convergence. The damped estimate α_1 is used rather than α_0' because the successive estimates α_j and α_j' oscillate about the maximum likelihood estimate, and damping substantially accelerates the rate of convergence.

The library macro WEIBULL (Aitkin and Francis, 1980) fits the Weibull distribution using this approach. The macro requires three arguments: the response variable, the censoring indicator, and a scalar which takes the value 1 if a Weibull fit is required (an exponential fit being the first iteration) or 0 if only the exponential fit is required. A macro called MODEL must also be specified containing the model to be fitted. Standard errors for the parameter estimates $\boldsymbol{\beta}$ are slightly underestimated because GLIM treats α as known in the estimation of $\boldsymbol{\beta}$. This can be corrected if necessary by additional computation outside GLIM. Aitkin and Clayton (1980) give details. See also Roger and Peacock (1982), and Roger (1985) for direct calculation in GLIM.

We illustrate with the GEHAN example.

```
$INPUT %PLC WEIBULL $
$INPUT 1 GEHAN $
$CALC %S = 1 $
$MACRO MODEL G $ENDMAC
$USE WEIBULL T W %S $
```

The deviance for the exponential model is 217.05, and that for the Weibull is 213.16, with an estimate $\hat{\alpha}$ of 1.365. The group difference estimate is now -1.731 with an approximate standard error of 0.398. The change in deviance of 3.89 is just on the 5% point of χ_1^2: there is marginal evidence to support a Weibull distribution rather than an exponential (but see Section 6.17 for further discussion of this example).

In Section 6.5 we compared the lognormal and exponential distributions on the FEIGL data, and found that the lognormal gave a slightly better fit. Does the extra shape parameter of the Weibull improve its fit compared to the lognormal?

```
$END
$INPUT 1 FEIGL $INPUT %PLC WEIBULL $
$CALC %S = 1 : W = 1 : LWBC = %LOG(WBC) $
$MACRO MODEL AG*LWBC $ENDMAC
$USE WEIBULL TIME W %S $
```

The deviance for the exponential fit is 291.32, exactly that calculated as the true value of $-2 \log L$ in Section 6.5. The deviance for the Weibull is 291.29, almost the same, and the shape parameter estimate is 0.979. Thus within the Weibull family the exponential provides almost the best fit. In Section 6.5 we found that the lognormal distribution for the same model gave a deviance of 288.38; this is 2.91 less and a non-significant difference.

6.12 The extreme value distribution

If T has a Weibull distribution with parameters α and μ as defined in Section 6.10, then $Y = \log T$ has the extreme value distribution with location parameter $\theta = 1/\alpha \log \mu$ and scale parameter $\sigma = 1/\alpha$. The density, survivor and hazard functions are

$$f(y) = \frac{1}{\sigma} \exp\left(\frac{y-\theta}{\sigma} - \exp\frac{y-\theta}{\sigma}\right) \qquad -\infty < y < \infty, -\infty < \theta < \infty, \sigma > 0$$

$$S(y) = \exp\left(-\exp\frac{y-\theta}{\sigma}\right)$$

$$h(y) = \frac{1}{\sigma} \exp\left(\frac{y-\theta}{\sigma}\right)$$

The hazard function increases exponentially, which limits the usefulness of the distribution as a model for survival data. The standard form of the density ($\theta = 0$, $\sigma = 1$) is plotted with the corresponding hazard function in Fig. 6.3.

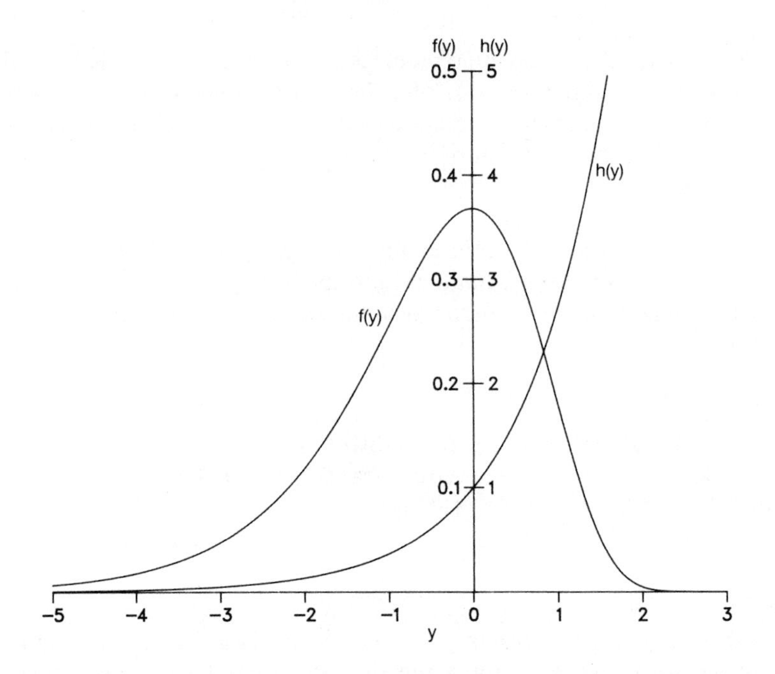

Fig. 6.3. Standard extreme value distribution: density and hazard functions

The density has a long left tail and short right tail: it has negative skew.

The extreme value distribution can be fitted using the Poisson representation of the likelihood as for the Weibull distribution. The likelihood is, for general location parameters θ_i

$$L(\theta_1,\ldots,\theta_n,\sigma) = \prod_{i=1}^{n} h(y_i)^{w_i} S(y_i)$$

$$= \prod_{i=1}^{n} H(y_i)^{w_i} e^{-H(y_i)} \prod_{i=1}^{n} \left[\frac{h(y_i)}{H(y_i)}\right]^{w_i}$$

where $H(y_i)$ is the integrated hazard function. The first part of the above expression is a Poisson likelihood with mean

$$H(y_i) = \exp\left(\frac{y_i - \theta_i}{\sigma}\right)$$

so that

$$\log H(y_i) = y_i/\sigma - \theta_i/\sigma$$

$$= \alpha y_i - \alpha\theta_i.$$

The second term in the likelihood expression is

$$\prod_{i=1}^{n} \frac{1}{\sigma^{w_i}} = \prod_{i=1}^{n} \alpha^{w_i} = \alpha^{n_1}$$

where n_1 is the number of uncensored observations.

The likelihood can be seen to be similar to that given in Section 6.11 for the Weibull distribution, with t_i being replaced by $y_i = \log(t_i)$. The parameter estimates from the Poisson fit correspond to the extreme value model

$$-\alpha\theta_i = \boldsymbol{\beta}'\mathbf{x}_i$$

or

$$\theta_i = -\boldsymbol{\beta}'\mathbf{x}_i/\alpha.$$

To fit the extreme value distribution in GLIM, we therefore need to replace $\log t_i$ in the offset by t_i. No other change is required, though for convenience a separate macro EXTVAL (not in the GLIM macro library) is used to fit the extreme value distribution.

We illustrate with the GEHAN example.

```
$END
$INPUT 1 EXTVAL GEHAN $
$MACRO MODEL G $ENDMAC
$USE EXTVAL T W $
```

The deviance is 240.01, much larger than the Weibull deviance of 213.16, and the estimate of alpha is 0.115 (corresponding to an estimate of σ of 8.70). This distribution is clearly inappropriate.

For the FEIGL data, it is found similarly that the deviance for the model AG*LWBC is 323.70 for the extreme value distribution, much larger than the value of 291.29 for the Weibull distribution. The extreme value distribution is again inappropriate.

6.13 The reversed extreme value distribution

The extreme value distribution is asymmetric, and is defined on the infinite range $(-\infty < y < \infty)$. The distribution of $-y$ is the *reversed* extreme value distribution, which has positive skew. The density, survivor and hazard functions are

$$f(y) = \frac{1}{\sigma}\exp\left(-\frac{y-\theta}{\sigma} - \exp\left(-\frac{y-\theta}{\sigma}\right)\right)$$

$$S(y) = 1 - \exp\left(-\exp\left(-\frac{y-\theta}{\sigma}\right)\right)$$

$$h(y) = f(y)/S(y)$$

The hazard function is monotone increasing and approaches an asymptote of $1/\sigma$. This form of the hazard function arises because for large y, $f(y)$ is essentially exponential, with constant hazard. The likelihood function for this distribution does not have the Poisson form, because the survivor function is not exponential. It is not easily fitted in GLIM and is not considered further in this book.

6.14 Survivor function plotting for the Weibull and extreme value distributions

In Section 6.8 we described the use of the Kaplan–Meier estimator of the survivor function to validate the exponential distribution. We can extend this procedure to assess the validity of the Weibull and extreme value distributions.

For the exponential distribution, we used the GLIM fitted values to construct the Kaplan–Meier estimate which was then plotted against u using a suitable scale. For the Weibull distribution,

$$S(t) = e^{-\lambda t^{\alpha}},$$

and the GLIM fitted values are

$$\hat{\theta}_i = \hat{\lambda}_i t_i^{\hat{\alpha}}$$

which are again approximately (censored or uncensored) standard exponential variables. Thus the same plotting procedures are used for these values based on the Weibull distribution.

In the same way for the extreme value distribution

$$S(y) = \exp\left(-\exp\{(y-\theta)/\sigma\}\right)$$

and the GLIM fitted values are $\exp\{(y-\hat{\theta})/\hat{\sigma}\}$, so again these have approximately standard exponential distributions, censored or uncensored.

We illustrate with the Gehan data

```
$INPUT %PLC WEIBULL $
$INPUT 1 GEHAN KAPLAN $
$MACRO MODEL G $ENDMAC
$CALC %S = 1 $USE WEIBULL T W %S $
$USE KAPLAN %FV W $
$CALC LS = -%LOG(ZA_) : LLS = %LOG(LS) :
      AS =  %ANG(ZA_) : AU  = %ANG(%EXP(-ZC_)):
      LU =  %LOG(ZC_) $

$PLOT LS ZC_ZC_ '+ *' :
     LLS LU_LU_'+ *' :
      AS AU AU '+ *' $
```

The plots are very closely straight lines: the curvature visible in the plots from the exponential model has been removed. Thus the assumption of a Weibull distribution seems well supported (but see Section 6.17).

6.15 The Cox proportional hazards model and the piecewise exponential distribution

A further application of the Poisson likelihood representation enables us to fit an important model developed by Cox (1972). In this *proportional hazards model* the hazard function is

$$h(t) = \lambda_0(t)e^{\boldsymbol{\beta}'\mathbf{x}}$$

where $\lambda_0(t)$ is an arbitrary, unspecified function of t—the base-line hazard when $\mathbf{x} = \mathbf{0}$. The likelihood function is

$$\begin{aligned} \mathrm{L}(\boldsymbol{\beta},\lambda_0) &= \prod_i \{h(t_i)\}^{w_i} S(t_i) \\ &= \prod_i \{h(t_i)\}^{w_i} e^{-H(t_i)} \\ &= \prod_i \{H(t_i)\}^{w_i} e^{-H(t_i)} \prod_i \{h(t_i)/H(t_i)\}^{w_i} \end{aligned}$$

where $H(t)$ is the integrated hazard function. Because of the proportional hazards assumption, the second product in the likelihood does not depend on $\boldsymbol{\beta}$, and the first is again a Poisson likelihood, with

$$\log H(t_i) = \log \Lambda_0(t_i) + \boldsymbol{\beta}'\mathbf{x}_i$$

where

$$\Lambda_0(t) = \int_0^t \Lambda_0(u)\mathrm{d}u$$

is the integrated baseline hazard. Since $\lambda_0(t)$ is arbitrary, it is not clear how to fit the model without parametrizing $\lambda_0(t)$ in some way. We use a semi-parametric approach based on the *piecewise exponential distribution*, due to Breslow (1974); see also Whitehead (1980), and Aitkin, Laird and Francis (1983).

As in the derivation of the Kaplan–Meier estimator in Section 6.8, we consider a set of discrete time-points $a_1 < a_2 < \ldots < a_N$. In each time interval $(a_{j-1}, a_j], j = 1, \ldots, N+1$, with $a_0 = 0$, $a_{N+1} = \infty$, we model the hazard function as *constant*, with parameter λ_j. That is, the survival time distribution $f(t)$ in $(a_{j-1},a_j]$ is exponential with mean $1/\lambda_j$, but the mean is different in each piece of the time axis—hence the name piecewise exponential.

This semi-parametric model has the same flexibility as the model in which $\lambda_0(t)$ is unspecified. We replace the assumption of an arbitrary hazard by the assumption of a step-function hazard with steps at $a_1,a_2,\ldots, a_N$.

We can thus write the baseline hazard as

$$\lambda_0(t)=\exp(\phi_j) \qquad a_{j-1}<t\leq a_j,\ j=1,\ldots N+1$$

where the ϕ_j are a set of $N+1$ time interval constants.

The proportional hazard assumption can then be written

$$h_j=\lambda_j=\exp(\phi_j+\boldsymbol{\beta}'\mathbf{x}).$$

To construct the likelihood function, we consider as in Section 6.8 the survival experience of each individual through time, shown in the table below.

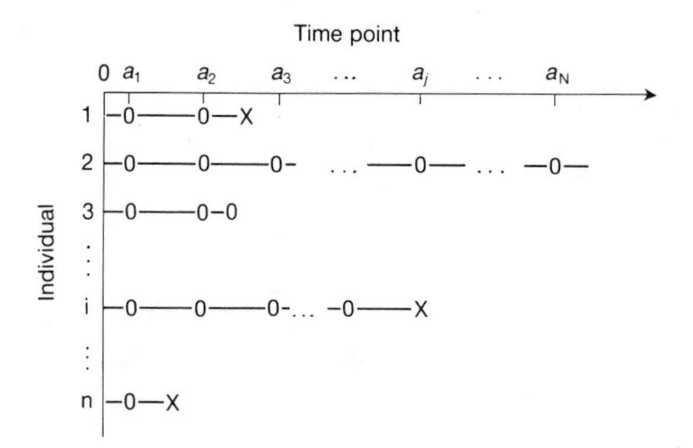

The circles represent censoring and the crosses death. The i-th individual experiences a sequence of censorings at $a_1,a_2,\ldots$ and either final censoring or death at t_i, defined to fall in the N_i-th interval, so that $a_{N_{i-1}} < t_i < a_{N_i}$. Let h_{ij} be the hazard function for the i-th individual in the j-th interval $(a_{j-1},a_j]$, with

$$h_{ij} = \exp(\phi_j + \boldsymbol{\beta}'\mathbf{x}_i)$$

and let w_{ij} be the censoring indicator for the i-th individual in the j-th interval. The survivor function for the i-th individual in the j-th interval is

$$\begin{aligned} S_{ij}(t) &= \exp\{-H_{ij}(t)\} \\ &= \exp(-h_{ij}t). \end{aligned}$$

Define e_{ij} to be the exposure time of the i-th individual in the j-th interval, so that

$$
\begin{aligned}
e_{ij} &= a_j - a_{j-1} \qquad && j = 1, \ldots, N_i - 1 \\
&= t_i - a_{N_i - 1} \qquad && j = N_i
\end{aligned}
$$

Then the contribution of the i-th individual to the likelihood is

$$
\mathrm{L}_i = \prod_{j=1}^{N_i} h_{ij}{}^{w_{ij}} S_{ij}(e_{ij}).
$$

The full likelihood is thus

$$
\begin{aligned}
L(\boldsymbol{\beta}, \phi_1 \ldots \phi_{N+1}) &= \prod_{i=1}^{n} \prod_{j=1}^{N_i} h_{ij}{}^{w_{ij}} \exp(-h_{ij} e_{ij}) \\
&= \prod_i \prod_j \theta_{ij}{}^{w_{ij}} \exp[-\theta_{ij}] / \prod_i \prod_j e_{ij}{}^{w_{ij}}
\end{aligned}
$$

with

$$
\begin{aligned}
\theta_{ij} &= h_{ij} e_{ij} \\
&= e_{ij} \exp(\phi_j + \boldsymbol{\beta}' \mathbf{x}_i).
\end{aligned}
$$

The denominator in the likelihood is a constant, and the numerator is again a Poisson likelihood, with a log-linear model for θ_{ij}:

$$
\log \theta_{ij} = \log e_{ij} + \phi_j + \boldsymbol{\beta}' \mathbf{x}_i.
$$

6.16 Maximum likelihood fitting of the piecewise exponential distribution

We now consider the maximization of the likelihood in the parameters $\boldsymbol{\beta}$, $\phi_1, \ldots, \phi_{N+1}$. We have

$$
\begin{aligned}
\ell = \log L &= \sum_{i=1}^{n} \sum_{j=1}^{N_i} (w_{ij} \log \theta_{ij} - \theta_{ij}) - \sum_{i=1}^{n} \sum_{j=1}^{N_i} w_{ij} \log e_{ij} \\
\frac{\partial \ell}{\partial \phi_j} &= \sum_{i \in R_j} (w_{ij} - \theta_{ij})
\end{aligned}
$$

where R_j is the risk set of individuals in the j-th interval. For a maximum of the likelihood,

$$
d_j = \sum_{i \in R_j} w_{ij} = \sum_{i \in R_j} \hat{\theta}_{ij} = \exp(\hat{\phi}_j) \sum_{i \in R_j} e_{ij} \exp(\hat{\boldsymbol{\beta}}' \mathbf{x}_i)
$$

where d_j is the number of deaths in the j-th interval. If there are no deaths in the j-th interval, then $d_j = 0$. Since $\hat{\theta}_{ij}$ is non-negative, this equality is only possible if $\exp(\hat{\theta}_j) = 0$, whatever the model $\boldsymbol{\beta}'\mathbf{x}$. Thus, in comparing two models using $\boldsymbol{\beta}_0$ and $\boldsymbol{\beta}_1$, intervals with no deaths contribute zero to the Poisson term in the maximized log-likelihood under both models. The second term in the log-likelihood depends only on t_i and the a_j and is the same under both models. Thus the comparison of two regression models by the likelihood ratio test is not affected by the inclusion or exclusion of intervals containing no deaths.

If $d_j \neq 0$, then we may solve explicitly for $\hat{\phi}_j$ in terms of $\hat{\boldsymbol{\beta}}$:

$$\exp(\hat{\phi}_j) = d_j / \{\sum_{i \in R_j} e_{ij} \exp(\hat{\boldsymbol{\beta}}'\mathbf{x}_i)\}.$$

The likelihood equation for $\hat{\boldsymbol{\beta}}$ is

$$\sum_i \sum_j (w_{ij} - \hat{\theta}_{ij})\mathbf{x}_i = \mathbf{0}$$

which is equivalent to

$$\sum_i (w_i - r_i \hat{\gamma}_i)\mathbf{x}_i = \mathbf{0},$$

that is,

$$\sum_i r_i (w_i / r_i - \hat{\gamma}_i)\mathbf{x}_i = \mathbf{0},$$

where

$$\hat{\gamma}_i = \exp(\hat{\boldsymbol{\beta}}'\mathbf{x}_i)$$

$$w_i = \sum_{j=1}^{N_i} w_{ij} = w_{iN_i}$$

$$r_i = \sum_{j=1}^{N_i} e_{ij} \exp(\hat{\phi}_j).$$

This likelihood equation can be recognized as that for a Poisson distribution with response variable w_i/r_i, prior weight variable r_i, regression model $\boldsymbol{\beta}'\mathbf{x}_i$ and log link function. The prior weight variable r_i is itself a function of the estimated hazard parameters ϕ_j; its effect in the analysis is to give heavier weight to individuals with a history of high hazard.

The model is fitted by a relaxation method (Green and Francis, 1986) similar to that used for the Weibull distribution in Section 6.11. We begin with an initial estimate of $\boldsymbol{\beta} = \mathbf{0}$ and obtain the first estimates of the ϕ_j. The prior weight variable r_i is then calculated from the $\hat{\phi}_j$ and the Poisson model is fitted. The fitted values $\hat{\boldsymbol{\beta}}'\mathbf{x}_i$ are then used to obtain a revised estimate of the ϕ_j, and these alternate steps continued until convergence. The standard error of $\hat{\boldsymbol{\beta}}$ is slightly underestimated because the ϕ_j are treated as known in each iteration; this effect is only slight because the estimates of the ϕ_j are nearly orthogonal to $\hat{\boldsymbol{\beta}}$. Standard errors for the $\hat{\phi}_j$ are not available using this model-fitting procedure. Correct standard errors for both $\hat{\boldsymbol{\beta}}$ and the $\hat{\phi}_j$ can be obtained if necessary by using the macro for time-dependent covariates described in Section 6.22.

The above derivation applies generally to any choice of cut-points a_j on the time scale. The model discussed by Breslow (1974) uses as cut-points the *distinct death times* in the sample. This choice of cut-points gives the MLE for $\boldsymbol{\beta}$ very close to that from the conditional likelihood approach originally used by Cox (1972), as can be shown if we consider the profile likelihood for $\boldsymbol{\beta}$ from the piecewise model. This is

$$PL(\boldsymbol{\beta}) = \prod_{i=1}^{n} \prod_{j=1}^{N_i} \hat{\theta}_{ij}(\boldsymbol{\beta})^{w_{ij}} \exp[-\hat{\theta}_{ij}(\boldsymbol{\beta})]$$

where

$$\hat{\theta}_{ij}(\beta) = \frac{e_{ij} d_j \exp(\boldsymbol{\beta}'\mathbf{x}_i)}{\sum_{i \in R_j} e_{ij} \exp(\boldsymbol{\beta}'\mathbf{x}_i)} = \frac{d_j \exp(\boldsymbol{\beta}'\mathbf{x}_i)}{\sum_{i \in R_j} \exp(\boldsymbol{\beta}'\mathbf{x}_i)}$$

since $e_{ij} = e_j$ for all i when death times occur only at cut-points. Interchanging i and j in the product, we have

$$PL(\boldsymbol{\beta}) = \prod_{j=1}^{N+1} \prod_{i \in R_j} \hat{\theta}_{ij}(\boldsymbol{\beta})^{w_{ij}} \exp[-\hat{\theta}_{ij}(\boldsymbol{\beta})]$$

Substituting for $\hat{\theta}_{ij}(\boldsymbol{\beta})$, we find

$$\prod_{i \in R_j} \hat{\theta}_{ij}(\boldsymbol{\beta})^{w_{ij}} = \left[\frac{d_j}{\sum_{i \in R_j} \exp(\boldsymbol{\beta}'\mathbf{x}_i)}\right]^{d_j} \exp\left(\boldsymbol{\beta}' \sum_{i \in R_j} w_{ij}\mathbf{x}_i\right)$$

$$\prod_{i \in R_j} \exp[-\hat{\theta}_{ij}(\boldsymbol{\beta})] = \exp(-d_j).$$

The profile likelihood, omitting constants, is thus

$$PL(\boldsymbol{\beta}) = \prod_{j=1}^{N+1} \frac{\exp\left(\boldsymbol{\beta}' \sum_{i \in R_j} w_{ij}\mathbf{x}_i\right)}{\left[\sum_{i \in R_j} \exp \boldsymbol{\beta}'\mathbf{x}_i\right]^{d_j}}.$$

If the cut-points a_j are the distinct death times, this likelihood is a very close approximation to the likelihood obtained by conditional or partial arguments; see Cox and Oakes (1984, pp. 91–98) for details. The direct maximization of this likelihood is not straightforward; see Whitehead (1980) and Clayton and Cuzick (1985) for alternative computational methods in GLIM.

The choice of interval cut-points determines the extent of *smoothing* of the estimated piecewise hazard function. If some intervals contain no deaths, the estimated hazard for these intervals is zero. Choosing the distinct death times as cut-points is equivalent to pooling the zero estimated hazards with the non-zero hazards from intervals containing at least one death, ensuring positive estimated hazards in all intervals.

The piecewise exponential model is fitted using macros in the subfile PIECEWISE.

6.17 Examples

We consider first the GEHAN data.

```
$INPUT 1 GEHAN PIECEWISE $
```

The PIECEWISE subfile contains a number of macros. CUTP sets up the vector of distinct death times which can be used for fitting the piecewise exponential distribution. It requires as arguments the survival time and the censoring indicator, and the name of the variable to contain the distinct death times.

```
$USE CUTP T W DDT $
```

INIC is an initializing macro which calculates the exposure times in each interval and other working vectors needed for the hazard parameter estimates. This must be used before fitting the model. It requires as arguments the survival time, the censoring indicator and the variable containing the cut-points defining the intervals for the piecewise model. This will normally be the distinct death times, but any set of specified times can be used.

```
$USE INIC T W DDT $
```

A macro called MODEL needs to be defined, as for the Weibull macro.

```
$MACRO MODEL G $ENDMAC
```

Finally, the model is fitted using the macro FITC. This requires as argument the censoring variate.

```
$USE FITC W $D E $
```

The output of FITC is only the deviance; as for an ordinary fit the parameter estimates have to be displayed explicitly. The group effect estimate of -1.521 is very similar to that from the exponential model, and noticeably smaller in magnitude than the Weibull estimate (-1.731). The deviance is 202.37 on 22 df, reflecting the estimation of 2 parameters in the model and 18 hazard parameters for the 17 distinct death times. The estimated hazard parameters $\exp(\hat{\phi}_j)$ are held in a vector HAZ_ and can be used in further calculations; a macro PHAZ provides a plot of the log hazard parameters $\hat{\phi}_j$ against the log of the death time variable, which has to be specified as the argument to the macro.

```
$USE PHAZ DDT $
```

```
               log hazard  v  log survival time
   -0.160 |                                     +
   -0.320 |
   -0.480 |
   -0.640 |
   -0.800 |
   -0.960 |
   -1.120 |
   -1.280 |                         +
   -1.440 |                              +
   -1.600 |                                  +
   -1.760 |                     +       +    +
   -1.920 |                               +
   -2.080 |
   -2.240 |                   +                  +
   -2.400 |        +       +
   -2.560 +                                 +
   -2.720 |                       +
   -2.880 |
   -3.040 |             +
   -3.200 |                            +
----------:---------:---------:---------:---------:---------:---------:
       0.000     0.800     1.600     2.400     3.200    4.000     4.800
```

The hazard for the last interval, for which there are no deaths, is suppressed in this plot. Standard errors for the $\hat{\phi}_j$ are not supplied by the FITC macro, but can be obtained if required at the cost of substantial computing time (see Section 6.22 for details). Good approximations to the standard errors can be based on standard results with a null model: if D is a Poisson variable with mean $\mu = e^{\phi}$, then $\hat{\phi} = \log d$ has large-sample variance $1/\mu \simeq 1/d$. The approximate standard error of $\hat{\phi}_j$ is thus $1/\sqrt{d_j}$ where d_j is the number of deaths in the j-th interval. (This approximation is close because the estimate $\hat{\boldsymbol{\beta}}$ is almost orthogonal to the hazard parameter estimates $\hat{\phi}_j$.)

The variation in the plot is consistent with a constant hazard, apart from the last interval where the hazard is considerably higher. This casts doubt on the Weibull model used in Section 6.11, despite the apparent better fit of the Weibull model implied by the marginally significant likelihood ratio test of $\alpha = 1$.

Thus the piecewise exponential distribution provides important information about the hazard function and may therefore suggest a suitable parametric distribution. See Aitkin, Davies and Francis (1986) for further discussion.

We repeat the analysis with the FEIGL data.

```
$END
$REINPUT 1 FEIGL PIECEWISE $
$CALC W = 1 $
$USE CUTP TIME W DDT $
$USE INIC TIME W DDT $
$CALC LWBC = %LOG(WBC) $
$MACRO MODEL AG*LWBC $ENDMAC
$USE FITC W $D E $
```

The deviance is 268.05 on 6 df.

The parameter estimates and standard errors are somewhat larger in magnitude than those for the exponential distribution, but less than those for the lognormal distribution (Section 6.5), and again the interaction can be omitted.

```
$MACRO MODEL AG+LWBC $ENDMAC
$USE FITC W $D E $
```

The deviance increases by 3.30 and the fitted model is

$$\widehat{\log h_j} = -1.275 + 1.020\ \mathrm{AG(2)} + 0.361\ \mathrm{LWBC} + \hat{\phi}_j$$

```
$USE PHAZ DDT $
```

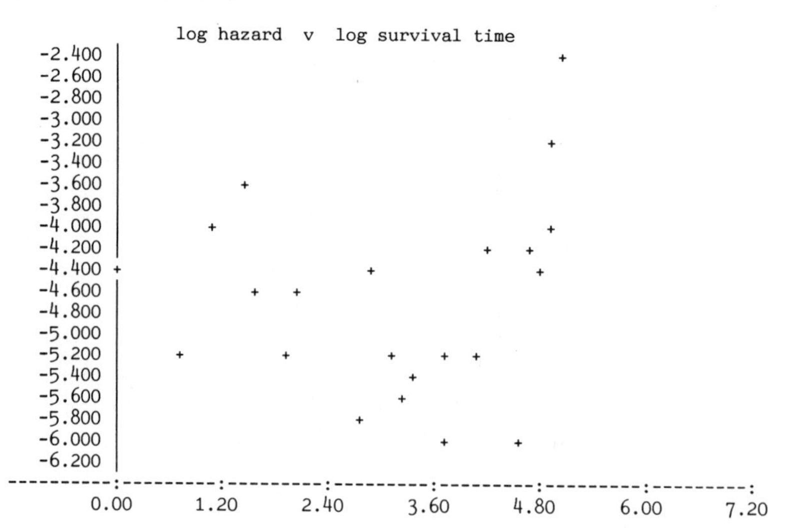

The plot shows a pronounced dip in the middle—a bath-tub shaped hazard inconsistent with the exponential distribution. Further discussion is given in Aitkin, Davies and Francis (1986), where the compound Rayleigh distribution is shown to provide a better fit.

The parameter estimates and their standard errors, and the deviance difference between the models, are closely equivalent to those reported in Cox and Oakes (1984, p. 100) obtained using the conditional likelihood, though Cox and Oakes use a different parametrization of the model, with AG reversed in sign and LWBC centred to an origin.

6.18 The logistic and log-logistic distributions

The logistic distribution is very similar to the normal but has slightly heavier tails. Its advantage in survival analysis in GLIM is that both right- *and* left-censoring can be easily handled, since the likelihood can be expressed as a special form of binomial likelihood (Bennett and Whitehead, 1981).

The density, survivor and hazard functions for the general location- and scale-parameter logistic distribution are

$$f(y) = \frac{1}{\sigma}\exp\left(\frac{y-\mu}{\sigma}\right)\bigg/\left\{1+\exp\left(\frac{y-\mu}{\sigma}\right)\right\}^2 \qquad -\infty < y < \infty$$

$$S(y) = 1\bigg/\left\{1+\exp\left(\frac{y-\mu}{\sigma}\right)\right\}$$

$$h(y) = \frac{1}{\sigma}\exp\left(\frac{y-\mu}{\sigma}\right)\bigg/\left\{1+\exp\left(\frac{y-\mu}{\sigma}\right)\right\};$$

the distribution mean and variance are μ and $\sigma^2\pi^2/3$ respectively.

The hazard function is monotone increasing; the distribution does not have a proportional hazard function under the usual model $\mu_i = \boldsymbol{\beta}'\mathbf{x}_i$, σ constant.

Now let w_i be the censoring indicator, with $w_i = 0$ for right-censoring, $w_i = 2$ for left-censoring and $w_i = 1$ for no censoring. Left censoring is uncommon, but may occur if individuals fail before the first measurement period occurs. For example, if survival is recorded in weeks and an individual dies within the first week, the value is left-censored at one week. (In general, recording in this way will lead to a *grouped* survival distribution, but this has little effect on inference unless the grouping is very coarse. In this case the discrete time approach of Section 6.23 should be used.)

For each individual, we record $(y_i, \mathbf{x}_i, w_i)$. Let c_i and b_i be dummy indicators for right- and left-censoring. The likelihood function is

$$L(\boldsymbol{\beta},\sigma) = \prod_i \{f(y_i)\}^{1-b_i-c_i}\{S(y_i)\}^{c_i}\{F(y_i)\}^{b_i}.$$

Write

$$F(y_i) = p_i;$$

then

$$S(y_i) = 1 - p_i$$

$$f(y_i) = p_i(1 - p_i)/\sigma$$

and

$$L(\boldsymbol{\beta},\sigma) = \frac{1}{\sigma^m} \prod_i p_i^{1-c_i}(1 - p_i)^{1-b_i}$$

where m is the number of uncensored observations. The product has the form of the likelihood from n binomial observations with $r_i = (1 - c_i)$ successes in $n_i = (2 - b_i - c_i)$ trials, and the success probability is

$$p_i = \exp\left(\frac{y_i - \mu_i}{\sigma}\right) \bigg/ \left\{1 + \exp\left(\frac{y_i - \mu_i}{\sigma}\right)\right\}$$

so that

$$\text{logit}\, p_i = (y_i - \mu_i)/\sigma = y_i/\sigma - \boldsymbol{\beta}'\mathbf{x}_i/\sigma.$$

(This parametrization is different from that in Bennett and Whitehead who absorb σ into β and use $1/\sigma$ as the scale parameter.)

Maximization of the likelihood can be achieved by a relaxation method using the binomial error model to estimate $\boldsymbol{\beta}$, and an additional step for the estimation of σ. For fixed σ, the logit model is fitted to the number of successes $1 - c_i$ with binomial denominator $2 - b_i - c_i$, using an offset of y_i/σ. The quantities $(y_i - \hat{\mu}_i)$ are obtained by multiplying the linear predictor by σ. A new estimate of σ is then obtained by solving the likelihood equation $\partial \log L/\partial\sigma = 0$, which reduces to

$$m\hat{\sigma} = \sum_i (n_i\hat{p}_i - r_i)(y_i - \hat{\mu}_i).$$

Alternate estimation of $\boldsymbol{\beta}$ and σ continues until convergence, with damping of successive estimates of σ. As with the Weibull macro, the standard errors of the estimated parameters in $\hat{\boldsymbol{\beta}}$ are slightly underestimated because σ is taken as known in each iteration. This can be corrected using the approach of Roger and Peacock (1982).

The macro LOGIST in the subfile CLOGISTIC fits the logistic distribution. The parameter estimates and standard errors differ from those of Bennett and Whitehead's macro by a scale factor σ, but are comparable with those for the normal distribution in Section 6.19.

Before considering an example, we describe the log–logistic distribution. If

Y is logistic, then $T=e^Y$ is log–logistic. The density, survivor and hazard functions are

$$f(t) = \frac{\alpha}{\theta} \frac{(t/\theta)^{\alpha-1}}{\{1 + (t/\theta)^{\alpha}\}^2} \qquad t,\theta,\alpha > 0$$

$$S(t) = \frac{1}{1 + (t/\theta)^{\alpha}}$$

$$h(t) = \frac{\alpha}{\theta} \frac{(t/\theta)^{\alpha-1}}{1 + (t/\theta)^{\alpha}}$$

where $\theta = e^{\mu}$, $\alpha = 1/\sigma$.

For $\alpha \leq 1$ the hazard is monotone decreasing, while for $\alpha > 1$ it has a single maximum. The log–logistic distribution is a convenient approximation to the lognormal since it possesses similar positive skew. It is easily fitted using the macro LLOGIST in the subfile CLOGISTIC, which gives estimates of $\boldsymbol{\beta}$ and σ, where $\log(\theta_i) = \boldsymbol{\beta}'\mathbf{x}_i$. The parameter estimates again differ from those in Bennett and Whitehead by the factor σ, and the deviance is different as Bennett and Whitehead omit the term $2\Sigma(1-b_i-c_i)\log t_i$ required to make the logistic and log–logistic comparable.

We illustrate with the GEHAN data. Both macros require as arguments the response variable and the censoring indicator. The fitted model is specified in a macro called MODEL, as for the Weibull distribution.

```
$INPUT 1 GEHAN CLOGISTIC $
$MACRO MODEL G $ENDMAC
$USE LOGIST T W $
```

The deviance is 236.9, with an estimated σ of 5.60.

```
$USE LLOGIST T W $
```

The deviance is 215.3, and the estimated group difference is 1.265, with standard error 0.319. the estimated σ is 0.546. The deviance is greater than that for the Weibull (213.2) in Section 6.11, and much lower than that for the logistic, as we should expect from the skewness of the survival times.

6.19 The normal and lognormal distributions

In Section 6.5 we compared the exponential and lognormal distributions for the FEIGL data, where there were no censored observations. Censoring complicates the fitting of the normal model because the survivor function does not have a simple analytic form. However, it is possible to fit the normal model with both left- and right-censoring; for simplicity we consider only right censoring (see Wolynetz, 1979 for the general case). For right censoring with censoring indicator w_i, we have

$$L(\boldsymbol{\beta},\sigma) = \prod_i \{f(y_i)\}^{w_i}\{S(y_i)\}^{1-w_i}$$

where

$$f(y_i) = \frac{1}{\sigma\sqrt{(2\pi)}} \exp\left\{-\frac{1}{2\sigma^2}(y_i - \mu_i)^2\right\}$$

$$\mu_i = \boldsymbol{\beta}'\mathbf{x}_i.$$

Maximization of the log likelihood can be achieved using an EM algorithm (Dempster, Laird and Rubin 1977, Wolynetz 1979, Aitkin 1981). The basis of the algorithm is the use of the log-likelihood for the complete data which has not been observed, in this case because some of the observations are censored. This complete data log-likelihood is replaced by its conditional expectation given the observed data and the current parameter estimates (the E-step of the algorithm). In the M-step, this expected log-likelihood is maximized as a function of the unknown parameters. Since the function being maximized now has the form of the log likelihood for a model without missing data, it is easily maximized using GLIM. The E- and M-steps are alternated until convergence.

In the normal regression model, the log-likelihood is linear in the sufficient statistics $\Sigma\mathbf{x}_i y_i$ and Σy_i^2. The alternate E- and M-steps of the algorithm consist of:

E-step For the censored observations, the values y_i and y_i^2 are replaced by their conditional expectations given the current parameter estimates.

M-step The likelihood equations are solved to give new parameter estimates using the conditional expectations obtained in the E-step as "real data".

For an observation right censored at a, the conditional density is

$$f(y|Y > a) = f(y)/S\left(\frac{a-\mu}{\sigma}\right) \qquad y > a.$$

The conditional expectation of Y is

$$E(Y|Y > a) = \int_a^\infty yf(y)\mathrm{d}y \bigg/ S\left(\frac{a-\mu}{\sigma}\right)$$

$$= \mu + \sigma h\left(\frac{a-\mu}{\sigma}\right)$$

and that of Y^2 is

$$E(Y^2|Y > a) = \int_a^\infty y^2 f(y)\mathrm{d}y \Big/ S\left(\frac{a-\mu}{\sigma}\right)$$

$$= \mu^2 + \sigma^2 + \sigma(\mu + a)h\left(\frac{a-\mu}{\sigma}\right)$$

where $h(y)$ is the hazard function.

Thus in each E-step, we construct from the current parameter estimates the expected observations

$$y_{1i} = w_i y_i + (1-w_i)\left\{\mu_i + \sigma h\left(\frac{y_i-\mu_i}{\sigma}\right)\right\}$$

$$y_{2i} = w_i y_i^2 + (1-w_i)\left\{\mu_i^2 + \sigma^2 + \sigma(\mu_i + y_i)h\left(\frac{y_i-\mu_i}{\sigma}\right)\right\}$$

with

$$\mu_i = \boldsymbol{\beta}'\mathbf{x}_i,$$

while in each M-step, we construct from the expected observations new parameter estimates using

$$\hat{\boldsymbol{\beta}} = \left(\sum_i \mathbf{x}_i\mathbf{x}_i'\right)^{-1} (\sum \mathbf{x}_i y_{1i})$$

$$\hat{\sigma}^2 = \sum_i (y_{2i} - 2\mu_i y_{1i} + \mu_i^2)/n$$

Both the normal and lognormal distributions can be fitted using the macros NORM and LNORM in the subfile CNORMAL. Both macros require as arguments the response variable and the censoring indicator. The fitted model is again specified in a macro called MODEL as for the Weibull distribution.

The standard errors printed out by GLIM based on $\hat{\sigma}^2(\Sigma \mathbf{x}_i\mathbf{x}_i')^{-1}$ may be seriously in error. The importance of variables in the model should therefore be assessed by omitting the variable and comparing the model deviances, not by relying on the "t-statistic" based on the standard error.

We illustrate with the GEHAN data.

```
$INPUT 1 GEHAN CNORMAL $
$MACRO MODEL G $ENDMAC
$USE NORM T W $
```

The deviance is 235.94, with an estimated σ of 9.55.

```
$USE LNORM T W $
```

The deviance is 213.41, and the estimated group difference is 1.349; the standard error attached is 0.273. The estimated σ is 0.924. The deviance for the lognormal is slightly greater than that for the Weibull (213.16) in Section 6.11, slightly less than for the log–logistic, and much lower than that for the normal, as we might expect.

```
$MACRO MODEL 1 $ENDMAC
$USE LNORM T W $
```

The deviance increases to 230.79, a change of 17.38, considerably less than $(1.349/0.273)^2 = 24.42$. The standard error is considerably underestimated (the likelihood in $\boldsymbol{\beta}$ is also non-normal when the censoring is heavy).

Note that the variance estimates for the lognormal and log–logistic distributions of t are very similar. The estimated variance of the log–logistic is $\hat{\sigma}^2\pi^2/3$, i.e. 0.981 with $\hat{\sigma} = 0.546$, while that of the lognormal is 0.854. It is very difficult to distinguish between these distributions in small samples.

The empirical survivor function of the standardized residuals from the normal or lognormal model can be constructed as in Section 6.8 using the vector SR_ as the first argument of KAPLAN; a Q–Q plot may then be constructed using the %ND transformation of 1–ZA_.

6.20 The choice between different survival distributions

We have seen in this chapter that a wide variety of survival distributions is available, however the Weibull, lognormal and log–logistic all have positive skew, and so it will be difficult to discriminate among them in small samples, despite the different behaviour of the hazard functions. Since the Weibull, gamma, lognormal and log–logistic are all special cases of the generalized F-distribution (see Cox and Oakes, 1984, p. 29) it is possible to compare the fit of these distributions using maximized log-likelihoods.

We conclude the discussion of different survival distributions with a complex example of medical survival data. The data are adapted from Prentice (1973); Prentice's data are reproduced in Kalbfleisch and Prentice (1980, pp. 223–4). They consist of survival times T in days of 137 lung cancer patients from a Veteran's Administration Lung Cancer trial, together with explanatory variables: performance STATUS, a measure of general medical status on a continuous scale 1–9.9, with 1–3 completely hospitalized, 4–6 partial confinement to hospital, 7–9.9 able to care for self; AGE in years; MFD, time in months from diagnosis to starting on the study; PRIOR therapy (1 no, 2 yes); TREATment (1 standard, 2 test) and tumour TYPE (1 squamous, 2 small, 3 adeno, 4 large). The censoring indicator is W.

We begin by fitting the piecewise exponential distribution with a main-effects regression model.

```
$INPUT 1 PRENTICE PIECEWISE $
$USE CUTP T W DDT $
$USE INIC T W DDT $
$MACRO MODEL STAT+MFD+AGE+PRIOR+TREAT+TYPE $ENDMAC
$USE FITC W $D E $
$USE PHAZ DDT $
```

```
                log hazard  v  log survival time
 0.900 |                                                 +
 0.600 |
 0.300 |
 0.000 |
-0.300 |
-0.600 |
-0.900 |                                         +   +
-1.200 |                                     +
-1.500 |                                 +    3      +
-1.800 |                                 +    2+ +   + +
-2.100 |             +           ++          +  ++   + +
-2.400 |                       2 +   2 ++   2
-2.700 |                    2 +      3   +2++    +
-3.000 +                      +++ + + 2 2+2     + ++
-3.300 |                    + 2      +  2  3 +2   + +
-3.600 |    +  + +  + +          22  + + +   +++2
-3.900 |                            +        +
-4.200 |                                                 +
-4.500 |                                   +
-4.800 |                            +
----------:---------:---------:---------:---------:---------:---------:
        0.00      1.60      3.20      4.80      6.40      8.00      9.60
```

The deviance is 1360.9 on 30 df, with 97 distinct death times. The main-effects model shows strong effects of status and type. The hazard plot appears to show an increasing log hazard with log time, suggesting a Weibull distribution.

```
$INPUT %PLC WEIBULL $
$CALC %S = 1 $
$USE WEIBULL T W %S $
```

The deviance is 1431.1 on 127 df, a change of 70.2 for the 97 parameters. The usual large-sample χ^2 distribution of the likelihood ratio test statistic is of doubtful validity here since the 98 hazard parameter estimates are each based on only one or a few (death) observations. The parameter estimates from the two models are very similar, apart from the intercept which is affected in the piecewise model by the hazard constants. The ML estimate $\hat{\alpha}$ is 1.077, strongly suggesting an exponential distribution, and the exponential deviance is 1432.3, only 1.2 larger than the Weibull. The apparent linear trend in the log hazard plot is misleading, as becomes evident if we put rough confidence limits for a constant hazard on the plot. Using a confidence interval of width $2/\sqrt{d}$, appropriately centred on the plot, with *d* taken as 1 for each point, we can see that all except 4 or 5 of the points are included.

We now fit the exponential.

```
$CALC %S = 0 $USE WEIBULL T W %S $
```

Parameter estimates are again very similar. We adopt the exponential distribution as a suitable probability model. Further examination of the regression model is made easier by using the explicit Poisson fit of the exponential model rather than the Weibull macro.

```
$CALC LT = %LOG(T) $OFFSET LT
$FIT . $D E $
```

The results are the same apart from the deviances, which differ because of a constant omitted from the likelihood in the explicit Poisson fit (this will have no effect on model selection as we need to look only at deviance differences). We extend the model to include all two-way interactions and then use backward elimination to remove the unnecessary terms. As GLIM does not allow interactions of variates to be declared in a model, the cross-products of variates must be explicitly calculated.

```
$CALC SA = STAT*AGE : SM = STAT*MFD : AM = AGE*MFD $
$FIT +SA+SM+AM+
(PRIO+TREA+TYPE).(STAT+AGE+MFD+PRIO+TREA+TYPE) $
$D E $
```

The change in deviance is 33.1 on 25 df. We proceed to backward elimination; to save space the terms omitted (not strictly in order of smallest *t*-value) and deviance change are summarized in the table below. Computing time can be reduced by using $RECYCLE.

Term omitted	df	Deviance change
AGE. TREA	1	0. 01
AGE. PRIO	1	0. 36
AM	1	0. 15
SM	1	0. 33
MFD. PRIO	1	0. 65
STAT. TYPE	3	1. 75
MFD. TREA	1	0. 95
TREA. TYPE	3	1. 31
PRIO. TYPE	3	1. 53
AGE. TYPE	3	2. 13

At this point the model contains main effects and the interactions SA, STAT.PRIO, STAT.TREA, PRIO.TREA and MFD.TYPE. Inspection of the parameter estimates shows that two df of the MFD.TYPE interaction can be omitted:

```
-0.0619   0.0197   MFD.TYPE(2)
-0.0398   0.0438   MFD.TYPE(3)
-0.0185   0.0433   MFD.TYPE(4)
```

```
$CALC MTY2 = MFD* %EQ(TYPE,2) $
$FIT - MFD.TYPE + MTY2 $D E $
```

The deviance increases by 0.95 on 2 df. Further eliminations are summarized below.

Term omitted	df	Deviance change
PRIO. TREA	1	3. 46
SA	1	3. 30
STAT. TREA	1	2. 19
AGE	1	0. 28

At this point the only remaining interactions are STAT.PRIO and MTY2. Main effects not marginal to these are now candidates for omission. The parameter estimates for MFD and MTY2 are 0.032 and −0.049, with standard errors 0.015 and 0.018, suggesting that for Type 2 the regression on MFD is negligible. To check this we define a common slope for Types 1, 3 and 4, and zero slope for Type 2:

```
$CALC MT134 = MFD - MTY2 $
$FIT + MT134 - MFD - MTY2 $D E $
```

The deviance increases by 2.87 on 1 df. Further eliminations proceed as follows.

Term omitted	df	Deviance change
TREA	1	2. 09
MT134	1	3. 56

All terms remaining are needed in the model. The treatment variable has been eliminated, showing that there are no convincing treatment differences. The

MFD regression for Types 1, 3 and 4 can also be omitted, despite a t-value of 2. The final model is shown below.

$$\widehat{\log h(t)} = \underset{(0.39)}{-3.75} \underset{(0.055)}{-0.236}\,\mathrm{STAT} + \underset{(0.65)}{1.76}\,\mathrm{PRIO(2)} \underset{(0.110)}{-0.287}\,\mathrm{STAT.PRIO(2)}$$

$$+\underset{(0.243)}{0.653}\,\mathrm{TYPE(2)} + \underset{(0.27)}{1.04}\,\mathrm{TYPE(3)} + \underset{(0.267)}{0.257}\,\mathrm{TYPE(4)}$$

Hazard declines with increasing performance status, as would be expected, but this decline is much more rapid for those patients with prior therapy, who have higher hazard for low performance status (the two regressions on performance status cross at STAT = 6.1). Type 2 and Type 3 tumours have higher hazard than Types 1 and 4. The test treatment appears to have no significantly different effect from the standard treatment.

Analyses of these data are given in Kalbfleisch and Prentice (1980, pp. 60–62 and 66–67). Results are very similar though slightly different models are used.

6.21 Competing risks

The survival time modelling of this chapter can be extended to an important class of processes in which failure may be from one of several causes. In the example considered below of heart transplantation patients, death of the patient may occur by rejection of the heart or from other causes not related to rejection (for example, from infections due to lowered resistance caused by immuno-suppressant drugs). If treatment or patient background variables affect the hazard differently for different causes of death, then an analysis which does not distinguish the different causes may misrepresent both the importance of the explanatory variables and the nature of the hazard function.

Consider in general the case of n individuals on whom we observe $(t_i, w_i, j_i, \mathbf{x}_i)$, for $i = 1, \ldots, n$. Here t_i, w_i and $\mathbf{x}_i$ are the survival time, censoring indicator and explanatory variables as in previous sections, and j_i is the cause of failure, taking one of the values $1, 2, \ldots k$ for the k possible causes of failure. The associated random variable will be denoted by J.

We define the *cause-specific hazard function* $h_j(t)$ by

$$h_j(t)\mathrm{d}t = \Pr(t < T \leq t + \mathrm{d}t, J = j | T > t)$$

That is, $h_j(t)$ is the instantaneous failure rate for failure from the j-th cause at time t, given survival to time t. The overall hazard function is then

$$h(t) = \sum_{j=1}^{k} h_j(t)$$

since failure must be from one of the k given causes. The survivor function is then

$$S(t) = e^{-H(t)}$$

with

$$H(t) = \int_0^t h(u)\mathrm{d}u = \sum_{j=1}^{k} H_j(t),$$

where $H_j(t)$ is the cause-specific integrated hazard function.

The *cause-specific density function* $f_j(t)$ for survival time for the j-th cause is then given by

$$\begin{aligned} f_j(t)\mathrm{d}t &= \Pr(t < T \leq t + \mathrm{d}t, J = j) \\ &= \Pr(t < T < t + \mathrm{d}t, J = j | T > t)\Pr(T > t) \\ &= h_j(t)\mathrm{d}t\, S(t) \end{aligned}$$

so that

$$f_j(t) = h_j(t)S(t) \qquad j = 1,\ldots,k.$$

To construct the likelihood function, define a set of k dummy indicators d_{ij} for the i-th individual by

$$\begin{aligned} d_{ij} &= 1 \text{ if failure for the } i\text{-th individual is from cause } j \\ &= 0 \text{ otherwise.} \qquad (j = 1,\ldots, k) \end{aligned}$$

Then

$$\sum_j d_{ij} = 1$$

for uncensored observations, and for censored observations $d_{ij} = 0$ for all j. The likelihood function can be written

$$\begin{aligned} L &= \prod_{i=1}^{n} \{f_{j_i}(t_i)\}^{w_i}\{S(t_i)\}^{1-w_i} \\ &= \prod_i \{h_{j_i}(t_i)\}^{w_i} S(t_i) \\ &= \prod_{i=1}^{n} \prod_{j=1}^{k} \{h_j(t_i)\}^{d_{ij}w_i} e^{-H_j(t_i)} \end{aligned}$$

Interchanging the products, we see that the likelihood is a product of k factors, the j-th being the likelihood obtained by treating death from cause j as the outcome, and deaths from any other cause as censoring. Since no assumptions have been made about $h_j(t)$, we can fit completely unrelated models to each cause of death very simply.

We illustrate with a much-analysed set of data from the Stanford Heart Transplantation programme (Crowley and Hu, 1977). For a discussion of the data and a detailed analysis see Aitkin, Laird and Francis (1983); slightly different data from the same study are presented and analysed in Kalbfleisch and Prentice (1980) and Cox and Oakes (1984).

Subfile STAN contains the data on 65 transplanted patients, consisting of the patient's AGE at transplantation, prior open-heart SURGery (1 = yes, 0 = no), a censoring indicator DIED (1 = yes, 0 = no), the SURVival time in days after transplant, a mismatch score MM representing the distance between the patient's and the donor's tissue type (values range from 0.00 to 3.05), and an indictor REJ for death by rejection (1 = yes, 0 = no). One zero survival time is recoded to 0.5. There are 41 deaths and 24 censored survivals, with 39 distinct death times. We begin by fitting the piecewise exponential distribution, without distinguishing the causes of death.

```
$INPUT 1 STAN PIECEWISE $
$USE CUTP SURV DIED DDT $
$USE INIC SURV DIED DDT $
$MACRO MODEL AGE+SURG+MM $ENDMAC
$USE FITC DIED $D E $
```

The parameter estimates and standard errors are shown below.

$$\underset{(0.0218)}{0.0559\,\text{AGE}} - \underset{(0.481)}{0.862\,\text{SURG}} + \underset{(0.281)}{0.427\,\text{MM}}$$

Age is clearly important: risk increases with age as would be expected. The importance of surgery and mismatch score are not clearly established (we obtain the same conclusions by looking at deviance changes due to omitting each variable in turn from the model). We now plot the log hazard function against log survival time.

```
$USE PHAZ DDT $
```

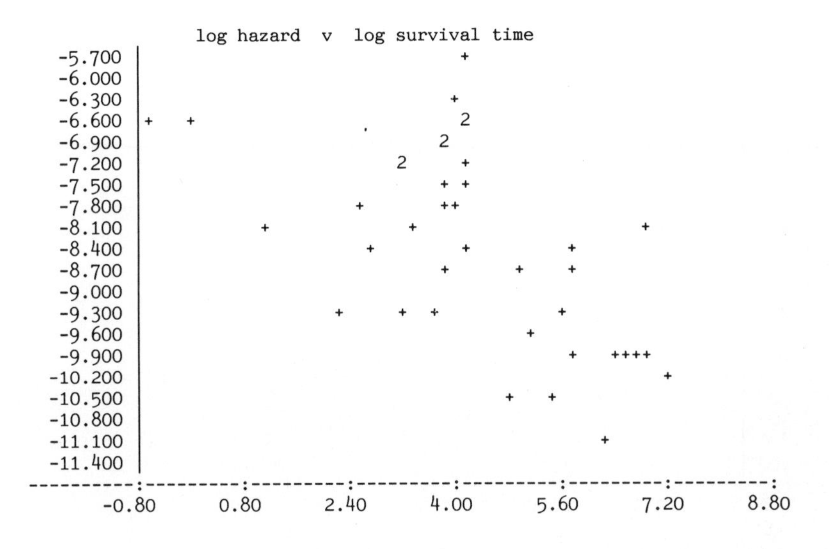

The plot shows a peculiar feature. While the hazard is generally decreasing with time, there is a rapid increase in hazard around 55 days (around 4 on the log scale). This is evident in the ordered death times, where there are 13 deaths between 44 and 68 days, and then a gap to 127 days (around 5 on the log scale). Of these 13 deaths, 11 are by rejection, compared with 29 out of 41 overall. Could this peculiar behaviour be due to different hazards for death by rejection and other causes? We first consider death by rejection.

```
$USE RESET DDT $
$CALC DREJ = REJ*DIED $
$USE CUTP SURV DREJ DRT $
$USE INIC SURV DREJ DRT $
$USE FITC DREJ $D E $
```

The macro RESET deletes the cut-point variate and other working variates, and so allows the piecewise macros to be reused with a different number of cut-points. The model parameter estimates and standard errors are shown below.

$$\begin{array}{lll} 0.0993\,\text{AGE} & -1.053\,\text{SURG} & +0.919\,\text{MM} \\ (0.0301) & (0.616) & (0.329) \end{array}$$

The standard errors have increased because the effective sample size—the number of deaths—has decreased. However, all three variables have increased in importance, and mismatch is now clearly significant.

```
$USE PHAZ DRT $
```

```
                    log hazard  v  log survival time
  -8.400 |                             +
  -8.700 |
  -9.000 |                          +
  -9.300 |                            ,+
  -9.600 |                            +
  -9.900 |                          +
 -10.200 |                            ++
 -10.500 |                                                           +
 -10.800 |                          ++
 -11.100 |                                          +
 -11.400 |                   +        +            +
 -11.700 |
 -12.000 |                        +             +
 -12.300 |                       +                             + +
 -12.600 |   +                                     +        +     +
 -12.900 |                +
 -13.200 |                                  +
 -13.500 |                                        +
 -13.800 |
 -14.100 |                                                   +
----------:---------:---------:---------:---------:---------:---------:
        2.00      3.00      4.00      5.00      6.00      7.00      8.00
```

There is now a rapid rise in the hazard from 10 to 65 days, followed by a much lower and fairly constant hazard from 130 days onward. This behaviour of the hazard does not correspond to any standard survival distribution. (The lognormal distribution does not provide an adequate fit, with a deviance of 407.67 compared with 359.42 for the piecewise exponential, a change of 48.24 for 27 df.)

We now analyse the deaths from other causes.

```
$USE RESET DRT $
$CALC NREJ = (1-REJ)*DIED $
$USE CUTP SURV NREJ DNRT $
$USE INIC SURV NREJ DNRT $
$USE FITC NREJ $D E $
```

The model parameter estimates and standard errors are shown below.

$$\underset{(0.032)}{-0.0081\,\text{AGE}} \; \underset{(0.782)}{-0.397\,\text{SURG}} \; \underset{(0.538)}{-0.449\,\text{MM}}$$

The standard errors have increased further because of the small number of deaths, and *none* of the variables appears important. The deviance is 165.55. We fit the null model.

```
$MACRO MODEL 1 $ENDMAC
$USE FITC NREJ $
```

The deviance changes by 1.04 on 3 df.

```
$USE PHAZ DNRT $
```

```
                log hazard  v  log survival time
 -3.300 |
 -3.600 |          +    +
 -3.900 |
 -4.200 |
 -4.500 |
 -4.800 |                          +                +
 -5.100 |                                      +
 -5.400 |                                                +
 -5.700 |                        .
 -6.000 |
 -6.300 |                                   +    +
 -6.600 |
 -6.900 |                                                +
 -7.200 |
 -7.500 |
 -7.800 |
 -8.100 |                                                         +    +
 -8.400 |
 -8.700 |
 -9.000 |                                                                    +
----------:---------:---------:---------:---------:---------:---------:
       -1.60      0.00      1.60      3.20      4.80      6.40      8.00
```

The log hazard function shows a steady decline with log time, suggesting a Weibull distribution for survival time for deaths from other causes. We fit the Weibull model.

```
$INPUT %PLC WEIBULL $
$CALC %S = 1
$USE WEIBULL SURV NREJ %S $
```

The Weibull deviance is 182.03, an increase of 15.44 on 12 df compared to the piecewise exponential: the Weibull distribution provides a satisfactory representation of survival time for deaths from other causes. The ML estimate of the shape parameter is 0.358: risk declines rapidly with time. Fitting the previous model with age, surgery and mismatch confirms their irrelevance for the Weibull distribution, with the same deviance change of 1.04 on 3 df.

Thus in this example the competing risk framework reveals the importance of mismatch and the very rapid increase in hazard in the first 60 days for deaths by rejection, and the rapidly declining hazard for deaths from other causes, which is unrelated to the explanatory variables.

6.22 Time-dependent explanatory variables

All of the analysis in this chapter has been based on the assumption that the explanatory variables **x** are *constant over time*, that is that they do not change during the lifetime of the individual. Many medical variables are of this type, but others may change their values during the individual's lifetime. For example, measures of physical or physiological status may be available during the course of treatment; these may be important predictors of survival time, and more relevant than the same measures taken before treatment begins. Variables of this type are called *time-dependent.*

Such variables can be incorporated into a proportional hazards model using the piecewise exponential distribution. We need first two notational changes. The time-varying explanatory variables may be changing their values continuously in time, but in practice they are measured only at follow-up or interview times, which will generally not correspond to the times a_j at which the hazard function changes. We assume that these variables can be taken as constant over the time intervals between measurements, and we extend the set of cut-points a_j to include the measurement times for each individual.

We now write $\mathbf{x}_{ij}$ for the value of the explanatory variables, measured at a_{j-1}, for the i-th individual in the j-th time interval $a_{j-1} < t \leq a_j$. In practice many, perhaps most, components of $\mathbf{x}_{ij}$ will be constant over time intervals.

The piecewise exponential model of Section 6.15 now applies with

$$h_{ij} = \exp(\phi_j + \boldsymbol{\beta}'\mathbf{x}_{ij}).$$

Maximization of the likelihood proceeds as in Section 6.16: we have

$$\frac{\partial \ell}{\partial \phi_j} = \sum_{i \in R_j} (w_{ij} - \theta_{ij})$$

giving as before

$$d_j = \sum_{i \in R_j} w_{ij} = \sum_{i \in R_j} \hat{\theta}_{ij} = \exp(\hat{\phi}_j) \sum_{i \in R_j} e_{ij} \exp(\hat{\boldsymbol{\beta}}'\mathbf{x}_{ij})$$

so that $\exp(\hat{\phi}_j) = 0$ if $d_j = 0$ as before. If $d_j \neq 0$,

$$\exp(\hat{\phi}_j) = d_j / \{\sum_{i \in R_j} e_{ij} \exp(\hat{\boldsymbol{\beta}}'\mathbf{x}_{ij})\}.$$

The likelihood equation for $\hat{\boldsymbol{\beta}}$ is

$$\sum_i \sum_j (w_{ij} - \hat{\theta}_{ij})\mathbf{x}_{ij} = \mathbf{0}$$

which is equivalent to

$$\sum_i \sum_j r_{ij} (w_{ij}/r_{ij} - \hat{\gamma}_{ij})\mathbf{x}_{ij} = \mathbf{0}$$

where

$$\hat{\gamma}_{ij} = \exp(\hat{\boldsymbol{\beta}}'\mathbf{x}_{ij})$$
$$r_{ij} = e_{ij} \exp(\hat{\phi}_j).$$

This likelihood equation is that for a Poisson distribution with response variable w_{ij}/r_{ij}, prior weight variable r_{ij}, regression model $\boldsymbol{\beta}'\mathbf{x}_{ij}$ and log link function. It should be noted that there are now

$$\sum_{i=1}^{n} N_i$$

observations, which may be a very large number if both n and the N_i are large. Since the estimated hazard is zero for intervals with no deaths, and such intervals do not contribute to the estimation of $\boldsymbol{\beta}$, it is computationally important to reduce the size of the data by eliminating from the data times a_j at which no deaths or changes in the time-varying explanatory variables occur (see discussion of this point in Section 6.16).

This model can again be fitted using a relaxation method as in Section 6.16. Macros to fit the model are available in the subfile PIECEWISE. The censoring vector w_{ij} and the time-constant explanatory variables have to be expanded to length ΣN_i. The macro INIB is used instead of INIC to form the vectors r_{ij} and w_{ij} and to initialize the fitting procedure—it takes the same arguments as macro INIC. Following this, the macro XPCC is used to expand the time-constant explanatory variables: it takes up to nine arguments, which are all names of the explanatory variables to be expanded. Models may now be fitted by declaring a macro called MODEL containing the model formula, and using the macro FITB (with the same argument as FITC) to fit the specified model. See Green and Francis (1986) for further details.

We noted in Section 6.16 that no standard errors are obtained for the $\hat{\phi}_j$ from the relaxation method used to fit the piecewise exponential model, and the standard errors of $\hat{\boldsymbol{\beta}}$ are slightly underestimated. Correct standard errors for both sets of parameters may be obtained using the above expansion method. We noted at the end of Section 6.15 that the w_{ij} could be treated as independent Poisson variables with means θ_{ij} given by a log-linear model

$$\log \theta_{ij} = \log e_{ij} + \phi_j + \boldsymbol{\beta}'\mathbf{x}_i.$$

This model can be fitted directly by expanding all the variables in $\mathbf{x}$ and the censoring indicator, using an offset $\log e_{ij}$, and a factor with N levels for the hazard parameters θ_j. The computation time for fitting this model is very considerable, since the data are of length

$$\sum_{i=1}^{n} N_i$$

and the hazard factor has N levels. With several hundred observations this fitting method is not practicable.

6.23 Discrete time models

The discussion so far in this chapter has assumed that time is measured on a continuous scale, or at least that the discreteness of the recording of time is small compared to the range of possible time values.

Time may, however, be grouped into quite broad categories; for example, when deaths of laboratory animals are recorded at relatively long intervals, or when human studies use follow-up periods which are long in comparison to the progress of the disease.

In such cases the likelihood construction of Section 6.15 is not appropriate, and a model for discrete time is needed. We follow the development in Prentice and Gloeckler (1978; see also Kalbfleisch and Prentice, 1980, p. 98 and Thompson, 1981).

Suppose the continuous distribution with density $f(t)$, survivor function $S(t)$ and hazard $h(t)$ is grouped into s intervals $A_j=(t_{j-1},t_j]$ $j=1,\ldots, s$, with $t_0=0$, $t_s=\infty$. Write f_j, s_j and h_j for the probability mass, survivor and hazard functions for the resulting discrete distribution.
Then

$$
\begin{aligned}
f_j &= \Pr(T\in A_j)\\
&= S(t_{j-1})-S(t_j)\\
&= s_j - s_{j+1},\\
h_j &= Pr(T\in A_j|T > t_{j-1})\\
&= f_j/(f_j+f_{j+1}+\ldots+f_s)\\
&= f_j/s_j
\end{aligned}
$$

so that

$$s_j = \prod_{k=1}^{j-1}(1-h_k)$$

as in Section 6.8. Consider the survival experience of each individual through time as in Section 6.15. The i-th individual experiences a sequence of censorings at $t_1,t_2,\ldots$ and either dies or is finally censored in the j_i-th interval. Let w_{ij} be the censoring indicator and h_{ij} the hazard for the i-th individual in the j-th interval; the contribution to the likelihood is then

$$
\begin{aligned}
L_i &= f_{ji}^{\,w_{i,j_i}} s_{ji+1}^{\,1-w_{i,j_i}}\\
&= \prod_{j=1}^{ji} h_{ij}^{\,w_{ij}}(1-h_{ij})^{1-w_{ij}}.
\end{aligned}
$$

For the proportional hazards model,

$$h(t,\mathbf{x}) = \lambda_0(t)e^{\boldsymbol{\beta}'\mathbf{x}}$$

$$S(t,\mathbf{x}) = \exp(-\Lambda_0(t)e^{\boldsymbol{\beta}'\mathbf{x}}).$$

Thus

$$s_j = \exp(-\Lambda_0(t_{j-1})e^{\boldsymbol{\beta}'\mathbf{x}})$$

$$h_j = 1 - s_{j+1}/s_j$$

$$= 1 - \exp[-e^{\boldsymbol{\beta}'\mathbf{x}}\{\Lambda_0(t_j) - \Lambda_0(t_{j-1})\}]$$

so that

$$h_{ij} = 1 - \exp(-e^{\boldsymbol{\beta}'\mathbf{x}_i + \psi_j})$$

where

$$\psi_j = \log\{\Lambda_0(t_j) - \Lambda_0(t_{j-1})\}.$$

Since nothing is assumed about $\lambda_0(t)$, the ψ_j are unrelated, unknown constants as in the proportional hazards model. The likelihood function over all individuals is then

$$L = \prod_{i=1}^{n} \prod_{j=1}^{j_i} h_{ij}^{w_{ij}}(1-h_{ij})^{1-w_{ij}} = \prod_{j=1}^{s} \prod_{i \in R_j} h_{ij}^{w_{ij}}(1-h_{ij})^{1-w_{ij}}$$

where R_j is the set of individuals at risk in the k-th time interval. This is a product of $\sum j_i$ Bernoulli likelihoods, one for each individual in each interval. In each interval the individuals currently at risk either die or survive to the next interval. The death probability in the j-th interval for the i-th individual is h_{ij}, and

$$\log\{-\log(1-h_{ij})\} = \boldsymbol{\beta}'\mathbf{x}_i + \psi_j.$$

Thus the discrete time proportional hazard model can be fitted by treating the observations in each time interval as independent across intervals, with the censoring indicator as the response variable with a Bernoulli distribution, and a complementary log–log link function (Section 4.2); the regression model contains the explanatory variables and the interval parameters ψ_j. If the explanatory variables are themselves categorical, the data have the form of a contingency table, and the response variable is the number of deaths in the interval, which has a binomial distribution with the number at risk in the interval as the binomial denominator.

Unlike the Poisson likelihood for the continuous time proportional hazards model, the binomial likelihood does not allow the interval parameters ψ_j to be estimated explicitly given a current estimate of $\boldsymbol{\beta}$; they must be estimated simultaneously by including a block factor for time interval in the model.

We illustrate with the GEHAN example. Suppose that remission times are recorded in months (units of 4 weeks) instead of weeks. The grouped data are presented below as a contingency table, with the number at risk and the number dying in each month for each treatment group. Censoring within an interval is treated as censoring at the end of the previous interval.

		Time					
		1-4	5-8	9-12	13-16	17-20	21-24
G1	d	7	6	4	1	1	2
	r	21	14	8	4	3	2
G2	d	0	4	1	2	0	2
	r	21	20	13	12	7	7

The analysis is straightforward.

```
$UNIT 12
$DATA D R
$READ
 7 21  6 14  4  8  1  4  1 3  2 2
 0 21  4 20  1 13  2 12  0 7  2 7
$FACTOR G 2 MONTH 6 $
$CALC G = %GL(2,6) : MONTH = %GL(6,1) $
$YVAR D $ERROR B R $LINK C
$FIT MONTH : + G $D E $
```

The deviance decreases from 24.88 to 6.80, and the estimate of G is -1.697 with s.e. 0.425. This value is between those for the exponential and Weibull distributions. Note that we have made no parametric assumption about the hazard.

An alternative to fitting the grouped time variable as a factor in the model is to construct a conditional likelihood which depends only on the common group difference. We condition on the total number at risk in each time interval, in the same way as for the 2×2 table in Section 4.6. Various tests based on the resulting conditional hypergeometric distribution are possible; the best known is the *log rank* test (Mantel, 1966).

It is not necessary to have categorical explanatory variables to fit the discrete time model. In general the model can be fitted using the macros in PIECEWISE. As for the proportional hazards model in continuous time with time dependent covariates, the data vector w_{ij} is expanded to length

$$\sum_i N_i,$$

as are the explantory variables $\mathbf{x}_i$. The initialising macro INIB can be used to set up the individuals at risk in each interval and expand the censoring indicator. In addition this macro provides us with a block factor BLOK for the time intervals which can be used explicitly in a direct GLIM fit. Before fitting the model it is also necessary to expand the explanatory variables to length ΣN_i (using the macro XPCC) and to declare the model as Bernoulli with the censoring indicator as the y-variable and complementary log–log link. Finally, the offset must be turned off as macro INIB sets this in readiness for the standard Poisson fit of the piecewise exponential distribution. The model may then be fitted with BLOK included as an explicit term in the model.

We illustrate with the FEIGL data, using the distinct death times as discrete cut-points.

```
$INPUT 1 FEIGL PIECEWISE $
$CALC W = 1 $
$USE CUTP TIME W DDT $
$USE INIB TIME W DDT $
$CALC LWBC = %LOG(WBC) $
$USE XPCC AG LWBC $
$CA N=1 $ERR B N $YVAR W $OFFSET $LINK C $
$FIT AG*LWBC + BLOK $D E $
```

The block hazard estimates are initially fairly constant but show a steady increase with block number from block 14 on. The behaviour of these estimates does not in general provide information about the form of the hazard function since it depends on both the spacing of the cut-points and on $\lambda(t)$. The fitted model estimates are given below with standard errors.

$$\underset{(1.06)}{-5.67} + \underset{(0.93)}{2.87}\,\mathrm{AG}(2) + \underset{(0.220)}{0.745}\,\mathrm{LWBC} - \underset{(0.280)}{0.596}\,\mathrm{AG}(2).\mathrm{LWBC}$$

The interaction is now significant (omitting it gives a deviance change of 4.67 on 1 df).

It is easily seen that if $t_j - t_{j-1}$ is small,

$$\begin{aligned}\psi_j &= \log\{\Lambda_0(t_j) - \Lambda_0(t_{j-1})\}\\ &\approx \log\{(t_j - t_{j-1})\Lambda_0(t_j)\}\\ &\approx \log(e_j) + \phi_j)\end{aligned}$$

from the piecewise exponential. As h_{ij} will also be very small.

$$\log\{-\log(1-h_{ij})\} \approx \log h_{ij}$$

so we obtain the previous piecewise exponential model as the Poisson limit of the binomial model for precise measurement in continuous time.

References

M. A. Aitkin, Simultaneous inference and choice of variable subsets in multiple regression *Technometrics*, **16,** 221–7 (1974).

M. A. Aitkin, The analysis of unbalanced cross-classifications (with Discussion) *J. Roy. Statist. Soc. A*, **141,** 195–223 (1978).

M. A. Aitkin, A note on the regression analysis of censored data *Technometrics*, **23,** 161–3 (1981).

M. Aitkin, D. Anderson and J. Hinde, Statistical modelling of data on teaching styles (with Discussion) *J. Roy. Statist. Soc. A*, **144,** 419–61 (1981).

M. Aitkin and D. Clayton, The fitting of exponential, Weibull and extreme value distributions to complex censored survival data using GLIM *Appl. Statist.*, **29,** 156–63 (1980).

M. Aitkin, R. B. Davies and B. J. Francis, Using the Cox proportional hazard model to choose a parametric survival distribution. (Centre for Applied Statistics Research Paper 85–15, Centre for Applied Statistics, University of Lancaster UK, 1986).

M. Aitkin and B. Francis, A GLIM macro for fitting the exponential or Weibull distribution to censored data *GLIM Newsletter*, June, 19–25 (1980).

M. Aitkin and B. Francis, Reader reaction: Interactive regression modelling *Biometrics*, **38,** 511–13 (1982).

M. Aitkin, N. Laird and B. Francis, A reanalysis of the Stanford heart transplant data (with Comments) *J. Amer. Statist. Assoc.*, **78,** 264–92 (1983).

M. Aitkin and N. T. Longford, Statistical modelling issues in school effectiveness studies *J. Roy. Statist. Soc. A*, **149,** 1–43 (1986).

D. M. Allen, Mean-square error of prediction as a criterion for selecting variables *Technometrics*, **13,** 465–75 (1971).

D. A. Anderson and M. Aitkin, Variance component models with binary response: interviewer variability *J. Roy. Statist. Soc. B*, **47,** 203–10 (1985).

J. A. Anderson, Regression and ordered categorical variables (with Discussion) *J. Roy. Statist. Soc. B*, **46,** 1–30 (1984).

F. J. Anscombe, Normal likelihood functions *Ann. Inst. Statist. Math.*, **26,** 1–19 (1964).

J. R. Ashford, An approach to the analysis of data from semi-quantal responses in biological response *Biometrics*, **15,** 573–81 (1959).

A. C. Atkinson, Robustness, transformations and two graphical displays for outlying and influential observations in regression *Biometrika*, **68,** 13–20 (1981).

A. C. Atkinson, Regression diagnostics, transformations and constructed variables (with Discussion) *J. Roy. Statist. Soc. B*, **44,** 1–36 (1982).

A. C. Atkinson, *Plots, transformations and regression. An introduction to graphical methods of diagnostic regression analysis* (*Oxford, 1985*).

C. Atkinson and J. Polivy, Effects of delay, attack and retaliation on state depression and hostility *J. Abnormal Psychology*, **85,** 370–76 (1976).

R. J. Baker, Glim 3.77 Reference guide. In *Glim 3.77 Reference Manual* (Numerical Algorithms Group, Oxford, 1985).

V. Barnett, Probability plotting methods and order statistics *Appl. Statist.*, **24,** 95–108 (1975).

V. Barnett and T. Lewis, *Outliers in statistical data* (Wiley, Chichester, 1978).

L. A. Baxter, S. M. Coutts and G. A. F. Ross, Applications of linear models in motor insurance, in *Proceedings of the 21st International Congress of Actuaries, Zurich*, 11–29 (1980).

D. A. Belsley, E. Kuh and R. E. Welsch, *Regression diagnostics: Identifying influential data and sources of collinearity* (Wiley, New York, 1980).

S. Bennett and J. Whitehead, Fitting logistic and log-logistic regression models to censored data using GLIM *GLIM Newsletter*, **4,** 12–19 (1981).

Y. M. M. Bishop, Full contingency tables, logits and split contingency tables *Biometrics*, **25,** 383–99 (1969).

Y. M. M. Bishop, S. E. Fienberg and P. W. Holland, *Discrete multivariate analysis: theory and practice* (M.I.T. Press, Cambridge, Mass, 1975).

R. D. Bock, *Multivariate statistical methods in behavioral research* (McGraw-Hill, New York, 1975).

R. D. Bock and G. Yates, *MULTIQUAL: Log-linear analysis of nominal or ordinal qualitative data by the method of maximum likelihood* (Chicago: National Educational Resources, 1973).

G. E. P. Box, Sampling and Bayes inference in scientific modelling and robustness (with Discussion) *J. Roy. Statist. Soc. A*, **143,** 383–430 (1980).

G. E. P. Box, An apology for ecumenism in statistics. In *Scientific inferences, data analysis and robustness*, Edited by G. E. P. Box, T. Leonard and C. F. Wu (Academic Press, New York, 1983).

G. E. P. Box and D. R. Cox, An analysis of transformations (with Discussion) *J. Roy. Statist. Soc. B*, **26,** 211–52 (1964).

G. E. P. Box and P. W. Tidwell, Transformations of the independent variable *Technometrics*, **4,** 531–50 (1962).

N. Breslow, Covariance analysis of censored survival data *Biometrics*, **30,** 89–99 (1974).

P. J. Brown, Multivariate calibration (with Discussion) *J. Roy. Statist. Soc. B*, **44,** 287–321 (1982).

P. J. Brown, J. Stone and C. Ord-Smith, Toxaemic signs during pregnancy. *Appl. Statist.*, **32,** 69–72 (1983).

C. Chatfield and G. J. Goodhardt, The beta-binomial model for consumer purchasing behaviour *Appl. Statist.*, **19,** 240–50 (1970).

D. Clayton and J. Cuzick, The EM algorithm for Cox's regression model using GLIM *Appl. Statist.*, **34,** 148–56 (1985).

R. D. Cook and S. Weisberg, *Residuals and influence in regression* (Chapman and Hall, London, 1982).

J. B. Copas, Regression, prediction and shrinkage (with Discussion) *J. Roy. Statist. Soc. B*, **45,** 311–54 (1983).

D. R. Cox, Regression models and life tables (with Discussion) *J. Roy. Statist. Soc. B*, **34,** 187–220 (1972).

D. R. Cox and D. Oakes, *Analysis of survival data* (Chapman and Hall, London, 1984).

M. J. Crowder, Beta-binomial ANOVA for proportions. *Appl. Statist.*, **27,** 34–7 (1978).

M. J. Crowder, Inference about the intraclass correlation in the beta-binomial ANOVA for proportions *J. Roy. Statist. Soc. B*, **41,** 230–4 (1979).

J. Crowley and M. Hu, Covariance analysis of heart transplant survival data *J. Amer. Statist. Assoc.*, **72,** 27–36 (1977).

A. P. Dempster, N. M. Laird and D. B. Rubin, Maximum likelihood from incomplete data via the EM algorithm (with Discussion) *J. Roy. Statist. Soc. B*, **39,** 1–38 (1977).

N. R. Draper and H. Smith, *Applied regression analysis.* 2nd Edn. (Wiley, New York, 1981).

A. W. F. Edwards, *Likelihood.* (Cambridge, 1972).

B. Efron and D. V. Hinkley, Assessing the accuracy of the maximum likelihood estimator: observed versus expected Fisher information *Biometrika*, **65,** 457–87 (1978).

B. H. Erickson and T. A. Nosanchuk, *Understanding data* (Open University Press, Milton Keynes, 1979).

P. Feigl and M. Zelen, Estimation of exponential probabilities with concomitant information *Biometrics*, **21,** 826–38 (1965).

J. J. Filliben, The probability plot correlation coefficient test for normality *Technometrics*, **17,** 111–17. Correction **17,** 520 (1975)

D. J. Finney, The estimation from original records of the relationship between dose and quantal response *Biometrika*, **34,** 320–34 (1947).

R. A. Fisher, On the mathematical foundations of theoretical statistics *Phil. Trans. Roy. Soc. London A*, **222,** 309–368 (1922). (Reprinted in *Contributions to Mathematical Statistics*, Wiley, New York, 1950).

W. A. Fuller and M. A. Hidiroglou, Regression estimation after correcting for attenuation *J. Amer. Statist. Assoc.*, **73,** 99–104 (1978).

E. A. Gehan, A generalized Wilcoxon test for comparing arbitrarily singly-censored samples. *Biometrika*, **52,** 203–23 (1965).

H. Goldstein, Multilevel mixed linear model analysis using iterative generalized least squares *Biometrika*, **73,** 43–56 (1986).

L. A. Goodman, Simple models for the analysis of ordered categorical data *J. Amer. Statist. Assoc.*, **74,** 537–52 (1979).

M. Green and B. Francis, Fitting the Cox proportional hazard model. Centre for Applied Statistics Research Paper 86–07, (Centre for Applied Statistics, University of Lancaster, UK 1986)

P. Green, Iteratively reweighted least squares for maximum likelihood estimation, and some robust and resistant alternatives (with discussion). *J. Roy. Statist. Soc. B*, **46,** 149–162 (1984).

D. A. Griffiths, Maximum likelihood estimation for the beta-binomial distribution and an application to the household distribution of the total number of cases of a disease *Biometrics*, **29,** 637–48 (1973).

S. J. Haberman, Log-linear models for frequency tables with ordered classifications *Biometrics*, **30,** 589–600 (1974).

H. O. Hartley, The maximum F-ratio as a short cut test for heterogeneity of variances *Biometrika*, **37,** 308–12 (1950).

H. V. Henderson and P. F. Velleman, Building multiple regression models interactively *Biometrics*, **37,** 391–411 (1981).

J. E. Higgins and G. G. Koch, Variable selection and generalized chi-square analysis of categorical data applied to a large cross-sectional occupational health survey *Int. Statist. Rev.*, **45**, 51–62 (1977).

J. Hinde, Compound Poisson regression models. In *GLIM82: Proceedings of the International Conference on Generalised Linear Models*. ed. R. Gilchrist, 109–21. (Springer-Verlag, New York, 1982).

J. P. Hinde and M. Aitkin, Canonical likelihoods: A new likelihood treatment of nuisance parameters *Biometrika*, **74**, 45–58 (1987).

D. C. Hoaglin and R. E. Welsch, The hat matrix in regression and ANOVA *The American Statistician*, **32**, 17–22. Corrigenda **32**, 146 (1978).

D. Holt, A. J. Scott and P. D. Ewings, Chi-squared tests with survey data *J. Roy. Statist. Soc. A*, **143**, 303–20 (1980).

P. O. Johnson and J. Neyman, Tests of certain linear hypotheses and their applications to some educational problems *Statist. Res. Mem.*, **1**, 57–93 (1936).

K. Jones, A geographical contribution to the aetiology of chronic bronchitis. Unpublished B.Sc. dissertation, University of Southampton (1975).

K. G. Jöreskog and D. Sörbom, *LISREL V. Analysis of linear structural relationships by maximum likelihood and least squares methods* (Department of Statistics, University of Uppsala 1981).

J. D. Kalbfleisch and R. L. Prentice, *The statistical analysis of failure time data* (Wiley, New York, 1980).

J. G. Kalbfleisch, *Probability and statistical inference II* New York: Springer-Verlag (1979).

E. L. Kaplan and P. Meier, Non-parametric estimation from incomplete observations *J. Amer. Statist. Assoc.*, **53**, 457–81 (1958).

H. H. Ku and S. Kullback, Loglinear models in contingency table analysis *The American Statistician*, **28**, 115–22 (1974).

D. V. Lindley and W. F. Scott, *New Cambridge elementary statistical tables* (Cambridge, 1984).

R. J. A. Little and M. D. Schluchter, Maximum likelihood estimation for mixed continuous and categorical data with missing values *Biometrika*, **72**, 497–512 (1985).

N. Longford, A fast scoring algorithm for maximum likelihood estimation in unbalanced mixed models with nested random effects *Biometrika*, **74**, 817–27 (1987).

P. McCullagh, Regression models for ordinal data (with Discussion) *J. Roy. Statist. Soc. B*, **42**, 109–42 (1980).

P. McCullagh and J. A. Nelder, *Generalized linear models* (Chapman and Hall, London, 1983).

N. Mantel, Evaluation of survival data and two new rank order statistics arising in its consideration *Cancer Chemotherapy Reports*, **50**, 163–70 (1966).

Ch. E. Minder and J. B. Whitney, A likelihood analysis of the linear calibration problem *Technometrics*, **17**, 463–71 (1975).

J. A. Nelder and R. W. M. Wedderburn, Generalized linear models *J. Roy. Statist. Soc. A*, **135**, 370–84 (1972).

Panel on Incomplete Data, Incomplete Data in Sample Surveys. Vols 1, 2 & 3 (Academic Press, New York, 1983).

E. S. Pearson and H. O. Hartley, *Biometrika tables for statisticians, Vol. 1* (3rd edn) (Cambridge, 1966).

K. Pearson (Ed.) *Tables of the incomplete beta function*. 2nd Edn (Cambridge, 1968).

R. L. Plackett, The marginal totals of a 2×2 table. *Biometrika*, **64,** 37–42 (1977).

R. F. Potthoff, On the Johnson–Neyman technique and some extensions thereof *Psychometrika*, **29,** 241–56 (1964).

D. Pregibon, Logistic regression diagnostics *Ann. Statist.*, **9,** 705–24 (1981).

R. L. Prentice, Exponential survivals with censoring and explanatory variables *Biometrika*, **60,** 279–88 (1973).

R. L. Prentice, and L. A. Gloeckler, Regression analysis of grouped survival data with application to breast cancer data *Biometrics*, **34,** 57–67 (1978).

S. Quine, Achievement orientation of aboriginals and white Australian adolescents. Unpublished Ph.D. Thesis. Australian National University (1975).

J. H. Roger and S. B. Peacock, Fitting the scale as a GLIM parameter for Weibull, extreme value, logistic and log-logistic regression models with censored data *GLIM Newsletter*, **6,** 30–37 (1982).

J. H. Roger, Using factors when fitting the scale parameter to Weibull and other survival regression models with censored data *GLIM Newsletter*, **11,** 14–15 (1985).

T. Ryan, B. Joiner and B. Ryan, *Minitab students handbook* (Duxbury Press, North Scituate, Mass 1976).

A. Scallan, R. Gilchrist and M. Green, Fitting parametric link functions in generalised linear models *Comp. Stat. and Data Anal.*, **2,** 37–49 (1984).

S. S. Shapiro and M. B. Wilk, An analysis of variance test for normality (complete samples) *Biometrika*, **52,** 591–611 (1965).

M. J. Silvapulle, On the existence of maximum likelihood estimators for the binomial response model *J. Roy. Statist. Soc. B*, **43,** 310–13 (1981).

T. M. F. Smith, The foundations of survey sampling: a review (with Discussion) *J. Roy. Statist. Soc A*, **139,** 183–204 (1976).

S. S. Stevens, On the theory of scales of measurement *Science*, **103,** 677–80 (1946).

M. Stone, Cross-validatory choice and assessment of statistical predictions (with Discussion). *J. Roy. Statist. Soc. B*, **36,** 111–47 (1974).

R. Thompson, Survival data and GLIM. Letter to the editor *Appl. Statist.*, **30,** 310 (1981).

R. Thompson, and R. J. Baker, Composite link functions in generalized linear models *Appl. Statist.*, **30,** 125–31 (1981).

D. Wechsler, *Wechsler intelligence scale for children* (The Psychological Corporation, New York, 1949).

J. Whitehead, Fitting Cox's regression model to survival data using GLIM *Appl. Statist.*, **29,** 268–75 (1980).

D. A. Williams, The analysis of binary responses from toxicological experiments involving reproduction and teratogenicity *Biometrics*, **31,** 949–52 (1975).

D. A. Williams, Extra-binomial variation in logistic linear models *Appl. Statist.*, **31,** 144–8 (1982).

M. S. Wolynetz, Maximum likelihood estimation in a linear model from confined and censored normal data. Algorithm AS139 *Appl. Statist.*, **28,** 195–206 (1979).

N. Wrigley, *Introduction to the use of logit models in geography*. CATMOG 10, University of East Anglia, Norwich: Geo. Abstracts Ltd (1976).

F. Yates, Contingency tables involving small numbers and the χ^2 test *J. Roy. Statist. Soc., Suppl.*, **1,** 217–35 (1934).

F. Yates, Tests of significance for 2×2 contingency tables (with Discussion) *J. Roy. Statist. Soc. A*, **147,** 426–63 (1984).

Appendix 1
Discussion

The section numbering used in this Appendix corresponds to the section numbering in the Chapter to which the Appendix section refers. For example the reference in Section 2.5 to the likelihood function when y is not measured with high precision in discussed fully in Section A1.2.5.

A1.2.5 The likelihood function and maximum likelihood estimation

We noted in Section 2.2 that "continuous" variables are in fact measured with finite precision, and can take only a finite number of discrete values. If the precision of measurement is low, this discreteness has to be allowed for explicitly in the model.

Suppose for example that a normal distribution $N(\mu,\sigma^2)$ is assumed for "true" height, but the value assigned to height can be measured or recorded only to the nearest 0.1 cm. The recorded value y defines the true value Y only to the interval $(y-0.05, y+0.05)$. In general if the measurement precision is δ, the recorded y corresponds to a true Y in the interval $(y-(\delta/2), y+(\delta/2))$. The probability of recording a value y is then

$$\Pr\{y-(\delta/2) < Y < y+(\delta/2)\} = \Phi\left\{\frac{y+(\delta/2)-\mu}{\sigma}\right\} - \Phi\left\{\frac{y-(\delta/2)-\mu}{\sigma}\right\}$$

and so, given observed values $y_1, \ldots, y_n$,

$$L(\mu,\sigma) = \prod_{i=1}^{n}\left[\Phi\left\{\frac{y_i+(\delta/2)-\mu}{\sigma}\right\} - \Phi\left\{\frac{y_i+(\delta/2)-\mu}{\sigma}\right\}\right]$$

where $\phi(x)$ is the normal cumulative distribution function. Maximization of this likelihood with respect to μ and σ using an EM algorithm is considerably more complicated than in Section 2.5 (Wolynetz, 1979). The Taylor expansion of

$$\Phi\left[\frac{y-\mu+\varepsilon}{\sigma}\right]$$

about $\varepsilon = 0$ shows that

$$\Pr\left\{y-\left(\frac{\delta}{2}\right) < Y < y+\left(\frac{\delta}{2}\right)\right\}$$
$$= (\delta/\sigma)\Phi\left(\frac{y-\mu}{\sigma}\right)\left[1+\frac{\delta^2}{24\sigma^2}\left\{\left(\frac{y-\mu}{\sigma}\right)^2-1\right\}+\mathrm{O}\left(\frac{\delta^4}{\sigma^4}\right)\right]$$

Thus if δ is small compared with σ—say $(\delta/\sigma) < 0.1$—the usual likelihood based on the normal density can be used, but if the standard deviation is not large compared to the measurement precision, the discrete representation of the probability distribution is necessary.

The usual representation of the likelihood for a continuous random variable is

$$L = \prod_{i=1}^{n} f(y_i)\mathrm{d}y_1\mathrm{d}y_2\ldots\mathrm{d}y_n$$

This effectively assumes that the measurement precision δ is very small, and can be represented by an infinitesimal differential element $\mathrm{d}y$, which can be omitted from the likelihood. In geometrical terms, the area under $f(y)$ from $y-(\delta/2)$ to $y+(\delta/2)$ is being approximated by the rectangle of height $f(y)$ and width δ.

A1.2.14 Maximum likelihood fitting of exponential family regression models

We present here the general iteratively reweighted least squares algorithm for exponential family regression models originally given by Nelder and Wedderburn (1972). The notation used is consistent with that in the GLIM manual and in McCullagh and Nelder (1983). See also Green (1984) for a relevant discussion on the IRLS algorithm.

We give first some standard results. Write the probability function for y in the exponential form

$$\log f(y) = (y\theta - b(\theta))/\phi + c(y,\phi).$$

We will see below that θ and ϕ are functions of the parameters of the distribution of y such that the mean of y involves only θ.

Differentiating w.r.t. θ,

$$\frac{\partial \log f}{\partial \theta} = \frac{1}{\phi}\{y - b'(\theta)\}$$

$$\frac{\partial^2 \log f}{\partial \theta^2} = -b''(\theta)/\phi.$$

Since

$$\int f(y)\mathrm{d}y = 1,$$

differentiating w.r.t. θ gives

$$0 = \int \frac{\partial f}{\partial \theta}\,\mathrm{d}y = \int \frac{\partial \log f}{\partial \theta}\cdot f\mathrm{d}y = E\left(\frac{\partial \log f}{\partial \theta}\right) = \frac{1}{\phi}\{E(y) - b'(\theta)\}$$

$$0 = \int \frac{\partial^2 f}{\partial \theta^2} \mathrm{d}y = \int \left\{ \frac{\partial^2 \log f}{\partial \theta^2} \cdot f + \left(\frac{\partial \log f}{\partial \theta} \right)^2 \cdot f \right\} \mathrm{d}y$$

$$= E\left(\frac{\partial^2 \log f}{\partial \theta^2} \right) + E\left(\frac{\partial \log f}{\partial \theta} \right)^2.$$

Thus

$$\mu = E(y) = b'(\theta)$$

$$\begin{aligned} \operatorname{var}(y) &= E\{y - b'(\theta)\}^2 \\ &= \phi^2 E\left(\frac{\partial \log f}{\partial \theta} \right)^2 \\ &= -\phi^2 E\left(\frac{\partial^2 \log f}{\partial \theta^2} \right) \\ &= \phi b''(\theta) \end{aligned}$$

Now consider the generalized linear model with

$$\eta_i = g(\mu_i) = \boldsymbol{\beta}' \mathbf{x}_i.$$

The log likelihood function is

$$\ell(\boldsymbol{\beta}) = \sum_i \{ y_i \theta_i - b(\theta_i) \} / \phi + \sum_i c(y_i, \phi)$$

and its derivative, the score function, is

$$\mathbf{s}(\boldsymbol{\beta}) = \frac{\partial \ell}{\partial \boldsymbol{\beta}} = \sum_i \{ y_i - b'(\theta_i) \} \cdot \frac{\partial \theta_i}{\partial \boldsymbol{\beta}} \Big/ \phi$$

Substitution of

$$\begin{aligned} \frac{\partial \theta_i}{\partial \boldsymbol{\beta}} &= \frac{\mathrm{d}\theta_i}{\mathrm{d}\mu_i} \cdot \frac{\mathrm{d}\mu_i}{\mathrm{d}\eta_i} \cdot \frac{\partial \eta_i}{\partial \boldsymbol{\beta}} \\ &= \frac{1}{b''(\theta_i)} \cdot \frac{1}{g'(\mu_i)} \cdot \mathbf{x}_i \end{aligned}$$

and

$$\phi b''(\theta_i) = \operatorname{var}(y_i) = V_i, \qquad b'(\theta_i) = \mu_i$$

gives

$$\mathbf{s}(\boldsymbol{\beta}) = \frac{\partial \ell}{\partial \boldsymbol{\beta}} = \sum_i (y_i - \mu_i) \mathbf{x}_i / V_i g'(\mu_i).$$

For the scoring algorithm used in GLIM, we need the expected second derivative of the log likelihood. A direct extension of the result above for a single parameter θ establishes that

$$E\left(\frac{\partial^2 \ell}{\partial \boldsymbol{\beta} \partial \boldsymbol{\beta}'}\right) = -E\left(\frac{\partial \ell}{\partial \boldsymbol{\beta}} \frac{\partial \ell}{\partial \boldsymbol{\beta}'}\right)$$

$$= -E\{\sum_i (y_i - \mu_i)\mathbf{x}_i / V_i g'(\mu_i) \quad \sum_j (y_j - \mu_j)\mathbf{x}_j' / V_j g'(\mu_j)\}$$

Since the y_i are independent, the expectations of $(y_i - \mu_i)(y_j - \mu_j)$ are zero for $i \neq j$. Thus

$$E\left(\frac{\partial^2 \ell}{\partial \boldsymbol{\beta} \partial \boldsymbol{\beta}'}\right) = -\Sigma \mathbf{x}_i \mathbf{x}_i' / V_i \{g'(\mu_i)\}^2$$

$$= -\Sigma w_i \mathbf{x}_i \mathbf{x}_i' = -X'WX$$

where W is a diagonal weight matrix with elements w_i given by

$$w_i^{-1} = V_i \{g'(\mu_i)\}^2.$$

In the r-th iteration of the scoring algorithm, the new estimate $\boldsymbol{\beta}_{r+1}$ of $\boldsymbol{\beta}$ is obtained from the previous estimate $\boldsymbol{\beta}_r$ by

$$\boldsymbol{\beta}_{r+1} = \boldsymbol{\beta}_r + (X'W_r X)^{-1} \mathbf{s}(\boldsymbol{\beta}_r)$$

where W_r is W evaluated at the estimate $\boldsymbol{\beta}_r$. Now

$$\mathbf{s}(\boldsymbol{\beta}) = \Sigma (y_i - \mu_i) g'(\mu_i) w_i \mathbf{x}_i$$

$$= X'W\mathbf{u}$$

with

$$u_i = (y_i - \mu_i) g'(\mu_i).$$

Thus

$$\boldsymbol{\beta}_{r+1} = \boldsymbol{\beta}_r + (X'W_r X)^{-1} X'W_r \mathbf{u}_r$$

$$= (X'W_r X)^{-1} (X'W_r X\boldsymbol{\beta}_r + X'W_r \mathbf{u}_r)$$

$$= (X'W_r X)^{-1} X'W_r \mathbf{z}_r$$

where

$$\mathbf{z}_r = X\boldsymbol{\beta}_r + \mathbf{u}_r$$

$$= \boldsymbol{\eta}_r + \mathbf{u}_r$$

where $\boldsymbol{\eta}_r$ is the vector of "fitted values" of the linear predictor $\boldsymbol{\eta}$ at the r-th iteration.

Thus each iteration of the scoring algorithm can be expressed as a weighted least squares regression of the "adjusted dependent variate" $\mathbf{z}$ on the explanatory variables $\mathbf{x}$ with weight vector (diagonal weight matrix) W.

A similar result holds for the Newton–Raphson algorithm using the observed rather than the expected information matrix at each iteration. The score function is as before

$$s(\boldsymbol{\beta}) = \frac{\partial \ell}{\partial \boldsymbol{\beta}} = \Sigma(y_i - b'(\theta_i)) \cdot \frac{\partial \theta_i}{\partial \boldsymbol{\beta}} \bigg/ \phi$$

with

$$\frac{\partial \theta_i}{\partial \boldsymbol{\beta}} = \phi \mathbf{x}_i / V_i g_i'(\mu_i).$$

The second derivative of the log likelihood can be written

$$\frac{\partial^2 \ell}{\partial \boldsymbol{\beta} \partial \boldsymbol{\beta}'} = \Sigma\,(-b''(\theta_i))\, \frac{\partial \theta_i}{\partial \boldsymbol{\beta}} \frac{\partial \theta_i}{\partial \boldsymbol{\beta}'} \bigg/ \phi + \Sigma(y_i - b'(\theta_i))\, \frac{\partial^2 \theta_i}{\partial \boldsymbol{\beta} \partial \boldsymbol{\beta}'} \bigg/ \phi.$$

Straightforward algebra gives

$$\begin{aligned} \frac{\partial^2 \ell}{\partial \boldsymbol{\beta} \partial \boldsymbol{\beta}'} &= -\Sigma \mathbf{x}_i \mathbf{x}_i' / V_i [g_i'(\mu_i)]^2 \\ &\quad -\Sigma(y_i - \mu_i)\mathbf{x}_i \mathbf{x}_i'(V_i g_i'' + V_i' g_i')/V_i^2 (g_i')^3 \\ &= -X' W^* X \end{aligned}$$

where W^* is a diagonal weight matrix with elements

$$w_i^* = w_i + (y_i - \mu_i)(V_i g_i'' + V_i' g_i')/V_i^2 (g_i')^3$$

and

$$V_i' = \frac{\mathrm{d}V_i}{\mathrm{d}\mu_i}, \qquad g_i'' = \frac{\mathrm{d}^2 g(\mu_i)}{\mathrm{d}\mu_i^2}.$$

Then each iteration of the Newton–Raphson algorithm can be expressed as

$$\begin{aligned} \boldsymbol{\beta}_{r+1} &= \boldsymbol{\beta}_r + (X' W_r^* X)^{-1} X' W_r \mathbf{u}_r \\ &= (X' W_r^* X)^{-1} (X' W_r^* \mathbf{n}_r + X' W_r \mathbf{u}_r) \\ &= (X' W_r^* X)^{-1} X' W_r^* \mathbf{z}_r^* \end{aligned}$$

with

$$\mathbf{z}^* = \boldsymbol{\eta} + W^{*-1} W \mathbf{u}.$$

The adjusted dependent variable in this case is more complicated. Note that $E(W^*) = W$.

To fit the logit model in Section 2.14 in which $\mu = p$,

$$\eta = g(p) = \log \frac{p}{1-p}, \qquad V = p(1-p)$$

$$g'(p) = \frac{1}{p(1-p)}.$$

The adjusted dependent variate is

$$z = \eta + \frac{y-p}{p(1-p)}$$

and the iterative weight for the GLIM scoring algorithm is

$$w = \frac{1}{p(1-p)}.$$

The scoring and Newton–Raphson algorithms are identical in this example and in general when

$$V_i g_i'' + V_i' g_i' = 0.$$

This is the case when

$$V(\mu_i) = \mathrm{k}_1 / g'(\mu_i)$$

for some arbitrary constant k_1.

Since

$$V(\mu) = \phi b''(\theta)$$

and

$$\mu = b'(\theta),$$

this condition is

$$\phi b''(\theta_i) = \mathrm{k}_1 b''(\theta_i) \Bigg/ \frac{\mathrm{d}g(\theta)}{\mathrm{d}\theta}$$

or

$$g(\theta) = \mathrm{k}_2 \theta$$

that is, the linear predictor is on the natural parameter scale of θ. Thus binomial logit, Poisson log-linear and gamma reciprocal link models all give the same results from the scoring and Newton–Raphson algorithms—the observed and estimated expected information matrices are the same. In all other cases the standard errors for model parameters will be different for the two algorithms. In large samples the differences are generally small but they may be quite large in small samples.

A1.2.15 Parameter transformations to normalize the likelihood

We noted in Section 2.15 the considerable difference in the shape of the log likelihood in p and in $\theta = \text{logit } p$. The use of the standard error from the information matrix is equivalent to the assumption that the log likelihood is quadratic in the parameter about the MLE. Since we may choose any monotonic function of the parameter to parametrize the model, we may choose that function which makes the log likelihood "closest" to quadratic.

Anscombe (1964) suggested using the function or transformation which makes the third derivative of the log likelihood zero at the MLE.

As in Section A1.2.14 we write the probability function for y in the exponential form

$$\log f(y) = y\theta - b(\theta) + c(y),$$

where for simplicity we take $\phi = 1$ since this does not affect the transformation. The log likelihood is

$$\ell(\theta) = \theta\Sigma y_i - nb(\theta) + \Sigma c(y_i).$$

Let the required parametrization be $\psi = g(\theta)$. Then

$$\ell'(\theta) = \ell'(\psi) \cdot g'(\theta) = \Sigma y_i - nb'(\theta)$$

$$\ell''(\theta) = \ell''(\psi)\{g'(\theta)\}^2 + \ell'(\psi)g''(\theta) = -nb''(\theta)$$

$$\ell'''(\theta) = \ell'''(\psi)\{g'(\theta)\}^3 + 3\ell''(\psi)g'(\theta)g''(\theta) + \ell'(\psi)g'''(\theta) = -nb'''(\theta).$$

At the MLE, we have $\ell'(\hat{\theta}) = \ell'(\hat{\psi}) = 0$ and require $\ell'''(\hat{\psi}) = 0$.

Then

$$3\ell''(\hat{\psi})g'(\hat{\theta})g''(\hat{\theta}) = -nb'''(\hat{\theta})$$

and

$$\ell''(\hat{\psi})[g'(\hat{\theta})]^2 = -nb''(\hat{\theta}).$$

Thus

$$\frac{3g''(\hat{\theta})}{g'(\hat{\theta})} = \frac{b'''(\hat{\theta})}{b''(\hat{\theta})}.$$

This holds if

$$\log g'(\theta) = (1/3) \log b''(\theta)$$

i.e. if

$$g(\theta) = \int^{\theta}\{b''(t)\}^{1/3}\mathrm{d}t.$$

For the binomial distribution with

$$P(r) = \binom{n}{r} p^r (1-p)^{n-r}$$

the natural parametrization is the logit, with $\theta = \log\{p/1-p)\}$. Then

$$\ell(\theta) = \log P(r) = \log\binom{n}{r} + r\theta - n\log(1 + e^{\theta})$$

with

$$b(\theta) = \log(1 + e^{\theta}),$$

$$b'(\theta) = e^{\theta}/(1 + e^{\theta})$$

$$b''(\theta) = e^{\theta}/(1 + e^{\theta})^2$$

Thus

$$g(\theta) = \int_{-\infty}^{\theta} e^{t/3}(1 + e^t)^{-2/3}\,dt$$

$$= \int_{0}^{e^\theta/(1+e^\theta)} x^{-2/3}(1-x)^{-2/3}dx$$

or in terms of the proportion p,

$$\psi = g(p) = I_p(\tfrac{1}{3},\tfrac{1}{3})$$

where $I_x(p,q)$ is the incomplete beta function (Pearson, 1968). Anscombe gave a table of ψ as a function of p.

For the other exponential family distributions in GLIM, the transformation gives $\mu^{1/3}$ for the Poisson mean μ, and $\sigma^{-2/3}$ for the normal standard deviation. These transformations are useful only for single-parameter models: regression models are not usually fitted on these scales, as they have no obvious physical interpretation.

A1.3.4 Profile likelihood for predicted values

We give here the profile likelihood for a predicted value y_0 for a new value $\mathbf{x}_0$ from the model

$$y \sim N(\boldsymbol{\beta}'\mathbf{x},\sigma^2)$$

given observations $(y_i,\mathbf{x}_i)$, $i = 1,\ldots, n$. Write

$$X^* = \begin{bmatrix} X \\ \mathbf{x}_0' \end{bmatrix}, \quad \mathbf{y}^* = \begin{bmatrix} y \\ y_0 \end{bmatrix}$$

where X is the design matrix for the n observations.

Treating y_0 as an unknown parameter of the model in addition to $\boldsymbol{\beta}$ and σ, we have for given y_0 the MLEs of $\boldsymbol{\beta}$ and σ^2 as functions of y_0:

$$\hat{\boldsymbol{\beta}}(y_0) = (X^{*\prime}X^*)^{-1}X^{*\prime}\mathbf{y}^*$$

$$\hat{\sigma}^2(y_0) = \text{RSS}(y_0)/(n+1)$$

We substitute these functions of y_0 into the likelihood to give the profile likelihood for y_0:

$$PL(y_0) = \frac{1}{\{\hat{\sigma}(y_0)\}^{n+1}}\,e^{-(n+1)/2}$$

where

$$\text{RSS}(y_0) = \mathbf{y}^{*\prime}\mathbf{y}^* - \mathbf{y}^{*\prime}X^*(X^{*\prime}X^*)^{-1}X^{*\prime}\mathbf{y}^*$$

is the residual sum of squares for the $n + 1$ observations including y_0 and is also a function of y_0.

The profile likelihood is maximized when $y_0 = \hat{\boldsymbol{\beta}}'\mathbf{x}_0$, where $\hat{\boldsymbol{\beta}} = (X'X)^{-1}X'\mathbf{y}$ is the MLE of $\boldsymbol{\beta}$ from the observed data, and so the profile relative likelihood in y_0 is

$$\begin{aligned} \mathrm{PRL}(y_0) &= PL(y_0)/PL(\hat{\boldsymbol{\beta}}'\mathbf{x}_0) \\ &= \left\{\frac{\mathrm{RSS}(\hat{\boldsymbol{\beta}}'\mathbf{x}_0)}{\mathrm{RSS}(y_0)}\right\}^{(n+1)/2}. \end{aligned}$$

To express this simply we need to calculate RSS ($\check{y}_0$) as an explicit function of y_0.

Now

$$\begin{aligned} X^{*\prime}X^* &= X'X + \mathbf{x}_0\mathbf{x}_0{}' \\ X^{*\prime}\mathbf{y}^* &= X'\mathbf{y} + \mathbf{x}_0 y_0 \\ \mathbf{y}^{*\prime}\mathbf{y}^* &= \mathbf{y}'\mathbf{y} + y_0{}^2, \end{aligned}$$

and we define

$$\mathbf{x}_0{}'(X'X)^{-1}\mathbf{x}_0 = h_0,$$

while

$$\mathbf{y}'X(X'X)^{-1}\mathbf{x}_0 = \hat{\boldsymbol{\beta}}'\mathbf{x}_0.$$

Then

$$(X^{*\prime}X^*)^{-1} = (X'X)^{-1} - (X'X)^{-1}\mathbf{x}_0\mathbf{x}_0{}'(X'X)^{-1}/(1 + h_0)$$

and

$$\mathbf{y}^{*\prime}X^*(X^{*\prime}X^*)^{-1}X^{*\prime}\mathbf{y}^* =$$
$$(\mathbf{y}'X + y_0\mathbf{x}_0{}')[(X'X)^{-1} - (X'X)^{-1}\mathbf{x}_0\mathbf{x}_0{}'(X'X)^{-1}/(1 + h_0)](X'\mathbf{y} + \mathbf{x}_0 y_0).$$

Simplifying, the RHS becomes

$$\mathbf{y}'H\mathbf{y} - y_0{}^2 - (y_0 - \hat{\boldsymbol{\beta}}'\mathbf{x}_0)^2/(1 + h_0)$$

and so

$$\mathrm{RSS}(y_0) = \mathbf{y}'(I - H)\mathbf{y} + (y_0 - \hat{\boldsymbol{\beta}}'\mathbf{x}_0)^2/(1 + h_0)$$

where H is the projection matrix

$$H = X(X'X)^{-1}X.$$

Finally,

$$\begin{aligned} PRL(y_0) &= \{1 + (y_0 - \hat{\boldsymbol{\beta}}'\mathbf{x}_0)^2/n\hat{\sigma}^2(1 + h_0)\} - (n + 1)/2 \\ &= \{1 + t^2/(n - p - 1)\} - (n + 1)/2 \end{aligned}$$

where $\hat{\sigma}^2$ is the MLE of σ^2 from the observed data and t is the t-statistic given in Section 3.4. An interval of values of y_0 based on the profile likelihood will then be equivalent to a t-interval.

A1.3.9 Profile likelihood for the calibration problem

The profile likelihood for the calibration problem can be constructed as in Section A1.3.4 for predicted values. We use the same notation, but now the new y_0 is observed and the new $\mathbf{x}_0$ is unknown (apart from the constant 1). Treating $\mathbf{x}_0$ as an unknown parameter of the model in addition to $\boldsymbol{\beta}$ and σ, we have for given $\mathbf{x}_0$,

$$\hat{\boldsymbol{\beta}}(\mathbf{x}_0) = (X^{*\prime}X^*)^{-1}X^{*\prime}\mathbf{y}^*$$

$$\hat{\sigma}^2(\mathbf{x}_0) = \mathrm{RSS}(\mathbf{x}_0)/(n+1)$$

$$PL(\mathbf{x}_0) = \frac{1}{\{\hat{\sigma}(\mathbf{x}_0)\}^{n+1}} e^{-(n+1)/2}$$

where

$$\mathrm{RSS}(\mathbf{x}_0) = \mathbf{y}^{*\prime}\mathbf{y}^* - \mathbf{y}^{*\prime}X^*(X^{*\prime}X^*)^{-1}X^*\mathbf{y}^*$$

is the residual sum of squares for the $n+1$ observations including the observed y_0.

As in Section A1.3.4 the profile likelihood is maximized when $\mathbf{x}_0{}'\hat{\boldsymbol{\beta}} = y_0$ (which may not uniquely define $\mathbf{x}_0$), and so the profile relative likelihood in $\mathbf{x}_0$ is

$$PRL(\mathbf{x}_0) = \{1 + (y_0 - \mathbf{x}_0{}'\hat{\boldsymbol{\beta}})^2/n\hat{\sigma}^2(1+h_0)\}^{-(n+1)/2}$$

where

$$h_0 = \mathbf{x}_0{}'(X'X)^{-1}\mathbf{x}_0.$$

A region of values of $\mathbf{x}_0$ with relative likelihood exceeding γ is the solution of

$$\mathbf{x}_0{}'(\hat{\boldsymbol{\beta}}\hat{\boldsymbol{\beta}}' - c_1(X'X)^{-1})\mathbf{x}_0 - 2y_0\hat{\boldsymbol{\beta}}'\mathbf{x}_0 + y_0{}^2 - c_1 < 0$$

where

$$c_1 = n\hat{\sigma}^2\{\gamma^{-2/(n+1)} - 1\}.$$

For the case $\mathbf{x}' = (1\ x)$ this is equivalent to the inequality $Q(x) < 0$ in Section 3.9, though c_1 is defined differently.

A1.4.3 Comparison of deviances from different link functions

The comparison of different link functions by their deviances in Section 4.3 is not straightforward because the different link functions cannot be represented by different values of a single parameter, as was the case with the identity, log and reciprocal transformations embedded in the Box–Cox transformation family in Chapter 3. Where this is possible, as with linear or log-linear models for the Poisson mean, the same approach can be used.

For the logit, probit and complementary log–log links, common sense dictates that a much smaller deviance for one of these links than the others indicates support for this link; how large a difference is required may be assessed by the usual likelihood ratio test, allowing one extra parameter for each additional link function considered. Thus two link functions $g_1(p)$ and $g_2(p)$ may be compared using

$$\theta(\phi) = \{g_1(p)\}^{\phi}\{g_2(p)\}^{1-\phi}$$

where only $\phi = 0$ and $\phi = 1$ are of interest. For three link functions, we may write

$$\theta(\phi_1,\phi_2) = \{g_1(p)\}^{\phi_1}\{g_2(p)\}^{\phi_2}\{g_3(p)\}^{1-\phi_1-\phi_2}$$

where the only values of interest are $\phi_1 = 0$, $\phi_2 = 0$, or $\phi_1 = 1, \phi_2 = 0$, or $\phi_1 = 0$, $\phi_2 = 1$.

A1.6.3 Observed and expected information matrices for the exponential distribution

As noted in Section A1.2.14, the observed and expected information matrices are different except when the "canonical link" is used, that is the linear predictor is on the natural parameter scale. For the exponential distribution with a log–linear model, we have

$$f(t_i/\mu_i) = \frac{1}{\mu_i}e^{-t_i/\mu_i}$$

and

$$\log \mu_i = \boldsymbol{\beta}'\mathbf{x}_i.$$

The log likelihood is

$$\ell(\boldsymbol{\beta}) = \Sigma(-\log \mu_i - t_i/\mu_i)$$

and

$$\frac{\partial \ell}{\partial \boldsymbol{\beta}} = \Sigma(-1 + t_i/\mu_i)\mathbf{x}_i$$

$$\frac{\partial^2 \ell}{\partial \boldsymbol{\beta}\partial \boldsymbol{\beta}'} = -\Sigma(t_i/\mu_i)\mathbf{x}_i\mathbf{x}_i'$$

The Newton–Raphson algorithm uses weights $(t_i/\hat{\mu}_i)$, while the scoring algorithm uses weights $E(t_i/\mu_i) = 1$, the identity weight matrix.

Appendix 2
GLIM Directives

The directive symbol is the first significant character of the directive. Directives may be typed in full, as just the symbol plus three characters or they may be further abbreviated to the portion in upper case in the list below.

In a description of a directive, *int* specifies an integer value and *number* a value that may contain a sign and a decimal point. *scalar* must be a scalar identifier. *macro* is a macro identifier, *vec* is a vector, *item* is a vector, scalar or macro identifier. Items in [ ] are optional.

The effect of each directive and its syntax is summarized below. Note that the description and syntax has been necessarily simplified for some directives—the GLIM 3.77 Reference Guide (Baker, 1985), should be consulted for full details.

$ACcuracy [*int*]
    number of digits output

$ALias
    switch to include/exclude intrinsically aliased parameters

$Argument *macro item* [*item*]*s* where *item* is *vector* or *scalar* or *macro* or %*digit* or *
    define up to 9 arguments for *macro*

$ASsign *vec* = *item* [*items*]
    where *item* is *number* or *vector* or *scalar*.
    Assigns to *vec* the concatenated list of values

$CAlculate *expression*
    evaluate and optionally print value of *expression*

$Comment *string*
    introduces comment in GLIM session

$CYcle [*int1* [*int2* [*number1* [*number2*]]]]
    controls the algorithmic aspects of model fitting
    int1 controls number of cycles
    int2 controls frequency of printing
    number1 controls convergence criterion
    number2 controls criterion for detecting aliasing

$DAta [*int*] *vec*(*s*)
define names for subsequent $READ or $DINPUT

$DElete *item*(*s*)
where *item* is *vector* or *macro*
Delete *item* and recover space

$DINput *channel* [*int*]
read data from file of width *int*

$Display *letter(s)*
display results after fitting a model. *letter* options are:

- A All parameter estimates (including aliased parameters)
- C Correlations of parameter estimates
- D (Scaled) deviance and degrees of freedom
- E Parameter estimates and s.e.s.
- L The current linear predictor
- M The current model specification
- R *y*-variate, fitted values and residuals
- S Standard errors of differences of estimates
- T Working matrix
- U Parameter estimates and s.e.s. (excluding aliased parameters)
- V Variance–covariance matrix of parameter estimates listed in U
- W As R, but excluding units for which the vector %RE is zero

$DUmp [*channel*]
save current state of program to file

$ECHO
reverses state of echoing of input to output

$EDit [*int1* [*int2*]] *vec*(*s*) *numbers*(*s*)
replace values of elements *int1*. . .*int2* of *vecs* with *numbers*

$End
end of job. Clears user space. Does not reset psuedo-random numbers or alter the state of any files

$Endmac
end of macro definition

$ENV [*channel*] *letters*
gives information on the current state of the program.
letter options are:

- C channels
- D directory contents
- E $PASS facilities

G graphical facilities
I implementation
P program control stack
R seed values for standard random number generator
S internal space used by system
U data space usage

$ERror *letter* [*vec*]
declares the error or probability distribution.
letter options are:
B binomial (needs *vec*)
G gamma
N Normal
P Poisson

$EXit [*int*]
pop program control stack *int* levels

$EXTract *vec(s)*
obtain values of vectors %VC, %PE or %VL

$FActor [*length*] [*vec int*]*s*
declare the vectors of length *length* to be factors with *int* levels

$FINish
$return and end-of-file marker after subfiles. Causes file to be rewound

$Fit [*model formula*]
carry out fitting algorithm for model specified in *model formula*

$FOrmat [*FREE*] or [*Fortran format specifiers*]
$GRAph
implementation dependent

$GROup [*vec2*=]*vec1* [Intervals[*]*vec3*[*]] [Values *vec4*]
group values of a vector into a factor using *vec3* as the lower bounds of the domain intervals. * indicates $\pm\infty$

$Help
turn on/off extended diagnostic messages following a fault

$HIstogram[(*option list*)] *vector*[/*weight*]]*s* ['*string*'] [*factor*]]
prints (nested) histogram(s) using *weight* as weights and characters of *string* as plotting symbols

$Input *channel* [*width*] [*subfile*]*s*
read GLIM directives from file on channel *channel* with width *width*. Optionally read *subfile*(*s*) only from file

$LAyout
implementation dependent

$LInk *letter* [*number*]
declare link function.
letter options are:
C complementary log log
E exponent (needs *number*)
G logit
I identity
L log
P probit
R reciprocal
S square root

$Look [(*option-list*)] [*int1* [*int2*]] *item*(*s*)
displays values of *item*(*s*) in parallel.
item(*s*) are either *vectors* or *scalars*, but not a mixture of the two

$LSeed [*int1* [*int2* [*int3*]]]
set local pseudo-random number generator seeds—implementation dependent

$Macro *macro string* $E
store *string* as *macro*

$MANual
implementation dependent

$MAP [*vec2*=]*vec1* [Intervals[*]*vec3*[*]] [Values *vec4*]
as $GROUP, but output vector is a variate

$Offset [*vec*]
declare *vec* as known component in linear predictor.
If *vec* absent no offset is set

$OUtput [*channel* [*width* [*height*]]]
directs output to *channel*; if *channel*=0, switches off output

$OWn *macro1 macro2 macro3 macro4*
declare the user's own generalised linear model.
macro1 produce %FV from %LP
macro2 produce %DR
macro3 produce %VA
macro4 produce %DI

$PAGe
turns pagination on/off

$PASs *int* [*vec* [*macro*]]
 pass facility, implementation dependent

$PAuse
 temporary return to operating system; implementation dependent

$PLOT (*option-list*) *y-vec*[*s*] *x-vec* [*'string'* [*factor*]]
 y-vectors plotted against *x-vector* with plotting characters specified in *string*

$PRint *items*(*s*)
 prints *items* on current output channel. *item* is '*string*' or *vector* or *scalar* or *macro* or *phrase* or / or ;

$Read *number*(*s*)
 read values of vectors named in $DATA

$RECycle [*int1* [*int2* [*number1* [*number2*]
 as $CYCLE, but use values of vector %FV as starting values for fitting procedure

$REInput *channel* [*width*][*subfiles*]
 has the same effect as
 $REWIND *channel* $INPUT *channel* [*width*] [*subfiles*]

$REStore [*channel*]
 reads a dump produced by $DUMP from file on channel *channel*

$RETurn
 pop input channel stack by 1 level

$REWind [*channel*]
 go to start of file

$SCale [*number*]
 if *number* > 0, use as scale factor; else estimate scale

$SEt *mode*
 mode is batch or interactive.
 Defines the mode of execution of GLIM

$SKip [*int*]
 pop program control stack *int* levels unless in a $WHILE loop

$Sort *vec1* [*vec2* or *int2* [*vec3* or *int3*]]
 sort *vec2* into *vec1* based on *vec3*. Use *int2* for obtaining ranks and *int3* for circular lags

$SSeed[*int1* [*int2* [*int3*]]]
 set pseudo-random number generator seeds

$STop
    end of session

$SUbfile *name*
    denotes the beginning of the subfile *name*

$SUSpend
    temporary reversion to primary input

$SWitch *scalar macro*(*s*)
    $use *scalar*th macro

$TAbulate [*phrases*]
    production of tables

  Input *phrases*:
    The *vec stat* [*number*]
      stat is one of / Mean / Total / Variance / Deviation / Smallest / Largest / Fifty / Interpolate / Percentile
      I and P require *number*
    With *vec*
      vec is input weights
    For *vec*[; *vec*]*s*
      each *vec* specifies a dimension of the table

  Output *phrases*:
    Into *vec* or [ ]
      store or print ([ ]) output table
    Using *vec*
      vec is output weights
    By *vec*[; *vec*]s or *scalar*[; *scalar*]s
      store the output classification

$TPrint *vec1*[; *vec1*]s *vec2*[; *vec2*] or *int*[; *int*]s
    prints values of *vec1*(s) as a table classified by *vec2*(s) or *int*(s)

$Transcript [*letter*]*s*
    determines which output is sent to transcript file
    letter options are:
      I input
      O ordinary output
      W warnings
      F faults
      H help messages
      V verification of macros

$UNits *int*
    define standard length of vectors (%NU)

$Use *macro* [*item*]s
    invoke macro passing arguments

$Variate [*int*] *vec*(s)
    declare variate(s) of length *int*

$VErify
    turn off/on macro verification

$WArning
    turn off/on warnings

$Weight [*vec*]
    declare or unset prior weights

$WHile *scalar macro*
    use macro repeatedly, while *scalar* is non-zero

$Yvariate *vec*
    declare *vec* as *y*-variate for subsequent fitting

The Macro Library

$ECHO $INPUT %PLC INFO $
    lists the contents

Appendix 3
System defined structures in GLIM

Scalars

%A %B ... %Z	ordinary scalars
%ACC	current accuracy setting (see $ACC)
%CC	convergence tolerance in $CYCLE/$RECYCLE
%CIC	the current input channel number
%CIL	the current input channel length
%COC	the current output channel number
%COH	the current output channel height
%COL	the current output channel length
%CYC	max. no. of cycles set in $CYCLE/$RECYCLE
%DF	degrees of freedom after each cycle of a fit
%DV	scaled deviance after each cycle of a fit
%ERR	indicator of current error setting (1 = Normal, 2 = Poisson, 3 = binomial, 4 = gamma, 9 = own model)
%JN	job number. Incremented by $END
%LIN	link function indicator (1 = identify, 2 = log, 3 = logit, 4 = reciprocal, 5 = square root, 6 = probit, 7 = comp–log–log, 8 = exponent, 9 = own model)
%ML	number of elements in (co)variance matrix of parameters. Length of vector %VC
%NU	standard length of vectors
%PDC	the primary dump channel number
%PI	π to machine accuracy
%PIC	the primary input channel number
%PIL	the primary input channel length
%PL	number of non-intrinsically aliased parameters. Length of vector %PE
%POC	the primary output channel number
%POH	the primary output channel height
%POL	the primary output channel length
%PRT	the printing frequency during iterative fits
%S1 . %S2 . %S3	current seeds for the standard pseudo-random-number generator

%SC	scale or mean deviance
%TOL	aliasing tolerance in $CYCLE/$RECYCLE
%TRA	amount of transcription
%X2	generalised Pearson's χ^2 after each cycle
%Z1 ... %Z9	local scalars in library macros

Indicators

The following system scalars have the value 0 if the corresponding item is not set, 1 otherwise.

%A1 ... %A9	in a macro, if the *i*th argument is set
%BDF	binomial denominator
%ECH	echo input
%HEL	extended error messages
%IM	interactive mode (0 if batch)
%OSF	offset
%PAG	pagination
%PWF	prior weight
%VER	echo directives in macros as they are executed
%WAR	warnings output
%YVF	*y*-variate

Vectors (Length given in brackets.)

%BD	binomial denominator (%NU)
%DI	deviance increment (%NU)
%DR	derivative $\partial(\eta)/\partial(\mu)$ (%NU)
%FV	fitted values (%NU)
%GM	grand mean (%NU)
%LP	linear predictors (%NU)
%OS	offset (%NU)
%PE	non-intrinsically aliased parameter estimates (%PL)
%PW	prior weights (%NU)
%RE	elements corresponding to zero elements are not PLOTted or DISPLAYed (%NU)
%VA	variance function values (%NU)
%VC	non-intrinsically aliased parameters (co)variance matrix (%ML)
%VL	variances of linear predictors (%NU)
%WT	iterative weights (%NU)
%WV	working vector for iterative models (%NU)
%YV	dependent variate values (%NU)

Functions

x, y, z are variates or scalars; k, n integer valued scalars.

%ANG(x)	arcsin(sqrt x)
%CU(x)	cumulative sums of x
%EQ(x,y)	1 if $x=y$, 0 otherwise
%EXP(x)	e^x
%GE(x,y)	1 if $x \geq y$, 0 otherwise
%GL(k,n)	generate factor levels 1 to k, blocks of n
%GT(x,y)	1 if $x > y$, 0 otherwise
%IF(x,y,z)	if x is non-zero then y, otherwise z
%LE(x,y)	1 if $x \leq y$, 0 otherwise
%LOG(x)	$\ln_e(x)$
%LR(n)	local pseudo-random integer on $[0,n]$
%LR(0)	local pseudo-random real on [0,1]
%LT(x,y)	1 if $x < y$, 0 otherwise
%ND(x)	normal deviate, inverse of %NP. $0 < x < 1$
%NE(x,y)	1 if $x \neq y$, 0 otherwise
%NP(x)	normal probability integral. $-\infty$ to x
%SIN(x)	sin x
%SR(0)	pseudo-random real on [0,1]
%SR(n)	pseudo-random integer on $[0,n]$
%SQRT(x)	square root
%TR(x)	integer x, truncated towards 0

Operators (used in $CALCULATE directive)

x and y are vectors, scalars, numbers, functions, suffixed vectors or expressions in matching brackets

Operator	Result
$x+y$	the sum $x+y$
$x-y$	the difference $x-y$
$x*y$	the product xy
x/y	the quotient x/y
$x**y$	x raised to the power y
$x>y$	1 if $x>y$, 0 otherwise
$x>=y$	1 if $x \geq y$, 0 otherwise
$x==y$	1 if $x=y$, 0 otherwise
$x<y$	1 if $x<y$, 0 otherwise
$x<=y$	1 if $x \leq y$, 0 otherwise
$x/=y$	1 if $x \neq y$, 0 otherwise
x & y	1 if x and y non zero, 0 otherwise
x ? y	1 if x or y non zero, 0 otherwise
$x=y$	y

Operator precedence:
highest: functions, monadic operators, brackets
**
* dyadic /
dyadic+ dyadic−
$<$ $<=$ $==$ $/=$ $>=$ $>$
&
?
lowest: =

All values are real numbers, in boolean expressions 1.0 indicates TRUE 0.0 indicates FALSE.

Appendix 4
Datasets and macros

This appendix contains a complete listing of the datasets and macros used in this book (excluding the macros found in the macro library). The contents of this appendix is available in machine-readable form from

GLIM Development Centre,
Centre for Applied Statistics,
Fylde College,
University of Lancaster,
Lancester LA1 4YF
U.K.

at a price of £8·00 for an IBM/PC floppy disk.

The price includes postage and packing.

```
!
!***********************************************************************
!
! GLIM DATA SETS AND MACROS FOR THE BOOK 'STATISTICAL MODELLING IN GLIM'
!
!***********************************************************************
$RETURN
!
$SUBFILE SOLV
!
! DATA FROM SCHOOL OF BEHAVIOURAL SCIENCES,MACQUARIE UNIVERSITY
!
$UNITS 24 $DATA TIME EFT$READ
317 59 464 33 525 49 298 69 491 65 196 26
268 29 372 62 370 31 739 139 430 74 410 31
342 48 222 23 219 9 513 128 295 44 285 49
408 87 543 43 298 55 494 58 317 113 407 7
!
$CALC GROUP=%GL(2,12)
$FACTOR GROUP 2$
!
!
! TIME - IS THE TIME TAKEN TO SOLVE FOUR BLOCK DESIGN PROBLEMS
!        BY 24 FIFTH-GRADE CHILDREN (12 BOYS + 12 GIRLS)
! EFT  - VALUE FOR THE EMBEDDED FIGURES TEST,MEASURE OF DIFFICULTY
!        IN ABSTRACTING LOGICAL STRUCTURE OF A PROBLEM FROM
!        ITS CONTEXT
! GROUP- CLASSIFICATION BY TYPE OF PROBLEMS PRESENTED FIRST
!        I.E. THOSE SOLVED BY ROW (GROUP 1) OR
!        FORMATION STRATEGY
!
!
$RETURN
!
!***********************************************************************
!
$SUBFILE BRONCHITIS
!
!    CARDIFF CHRONIC BRONCHITIS DATA
!    REPRODUCED FROM
!      'INTRODUCTION TO THE USE OF LOGIT MODELS IN GEOGRAPHY'
!       BY NEIL WRIGLEY (1976)
!
$UNITS 212
$DATA R CIG POLL$
$READ
0  5.15  67.1   1  0.0   66.9    0  2.5   66.7    0  1.75  65.8
0  6.75  64.4   0  0.0   64.4    1  0.0   65.1    1  9.5   66.2
0  0.0   65.9   0  0.75  67.1    0  5.25  67.9    1  8     68.1
1  5.15  67     1 30.0   66.3    0  0.0   65.7    0  0.0   65.2
0  5.25  64.2   0 10.05  64.6    0  0.0   63.5    1  3.4   63.0
0  0.0   62.7   0   .55  62.7    1  9.5   62.1    1 12.5   63.7
0  0.0   63.1   0  3.4   63      0  2.2   62.7    0  6.7   63.1
0  1.1   62.4   0  1.8   64.4    0  0.0   64.2    1  3.6   64.2
0  1.6   63     0  6.2   62.2    0 14.75  62.3    0   .35  63.7
1 13.75  63.8   0  0.0   63.1    1  7.5   62.7    0  1.0   62.9
0  0.0   62.5   1 14.8   61.7    1  3.5   61.6    0  0.0   61.6
0  0.0   61.4   0   .25  61.4    0  1.55  62      1  0.0   61.8
0  0.0   60.9   0  5.9   60.8    0 16.45  60.6    0  2.65  62.9
1 12.5   62.6   0  0.0   62.1    0 14.55  61.7    1 11     61
1  6.75  62.7   0  0.0   62.7    1  0.0   61.7    0  1.75  60.9
0  2.4   60.6   0 10.05  60.4    1 12.75  61.7    0  0.0   61.9
0  5     61.3   0   .6   60.7    0  0.0   60.8    0   .85  60.5
0   .9   59.7   0   .0   59.5    1  8.75  59.6    0   .8   59.1
1  6.6   59.4   0  1.0   58.5    0  0.0   60      1  8.15  59.8
0  0.0   59.7   1  5     59.4    0  2.55  59.2    0  1.2   58.6
```

```
0  0.0   60.8   1 11.25  60.4     0  0.0  60.2     0  2     60
0  1.9   59.4   0    .45 59.8     1  0.0  59.7     0  0.0   59.0
1  6.9   59.0   0  2.35  58.6     0  3.95 59.7     0   .6   59.6
1 15     59.4   0  0.0   59.4     0   .95 59.4     0  0.0   59.3
0  1.4   54.2   0   .5   54.0     0   .6  53.8     0  0.0   53.7
0  2.45  53.7   0  1.75  53.1     0  0.0  54.4     0  3.1   54.2
0 10.05  53.9   0   .55  53.2     0   .85 53.2     0  1.1   54.9
0  0.0   54.9   0  0.0   54.5     0  1.45 54.2     0  2.05  54.2
1  10.5  54     0   .5   55.8     1  9.2  55.5     0   .55  55.6
0  0.0   55.5   0  0.96  54.9     0  1    54.6     0  0.0   56.9
0  5.25  56.4   1  0     55.9     0  9    55.8     0  1.6   55.6
1  10.9  57.6   0  0.0   57.7     0  0    57.6     0  2.25  57.8
0  2.65  57.8   0   .55  58.4     0  0    58.2     1  4.5   58
0 15     58.1   0  0     57.9     0  0    57.3     0  4.2   58.3
0   .55  58.1   1 10     57.9     0  0    57.6     0  7.1   57.3
0  3.2   57.1   1  0     58.9     1  6.8  58.6     0  0     58.7
0  0     57.5   0  2.35  57.2     0 24.9  58       0  2.65  57.9
1  3.7   57.2   0 17.1   57.3     0  0    57.5     0   .95  57.2
0 10.05  53.1   0  1.15  53       1 18.25 53       0 10     52.9
0   .75  52.6   0  0     53.1     0  4.2  53       0   .8   52.9
0   .55  52.7   0   .95  52.6     0  0    52.1     0  3.1   54.1
0   .8   53.7   0  1.55  53.1     0   .4  53.3     0  6.2   53
0   .6   53     0   .4   53.9     1  7.5  53.7     0  7.15  53.4
0   .25  53.2   0  3.6   53.4     0   .95 53.2     0  2.8   54.9
1 20.25  54.9   0   .95  54.6     0  4.25 54.1     0  4.15  54.2
0 10     57.4   0  3.4   57.3     0  0.0  57.3     0  3.6   56.7
0   .9   56.5   0  0.0   56.8     0  0    56.6     1  6.4   56.5
0   .95  56.3   0  1.06  56.3     0 13.3  56.2     0  1.1   56.6
0  17.2  55.9   0  1.65  56       1  5    55.8     0  2.1   55.7
0    .6  57     1  8.25  56.7     0   .9  56.4     0  0.0   56.5
1  12.3  55.2   0  1.15  56.9     0  2.2  56.7     0  3.6   56
1  10    55.5   0  0.6   55.3     0  9.5  56.5     0   .7   56.3
1  9     56.1   0  0     55.9     0   .5  55.5     0   .9   55.4
$CALC N=1
!     VARIABLES
!      R          CHRONIC BRONCHITIS (1=YES,0=NO)
!      N         (BINOMIAL DENOMINATOR..1 FOR ALL RESPONDENTS)
!      CIG        CIGARETTE CONSUMPTION
!      POLL       SMOKE LEVEL OF LOCALITY OF RESPONDENTS HOME
$RETURN
!
!**********************************************************************
!
$SUBFILE GHQ
!
!..................................................................
! Reproduced by permission of the Royal Statistical Society
! Silvapulle, M J (1981), JRSS B, 43, 310-313
!..................................................................
$UNIT 17
$DATA GHQ C NC$
$READ
0  0  18     1  0   8     2  1   1     4  1   0
5  3   0     7  2   0    10  1   0     0  2  42
1  2  14     2  4   5     3  3   1     4  2   1
5  3   0     6  1   0     7  1   0     8  3   0
9  1   0
$CALC I=%GL(17,1) : SEX=%IF(%LE(I,7),1,2) $FACTOR SEX 2$
$CALC N=C+NC$
$RETURN
!
!**********************************************************************
!
$SUBFILE VASO
!..................................................................
! Reproduced by permission of the Biometrika Trustees
```

```
! Finney, D.J. (1947) Biometrika,34,320-334
!.................................................................
!
$UNIT 39$DATA VOL RATE RESP
$READ
3.7   .825 1    3.5  1.09  1    1.25 2.5    1     .75 1.5  1
 .8  3.2   1     .7  3.5   1     .6   .75   0    1.1  1.7  0
.9    .75  0     .9   .45  0     .8   .57   0     .55 2.75 0
.6   3.0   0    1.4  2.33  1     .75 3.75   1    2.3  1.64 1
3.2  1.6   1     .85 1.415 1    1.7  1.06   0    1.8  1.8  1
 .4  2.0   0     .95 1.36  0    1.35 1.35   0    1.5  1.36 0
1.6  1.78  1     .6  1.5   0    1.8  1.5    1     .95 1.9  0
1.9   .95  1    1.6   .4   0    2.7   .75   1    2.35  .03 0
1.1  1.83  0    1.1  2.2   1    1.2  2.0    1    0.8  3.33 1
 .95 1.9   0     .75 1.9   0    1.3  1.625  1
!
!       VARIABLES:
!         VOL          VOLUME OF AIR INSPIRED
!        RATE          RATE OF AIR INSPIRED
!        RESP          RESPONSE (1= OCCURENCE OF VASO-CONSTRICTION
!                                2=NON-OCCURENCE OF  "      ""  )
$RETURN
!
!*******************************************************************
!
$SUBFILE TREES
!
!          USABLE WOOD IN CHERRY TREES.
!.................................................................
! Reproduced by permission of the Duxbury Press
! Ryan, T., Joiner, B. and Ryan B. (1976)
! Minitab Student Handbook
!.................................................................
!
$UNIT 31
$DATA D H V
$READ
 8.3 70 10.3     8.6 65 10.3     8.8 63 10.2    10.5 72 16.4
10.7 81 18.8    10.8 83 19.7    11.0 66 15.6    11.0 75 18.2
11.1 80 22.6    11.2 75 19.9    11.3 79 24.2    11.4 76 21.0
11.4 76 21.4    11.7 69 21.3    12.0 75 19.1    12.9 74 22.2
12.9 85 33.8    13.3 86 27.4    13.7 71 25.7    13.8 64 24.9
14.0 78 34.5    14.2 80 31.7    14.5 74 36.3    16.0 72 38.3
16.3 77 42.6    17.3 81 55.4    17.5 82 55.7    17.9 80 58.3
18.0 80 51.5    18.0 80 51.0    20.6 87 77.0
!
!      VARIABLES:
!           D-    DIAMETER (IN INCHES) OF 31 CHERRY TREES
!                  AT A HEIGHT OF 4.5 FEET FROM THE GROUND
!
!           H-    HEIGHT IN FEET OF THE TREES
!
!           V-    VOLUME OF USEABLE WOOD IN CUBIC FEET
!
$RETURN
!
!*******************************************************************
!
$SUBFILE POIS
!
!  SURVIVAL TIMES OF ANIMALS AFTER POISONING TREATMENT
!.................................................................
! Reproduced by permission of the Royal Statistical Society
! Box, G.E.P and Cox, D.R. (1964), JRSS B, 26, 211-252
!.................................................................
!
```

```
$UNITS 48
$DATA TIME TYPE TREAT
$FACTOR TYPE 3 TREAT 4
$READ
0.31  1 1    0.45  1 1    0.46  1 1    0.43  1 1
0.82  1 2    1.1   1 2    0.88  1 2    0.72  1 2
0.43  1 3    0.45  1 3    0.63  1 3    0.76  1 3
0.45  1 4    0.71  1 4    0.66  1 4    0.62  1 4
0.36  2 1    0.29  2 1    0.4.  2 1    0.23  2 1
0.92  2 2    0.61  2 2    0.49  2 2    1.24  2 2
0.44  2 3    0.35  2 3    0.31  2 3    0.4   2 3
0.56  2 4    1.02  2 4    0.71  2 4    0.38  2 4
0.22  3 1    0.21  3 1    0.18  3 1    0.23  3 1
0.3   3 2    0.37  3 2    0.38  3 2    0.29  3 2
0.23  3 3    0.25  3 3    0.24  3 3    0.22  3 3
0.3   3 4    0.36  3 4    0.31  3 4    0.33  3 4
!    VARIABLES:
!         TIME    SURVIVAL TIME OF RATS
!         TREAT   TREATMENT (1-4)
!         TYPE    TYPE OF POISON
$RETURN
!
!*********************************************************************
!
$SUBFILE CAR1
!           PETROL CONSUMPTION DATA
!..................................................................
! Reproduced from:
!  M.V.Henderson and P.F. Velleman, "Building Multiple Regression Models
!  Interactively". BIOMETRICS 37: 391-411. 1981
!  With permission from the Biometric Society.
!..................................................................
!
$UNITS 32$
$DATA S C T G  DISP HP CB DRAT WT QMT   MPG $
$READ
0 6 1 4  160.0 110 4 3.90 2620 16.46   21.0      !MAZDA RX-4
0 6 1 4  160.0 110 4 3.90 2875 17.02   21.0      !MAZDA RX-4 WAGON
1 4 1 4  108.0  93 1 3.85 2320 18.61   22.8      !DATSUN 710
1 4 0 3  258.0 110 1 3.08 3215 19.44   21.4      !HORNET 4 DRIVE
0 8 0 3  360.0 175 2 3.15 3440 17.02   18.7      !HORNET SPORTABOUT
1 6 0 3  225.0 105 1 2.76 3460 20.22   18.1      !VALIANT
0 8 0 3  360.0 245 4 3.21 3570 15.84   14.3      !DUSTER 360
1 4 0 4  146.7  62 2 3.69 3190 20.00   24.4      !MERCEDES 240D
1 4 0 4  140.8  95 2 3.92 3150 22.90   22.8      !MERCEDES 230
1 6 0 4  167.6 123 4 3.92 3440 18.30   19.2      !MERCEDES 280
1 6 0 4  167.6 123 4 3.92 3440 18.90   17.8      !MERCEDES 280C
0 8 0 3  275.8 180 3 3.07 4070 17.40   16.4      !MERCEDES 450SE
0 8 0 3  275.8 180 3 3.07 3730 17.60   17.3      !MERCEDES 450SL
0 8 0 3  275.8 180 3 3.07 3780 18.00   15.2      !MERCEDES 450SLC
0 8 0 3  472.0 205 4 2.93 5250 17.98   10.4      !CADILLAC FLEETWOOD
0 8 0 3  460.0 215 4 3.00 5425 17.82   10.4      !LINCOLN CONTINENTAL
0 8 0 3  440.0 230 4 3.23 5345 17.42   14.7      !IMPERIAL
1 4 1 4   78.7  66 1 4.08 2200 19.47   32.4      !FIAT 128
1 4 1 4   75.7  52 2 4.93 1615 18.52   30.4      !HONDA CIVIC
1 4 1 4   71.1  65 1 4.22 1835 19.90   33.9      !TOYOTA COROLLA
1 4 0 3  120.1  97 1 3.70 2465 20.01   21.5      !TOYOTA CORONA
0 8 0 3  318.0 150 2 2.76 3520 16.87   15.5      !DODGE CHALLENGER
0 8 0 3  304.0 150 2 3.15 3435 17.30   15.2      !AMC JAVELIN
0 8 0 3  350.0 245 4 3.73 3840 15.41   13.3      !CHEVROLET CAMARO Z-28
0 8 0 3  400.0 175 2 3.08 3845 17.05   19.2      !PONTIAC FIREBIRD
1 4 1 4   79.0  66 1 4.08 1935 18.90   27.3      !FIAT X1-9
0 4 1 5  120.3  91 2 4.43 2140 16.70   26.0      !PORSCHE 914-2
1 4 1 5   95.1 113 2 3.77 1513 16.90   30.4      !LOTUS EUROPA
0 8 1 5  351.0 264 4 4.22 3170 14.50   15.8      !FORD PANTERA L
0 6 1 5  145.0 175 6 3.62 2770 15.50   19.7      !FERRARI DINO 1973
```

```
0 8 1 5  301.0 335 8 3.54 3570 14.60   15.0        !MASERATI BORA
1 4 1 4  121.0 109 2 4.11 2780 18.60   21.4        !VOLVO 142E
$CA WT=WT/1000
!  S      SHAPE OF ENGINE   (1 = STRAIGHT,0 = VEE)
!  C      NO. OF CYLINDERS
!  T      TRANSMISSION TYPE (0= AUTOMATIC,1=MANUAL)
!  G       NO. OF GEARS
!  DISP   ENGINE DISPLACEMENT IN CUBIC INCHES
!  HP     HORSEPOWER OF CAR
!  CB     NO. OF CARBURETTOR BARRELS
!  DRAT   DRIVE RATIO
!  WT     WEIGHT OF CAR/1000
!  QMT    QUARTER-MILE TIME
!  MPG    PETROL CONSUMPTION IN MILES PER GALLON
$RETURN
!
!*********************************************************************
!
$SUBFILE KULLBACK
!
!  CORONARY HEART DISEASE STUDY
!.................................................................
! Reproduced by permission of the American Statistical Association
! from Ku,H.H. and Kullback,S. ,American Statistician,1974,28,p117
!.................................................................
!
!
$UNITS 16
$DATA R N$
$READ
2 119 3 124 3 50 4 26  3 88 2 100 0 43 3 23
8 127 11 220 6 74 6 49 7 74 12 111 11 57 11 44
$FACTOR CHOL 4 BP 4
$CA CHOL=%GL(4,4) : BP=%GL(4,1)
! VARIABLES:
!            N     NO. OF SUBJECTS IN EACH GROUP
!            R     NO. OF SUBJECTS DIAGNOSED AS HAVING CORONARY HEART
!                           DISEASE
!           CHOL   SERUM CHOLESTEROL IN MG/100CC
!                   1= <200, 2= 200-219,3=220-259,4=260+
!           BP     BLOOD PRESSURE IN MM MERCURY
!                   1= <127, 2=127-146,3=147-166,4=167+
$RETURN
!
!*********************************************************************
!
$SUBFILE BYSSINOSIS
!   BYSSINOSIS DATA ....
!   NO. OF COTTON WORKERS SUFFERING FROM BYSSINOSIS
!.................................................................
! Reproduced by permission of the International Statistical Inst.
! Higgins,J.E. and Koch,G.G.,International Statist. Review,45,p51-62
!.................................................................
!
$UNITS 72
$DATA YES NO$
$READ
3     37    0     74    2     258    25    139    0    88    3    242
0     5     1     93    3     180    2     22     2    145   3    260
0     16    0     35    0     134    6     75     1    47    1    122
0     4     1     54    2     169    1     24     3    142   4    301
8     21    1     50    1     187    8     30     0     5    0     33
0     0     1     33    2     94     0     0      0    4     0    3
2     8     1     16    0     58     1     9      0    0     0    7
0     0     0     30    1     90     0     0      0    4     0    4
31    77    1     141   12    495    10    31     0    1     0    45
```

```
0    1    3    91    3    176    0    1    0    0    0    2
5    47   0    39    3    182    3    15   0    1    0    23
0    2    3    187   2    340    0    0    0    2    0    3
$FACT DUST 3 RACE 2 SEX 2 SMOK 2 EMP 3$
$CA DUST=%GL(3,1):RACE=%GL(2,3):SEX=%GL(2,6):SMOK=%GL(2,12)
:EMP=%GL(3,24)$
!
!   VARIABLES:
!        YES    NO. OF COTTON WORKERS SUFFERING FROM BYSSINOSIS
!         NO    NO. OF COTTON WORKERS NOT SUFFERING FROM BYSSINOSIS
!       DUST    DUSTINESS OF WORKPLACE(1=HIGH,2=MEDIUM,3=LOW)
!       RACE    RACE OF WORKER(1=WHITE,2=OTHER)
!        SEX    SEX OF WORKER(1=MALE,2=FEMALE)
!       SMOK   SMOKING HABIT(1=SMOKER,2=NON-SMOKER)
!        EMP    LENGTH OF EMPLOYMENT IN YEARS(1= <10,2=10-20,3= >20YRS)
!
$RETURN
!
!*********************************************************************
!
$SUBFILE VIETNAM
!
! SURVEY OF STUDENT OPINION ON THE VIETNAM WAR, UNIVERSITY OF
! NORTH CAROLINA AT CHAPEL HILL, MAY 1967
! POLICIES :
!            A- DEFEAT POWER OF VIETNAM BY WIDESPREAD BOMBING
!               AND LAND INVASION
!            B- FOLLOW THE PRESENT POLICY
!
!            C- WITHDRAW TROOPS TO STRONG POINTS AND OPEN
!               NEGOTIATIONS ON ELECTIONS INVOLVING THE VIETCONG
!            D- IMMEDIATE WITHDRAWAL OF ALL U.S. TROOPS
!
! EXPLANATORY VARIABLES : SEX=1 MALE
!                             =2 FEMALE
!                        YEAR=1,2,3,4 UNDERGRADUATE
!                             =5 GRADUATE
!
$UNIT 10          !
$DATA A B C D     !
$READ             !
175 116 131  17   !
160 126 135  21   !
132 120 154  29   !
145  95 185  44   !
118 176 345 141   !
!
 13  19  40   5   !
  5   9  33   3   !
 22  29 110   6   !
 12  21  58  10   !
 19  27 128  13   !
!
$FACTOR SEX 2 YEAR 5 $
$CALC SEX=%GL(2,5) : YEAR=%GL(5,1) $
$RETURN
!
!*********************************************************************
!
$SUBFILE MINERS
!
! DATA ON PNEUMOCONIOSIS FOR COALMINERS
!...............................................................
! Reproduced from
! J.A.Ashford, "An approach to the Analysis of Data for Semi-quantal
! Responses in Biological Response". BIOMETRICS 15: 573-581 1959
```

```
! With permission from the Biometric Society.
!..................................................................
!
! N=NORMAL, M=MILD, S=SEVERE PNEUMOCONIOSIS
!
$UNIT 8
$DATA N M S
$READ
98  0  0    51  2  1    34  6  3    35  5  8
32 10  9    23  7  8    12  6 10     4  2  5
$FACTOR P 8 $CALC P=%GL(8,1) :
YEARS=5.8*%EQ(P,1)+15*%EQ(P,2)+21.5*%EQ(P,3)+27.5*%EQ(P,4)+
     33.5*%EQ(P,5)+39.5*%EQ(P,6)+46*%EQ(P,7)+51.5*%EQ(P,8) $
$RETURN
!
!************************************************************************
!
$SUBFILE TOXAEMIA
!..................................................................
! Reproduced by permission of the Royal Statistical Society
! Brown,P.J.,Stone,J. and Ord-Smith,C(1983),Appl. Statist.,32,69-72
!..................................................................
!
$UNITS 15
$DATA HU HN NU NN$READ
  28    82   21   286      5    24    5    71      1    3    0    13
  50   266   34   785     13    92   17   284      0   15    3    34
 278  1101  164  3160    120   492  142  2300     16   92   32   383
  63   213   52   656     35   129   46   649      7   40   12   163
  20    78   23   245     22    74   34   321      7   14    4    65
$FAC CLASS 5 SMOK 3$
$CAL CLASS=%GL(5,3):SMOK=%GL(3,1)$
!
!  Variables:
!
!       SMOK    Number of cigarettes smoked per day by mother during
!               pregnancy
!       CLASS   Social class of mother (I to V)
!
!  Toxaemic signs exhibited by mother during pregnancy
!
!        HU     Hypertension and protein urea
!        NU     protein urea only
!        HN     Hypertension only
!        NN     neither sign exhibited
!
$RETURN
!
!************************************************************************
!
$SUBFILE CLAIMS
!
! CLAIM FREQUENCY DATA FROM BAXTER ET AL (1980)
!  Proceedings of the 21st International Congress of Actuaries,11-29
!
$UNIT 64
$DATA N C $
$READ
197 38   264 35   246 20   1680 156
284 63   536 84   696 89   3582 400
133 19   286 52   355 74   1640 233
 24  4    71 18    99 19    452  77
 85 22   139 19   151 22    931  87
149 25   313 51   419 49   2443 290
 66 14   175 46   221 39   1110 143
  9  4    48 15    72 12    322  53
```

```
 35  5    73 11    89 10    648  67
 53 10   155 24   240 37   1635 187
 24  8    78 19   121 24    692 101
  7  3    29  2    43  8    245  37
 20  2    33  5    40  4    316  36
 31  7    81 10   122 22    724 102
 18  5    39  7    68 16    344  63
  3  0    16  6    25  8    114  33
$FACTOR DISTRICT 4 CAR 4 AGE 4 $
$CA DIST=%GL(4,16) : CAR=%GL(4,4) : AGE=%GL(4,1) $
!
! AGE IS POLICYHOLDER'S AGE
!
$RETURN
!
!***********************************************************************
!
$SUBFILE QUINE
!
$UNITS 146
$DATA DAYS C S A L $
$FACT C 2 S 2 A 4 L 2$
$READ
  2 1 1 1 1    11 1 1 1 1    14 1 1 1 1     5 1 1 1 2
  5 1 1 1 2    13 1 1 1 2    20 1 1 1 2    22 1 1 1 2
  6 1 1 2 1     6 1 1 2 1    15 1 1 2 1     7 1 1 2 2
 14 1 1 2 2     6 1 1 3 1    32 1 1 3 1    53 1 1 3 1
 57 1 1 3 1    14 1 1 3 2    16 1 1 3 2    16 1 1 3 2
 17 1 1 3 2    40 1 1 3 2    43 1 1 3 2    46 1 1 3 2
  8 1 1 4 2    23 1 1 4 2    23 1 1 4 2    28 1 1 4 2
 34 1 1 4 2    36 1 1 4 2    38 1 1 4 2     3 1 2 1 1
  5 1 2 1 2    11 1 2 1 2    24 1 2 1 2    45 1 2 1 2
  5 1 2 2 1     6 1 2 2 1     6 1 2 2 1     9 1 2 2 1
 13 1 2 2 1    23 1 2 2 1    25 1 2 2 1    32 1 2 2 1
 53 1 2 2 1    54 1 2 2 1     5 1 2 2 2     5 1 2 2 2
 11 1 2 2 2    17 1 2 2 2    19 1 2 2 2     8 1 2 3 1
 13 1 2 3 1    14 1 2 3 1    20 1 2 3 1    47 1 2 3 1
 48 1 2 3 1    60 1 2 3 1    81 1 2 3 1     2 1 2 3 2
  0 1 2 4 2     2 1 2 4 2     3 1 2 4 2     5 1 2 4 2
 10 1 2 4 2    14 1 2 4 2    21 1 2 4 2    36 1 2 4 2
 40 1 2 4 2     6 2 1 1 1    17 2 1 1 1    67 2 1 1 1
  0 2 1 1 2     0 2 1 1 2     2 2 1 1 2     7 2 1 1 2
 11 2 1 1 2    12 2 1 1 2     0 2 1 2 1     0 2 1 2 1
  5 2 1 2 1     5 2 1 2 1     5 2 1 2 1    11 2 1 2 1
 17 2 1 2 1     3 2 1 2 2     4 2 1 2 2    22 2 1 3 1
 30 2 1 3 1    36 2 1 3 1     0 2 1 3 2     1 2 1 3 2
  5 2 1 3 2     7 2 1 3 2     8 2 1 3 2    16 2 1 3 2
 27 2 1 3 2     0 2 1 4 2    10 2 1 4 2    14 2 1 4 2
 27 2 1 4 2    30 2 1 4 2    41 2 1 4 2    69 2 1 4 2
 25 2 2 1 1    10 2 2 1 2    11 2 2 1 2    20 2 2 1 2
 33 2 2 1 2     0 2 2 2 1     1 2 2 2 1     5 2 2 2 1
  5 2 2 2 1     5 2 2 2 1     5 2 2 2 1     5 2 2 2 1
  7 2 2 2 1     7 2 2 2 1    11 2 2 2 1    15 2 2 2 1
  5 2 2 2 2     6 2 2 2 2     6 2 2 2 2     7 2 2 2 2
 14 2 2 2 2    28 2 2 2 2     0 2 2 3 1     2 2 2 3 1
  2 2 2 3 1     3 2 2 3 1     5 2 2 3 1     8 2 2 3 1
 10 2 2 3 1    12 2 2 3 1    14 2 2 3 1     1 2 2 3 2
  1 2 2 4 2     3 2 2 4 2     3 2 2 4 2     5 2 2 4 2
  9 2 2 4 2    15 2 2 4 2    18 2 2 4 2    22 2 2 4 2
 22 2 2 4 2    37 2 2 4 2
$RETURN
!
!***********************************************************************
!
$SUBFILE STAN
!
```

```
!    STANFORD HEART TRANSPLANT DATA.. AITKIN,LAIRD & FRANCIS(1983)
!...............................................................
! Reproduced by permission of the American Statistical Association
! Crowley, J and Hu,M. (1977)
! Journal of the American Statistical Association, 72, 27-36
!...............................................................
!
$UNIT 65 $DATA ID ZA ZB AGE SURG ACC DIED SURV NMM HLA MM REJ
$READ
  3    0    1    54    0    0    1    15    2    0    1.11    0
  4    0    1    40    0   35    1     3    3    0    1.66    0
  7    0    1    50    0   50    1   624    4    0    1.32    1
 10    0    1    42    0   11    1    46    2    0     .61    1
 11    0    1    47    0   25    1   127    1    0     .36    0
 13    0    1    54    0   16    1    61    3    0    1.89    1
 14    0    1    54    0   36    1  1350    1    0     .87    1
 16    0    1    49    0   27    1   312    2    0    1.12    1
 18    0    1    56    0   19    1    24    3    0    2.05    0
 20    0    1    55    0   17    1    10    3    1    2.76    1
 21    0    1    43    0    7    1  1024    2    0    1.13    1
 22    0    1    42    0   11    1    39    3    0    1.38    1
 23    0    1    42    0    2    1   730    3    0     .96    1
 24    0    1    58    0   82    1   136    3    1    1.62    1
 25    0    1    30    0   24    0  1775    2    0    1.06    0
 28    0    1    54    0   70    1     1    2    0     .47    0
 30    0    1    44    0   15    1   836    4    0    1.58    1
 32    0    1    64    0   16    1    60    4    0     .69    1
 33    0    1    48    0   50    0  1536    3    0     .91    0
 34    0    1    40    0   22    0  1549    2    0     .38    0
 36    0    1    48    0   45    1    54    2    0    2.09    1
 37    0    1    61    0   18    1    47    3    1     .87    1
 38    0    1    41    0    4    1     0    3    0     .87    0
 40    0    1    48    1   40    0  1367    4    0     .75    0
 41    0    1    45    1   57    0  1264    2    0     .98    0
 45    0    1    36    0    0    1    44    1    0    0.00    0
 46    0    1    48    1    1    1   993    2    0     .81    1
 47    0    1    47    0   20    1    51    3    0    1.38    1
 49    0    1    36    1   35    0  1106    4    0    1.35    0
 51    0    1    48    0   31    1   253    4    1    1.08    1
 55    0    1    52    0    9    1    51    2    0    1.51    1
 56    0    1    38    0   66    0   875    4    0     .98    0
 58    0    1    48    1   20    1   322    2    1    1.82    1
 59    0    1    41    1   77    0   838    2    0     .19    0
 60    0    1    49    0    2    1    65    3    0     .66    1
 63    0    1    32    0   26    0   815    3    1    1.93    0
 64    0    1    48    1   32    1   551    1    0     .12    0
 65    0    1    51    0   13    1    64    2    0    1.12    1
 67    0    1    19    0   56    1   228    3    0    1.02    0
 68    0    1    45    0    2    1    65    3    1    1.68    1
 69    0    1    47    0    9    0   660    2    0    1.20    0
 70    0    1    53    0    4    1    25    3    1    1.68    1
 71    0    1    47    0   30    0   954    3    0    0.97    0
 72    0    1    26    0    3    0   591    3    1    1.46    0
 73    0    1    56    0   26    1    63    3    1    2.16    1
 74    0    1    29    0    4    1    12    1    0    0.61    0
 76    0    1    52    1   45    0   498    3    1    1.70    0
 78    0    1    48    0  209    0   304    3    0    0.81    0
 79    0    1    53    0   66    1    29    2    0    1.08    1
 80    0    1    46    1   25    0   455    3    0    1.41    0
 81    0    1    52    0    5    0   438    4    1    1.94    0
 83    0    1    53    0   31    1    48    4    0    3.05    0
 84    0    1    42    0   36    1   297    4    0    0.60    1
 86    0    1    48    0    7    0   388    3    1    1.44    0
 87    0    1    46    0   59    1    50    2    0    2.25    1
 88    0    1    54    0   30    0   338    3    0    0.68    0
 89    0    1    51    0  138    1    68    4    1    1.33    1
```

```
 90   0   1   52   1  159   1   26   3   1   0.82   0
 92   0   1   44   0  309   0   29   1   0   0.16   0
 93   0   1   47   0   27   0  236   2   0   0.33   0
 94   0   1   43   1    3   1  161   3   0   1.20   1
 96   0   1   26   0   12   0  166   2   0   0.46   0
 97   0   1   23   0   20   0  109   3   1   1.78   0
 98   0   1   28   0   95   0   13   4   1   0.77   0
100   0   1   35   1   37   0    1   3   0   0.67   0
!
!
!      ID    ID number of patient
!      ZA    Censor variate for pretransplant survival.
!                           1=died before transplant
!                           0=transplanted or still waiting for transplant
!      ZB    Censor variate for waiting time to transplant
!                           1=transplanted
!                           0=died before transplant or still waiting for tplnt
!     AGE    Age at acceptance into program in years
!    SURG    Prior surgery (0=none,1=previous open heart surgery)
!     ACC    Days since January 1st,1967 to acceptance into program
!    DIED    Censor variate(0=Alive,1=dead)
!    SURV    Survival time of patient in days
!     NMM    Number of mismatches
!     HLA    HLA score
!      MM    Mismatch score
!     REJ    Rejection
!
! Recode zero survival times to 0.5
!
$CA SURV=SURV+0.5*%EQ(SURV,0)$
!
$RETURN
!
!************************************************************************
!
$SUBFILE GEHAN
!...............................................................
! Reproduced by permission of the Biometrika Trustees
! Gehan, E.A (1965), Biometrika,52,203-223
!...............................................................
$UNITS 42
!     REMISSION TIMES IN ACUTE LEUKEMIA
$DATA T $READ
1 1 2 2 3 4 4 5 5  8 8 8 8 11 11 12 12 15 17 22 23
6 6 6 6 7 9 10 10 11 13 16 17 19 20 22 23 25 32 32 34 35
$DATA W $READ
1 1 1 1 1 1 1 1 1 1 1 1 1 1 1 1 1 1 1 1 1
1 1 1 0 1 0 1 0 0 1 1 0 0 0 1 1 0 0 0 0 0
!
$CA G=%GL(2,21)$FACTOR G 2 $
!     VARIABLES:
!          T  REMISSION TIMES IN WEEKS
!          W  CENSOR VARIATE (1=UNCENSORED OBSN, 0=CENSORED OBSN)
!          G  GROUP TREATMENT(1=PLACEBO,2=6-MERCAPTOPURINE(6-MP))
!
$RETURN
!
!************************************************************************
!
$SUBFILE FEIGL
!
!...............................................................
! Reproduced from:
!  P. Feigl and M. Zelen, "Estimation of Exponential Probabilities
!  with Concomitant Information". BIOMETRICS 21: 826-838 1965
!  With permission from the Biometric Society.
```

```
!..................................................................
!
! SURVIVAL TIMES IN WEEKS OF PATIENTS WITH ACUTE MYELOGENEOUS
! LEUKAEMIA. COVARIATES WBC - WHITE BLOOD CELL COUNT IN THOUSANDS
! AND AG-FACTOR (1=POS, 2=NEG)
$UNIT 33
$DATA TIME WBC
$READ
 65     2.3  156   0.75  100    4.3  134    2.6   16    6.0
108    10.5  121  10.0     4   17.0   39    5.4  143    7.0
 56     9.4   26  32.0    22   35.0    1  100.0    1  100.0
  5    52.0   65 100.0    56    4.4   65    3.0   17    4.0
  7     1.5   16   9.0    22    5.3    3   10.0    4   19.0
  2    27.0    3  28.0     8   31.0    4   26.0    3   21.0
 30    79.0    4 100.0    43  100.0
!
$CALC AG=1+%GE(%CU(1),18) $FACTOR AG 2 $
$RETURN
!
!*************************************************************************
!
$SUBFILE PRENTICE
!
!  PRENTICE LUNG CANCER DATA
!..................................................................
! Reproduced by permission of the Biometrika Trustees
! Prentice, R.L (1973), Biometrika,60,279-288
!..................................................................
!
$UNIT 137!
$C THE CENSORED OBSERVATIONS APPEAR FIRST!
$DATA T W MFD AGE PRIO TREA TYPE STAT
$READ
 100 0  6 70 1 1 1 7.0       25 0  9 52 2 1 1 8.0       123 0  3 55 1 1 2 4.0
  97 0  5 67 1 1 2 6.0      182 0  2 62 1 1 4 9.0        87 0  3 48 1 2 1 8.0
 231 0  8 52 2 2 1 5.0      103 0 22 36 2 2 2 7.0        83 0  3 57 1 2 3 9.9
  72 1  7 69 1 1 1 6.0      411 1  5 64 2 1 1 7.0       228 1  3 38 1 1 1 6.0
 126 1  9 63 2 1 1 6.0      118 1 11 65 2 1 1 7.0        10 1  5 49 1 1 1 2.0
  82 1 10 69 2 1 1 4.0      110 1 29 68 1 1 1 8.0       314 1 18 43 1 1 1 5.0
  42 1  4 81 1 1 1 6.0        8 1 58 63 2 1 1 4.0       144 1  4 63 1 1 1 3.0
  11 1 11 48 2 1 1 7.0       30 1  3 61 1 1 2 6.0       384 1  9 42 1 1 2 6.0
   4 1  2 35 1 1 2 4.0       54 1  4 63 2 1 2 8.0        13 1  4 56 1 1 2 6.0
 153 1 14 63 2 1 2 6.0       59 1  2 65 1 1 2 3.0       117 1  3 46 1 1 2 8.0
  16 1  4 53 2 1 2 3.0      151 1 12 69 1 1 2 5.0        22 1  4 68 1 1 2 6.0
  56 1 12 43 2 1 2 8.0       21 1  2 55 2 1 2 4.0        18 1 15 42 1 1 2 2.0
 139 1  2 64 1 1 2 8.0       20 1  5 65 1 1 2 3.0        31 1  3 65 1 1 2 7.5
  52 1  2 55 1 1 2 7.0      287 1 25 66 2 1 2 6.0        18 1  4 60 1 1 2 3.0
  51 1  1 67 1 1 2 6.0      122 1 28 53 1 1 2 8.0        27 1  8 62 1 1 2 6.0
  54 1  1 67 1 1 2 7.0        7 1  7 72 1 1 2 5.0        63 1 11 48 1 1 2 5.0
 392 1  4 68 1 1 2 4.0       10 1 23 67 2 1 2 4.0         8 1 19 61 2 1 3 2.0
  92 1 10 60 1 1 3 7.0       35 1  6 62 1 1 3 4.0       117 1  2 38 1 1 3 8.0
 132 1  5 50 1 1 3 8.0       12 1  4 63 2 1 3 5.0       162 1  5 64 1 1 3 8.0
   3 1  3 43 1 1 3 3.0       95 1  4 34 1 1 3 8.0       177 1 16 66 2 1 4 5.0
 162 1  5 62 1 1 4 8.0      216 1 15 52 1 1 4 5.0       553 1  2 47 1 1 4 7.0
 278 1 12 63 1 1 4 6.0       12 1 12 68 2 1 4 4.0       260 1  5 45 1 1 4 8.0
 200 1 12 41 2 1 4 8.0      156 1  2 66 1 1 4 7.0       143 1  8 60 1 1 4 9.0
 105 1 11 66 1 1 4 8.0      103 1  5 38 1 1 4 8.0       250 1  8 53 2 1 4 7.0
 100 1 13 37 2 1 4 6.0      999 1 12 54 2 2 1 9.0       112 1  6 60 1 2 1 8.0
 242 1  1 70 1 2 1 5.0      991 1  7 50 2 2 1 7.0       111 1  3 62 1 2 1 7.0
   1 1 21 65 2 2 1 2.0      587 1  3 58 1 2 1 6.0       389 1  2 62 1 2 1 9.0
  33 1  6 64 1 2 1 3.0       25 1 36 63 1 2 1 2.0       357 1 13 58 1 2 1 7.0
 467 1  2 64 1 2 1 9.0      201 1 28 52 2 2 1 8.0         1 1  7 35 1 2 1 5.0
  30 1 11 63 1 2 1 7.0       44 1 13 70 2 2 1 6.0       283 1  2 51 1 2 1 9.0
  15 1 13 40 2 2 1 5.0       25 1  2 69 1 2 2 3.0        21 1  4 71 1 2 2 2.0
  13 1  2 62 1 2 2 3.0       87 1  2 60 1 2 2 6.0         2 1 36 44 2 2 2 4.0
  20 1  9 54 2 2 2 3.0        7 1 11 66 1 2 2 2.0        24 1  8 49 1 2 2 6.0
```

```
  99 1  3 72 1 2 2 7.0        8 1  2 68 1 2 2 8.0       99 1  4 62 1 2 2 8.5
  61 1  2 71 1 2 2 7.0       25 1  2 70 1 2 2 7.0       95 1  1 61 1 2 2 7.0
  80 1 17 71 1 2 2 5.0       51 1 87 59 2 2 2 3.0       29 1  8 67 1 2 2 4.0
  24 1  2 60 1 2 3 4.0       18 1  5 69 2 2 3 4.0       31 1  3 39 1 2 3 8.0
  51 1  5 62 1 2 3 6.0       90 1 22 50 2 2 3 6.0       52 1  3 43 1 2 3 6.0
  73 1  3 70 1 2 3 6.0        8 1  5 66 1 2 3 5.0       36 1  8 61 1 2 3 7.0
  48 1  4 81 1 2 3 1.0        7 1  4 58 1 2 3 4.0      140 1  3 63 1 2 3 7.0
 186 1  3 60 1 2 3 9.0       84 1  4 62 2 2 3 8.0       19 1 10 42 1 2 3 5.0
  45 1  3 69 1 2 3 4.0       80 1  4 63 1 2 3 4.0       52 1  4 45 1 2 4 6.0
 164 1 15 68 2 2 4 7.0       19 1  4 39 2 2 4 3.0       53 1 12 66 1 2 4 6.0
  15 1  5 63 1 2 4 3.0       43 1 11 49 2 2 4 6.0      340 1 10 64 2 2 4 8.0
 133 1  1 65 1 2 4 7.5      111 1  5 64 1 2 4 6.0      231 1 18 67 2 2 4 7.0
 378 1  4 65 1 2 4 8.0       49 1  3 37 1 2 4 3.0
!
$FACTOR PRIOR 2 TREAT 2 TYPE 4 $!
!   VARIABLES:
!         T      SURVIVAL TIME IN DAYS
!      STATUS    PERFORMANCE STATUS
!                  1,2,3 ... COMPLETELY HOSPITALIZED
!                  4,5,6 ... PARTIAL CONFINEMENT
!                  7,8,9 ... ABLE TO CARE FOR SELF
!        MFD     MONTHS FROM DIAGNOSIS TO STARTING ON STUDY
!        AGE     AGE IN YEARS
!       PRIOR    PRIOR THERAPY (1=NO,2=YES)
!       TREAT    TYPE OF TREATMENT(1=STANDARD,2=TEST)
!       TYPE     TYPE OF CANCER  (1=SQUAMOUS,2=SMALL,
!                                 3=ADENO,4=LARGE)
!        W       CENSOR VARIATE (1=UNCENSORED,0=CENSORED)
!
$RETURN
!
!***************************************************************************
!
$SUBFILE HOST
!
!.......................................................................
! Reproduced by permission of McGraw Hill Ryerson Ltd. and
!     Open University Educational Enterprises
! Erickson, B.H. and Nosanchuk, T.A. (1977) Understanding Data
!  Open University Press, Milton Keynes
!.......................................................................
!
$UNITS 19
$DATA HBEF HAFT SEX$FACTOR SEX 2$READ
51 58 1 54 65 1 61 86 1 54 77 1 49 74 1
54 59 1 46 46 1 47 50 1 43 37 1 86 82 2
28 37 2 45 51 2 59 56 2 49 53 2 56 90 2
69 80 2 51 71 2 74 88 2 42 43 2
$RETURN
!
!***************************************************************************
!
$SUBFILE BINOMIAL
!
! FILE WITH VARIOUS BINOMIAL LIKELIHOOD MACROS
!
!
$MAC RPROB
$DEL LP_ DV_ P_ OF_ X_!
$PR 'Profile deviance of ' *N %1 ' = ' %X ' from p = '%a' to '%z $!
$CA %G=1+(%Z-%A)/%I : %C=1 : %F=%G$!
$VAR %G P_ LP_ DV_$!
$CA P_=%A+%CU(%I)-%I : LP_=%LOG(P_/(1-P_)) : X_=%1-%X $OFF OF_$!
$OUT$TRA$!
$WHILE %F RPRA$!
$OUT %POC $TRA I O W F H$!
```

```
$PLOT DV_ P_ '+' $!
$LOOK DV_ P_ $!
$OFF$DEL LP_ DV_ P_ OF_$!
$$E
!
$MAC RPRA
$CA OF_=LP_(%C)$FIT X_-1$CA DV_(%C)=%DV : %C=%C+1 : %F=%LE(%C,%G) $!
$E
!
!
$MACRO LD
$DEL DV_ LD_ X_ THO_!
$PRINT 'Profile deviance of LD value (p =' %p ') for ' *n %1 ' from '!
       %a  ' to ' %z$!
$CA %G= 1 +(%Z-%A)/%I : %C=1 : %F=%G : %Z1=%A$!
$VAR %G DV_ LD_$!
$CA THO_=%LOG(%P/(1-%P)) : X_=%1-%A!
$OFF THO_!
$OUT $TRA $!
$WHILE %F LDA$!
$OUT %POC$TRA I O F W H$!
$PLOT DV_ LD_ '+'$!
$LOOK DV_ LD_$!
$DEL DV_ LD_ X_ THO_$!
$E
!
$M LDA
$FIT X_ -1 $CA LD_(%C)=%Z1 : DV_(%C)=%DV : %C=%C+1 : %F=%LE(%C,%G) !
: %Z1=%Z1+%I : X_=X_-%I$!
$E
!
!
$MACRO RELPOT
$DEL DV_ TH_ X_$!
$PRINT 'Profile deviance of relative potency of '*N %2 ' for '!
 *N %1 ' from ' %A  ' to ' %Z$!
$CA %G= 1 +(%Z-%A)/%I : %C=1 : %F=%G : %Z1=%A$!
$VAR %G DV_ TH_$!
$ARG RELA %1 %2$!
$OUT $TRA $!
$WHILE %F RELA$!
$OUT %POC$TRA I O F W H$!
$PLOT DV_ TH_ '+'$!
$LOOK DV_ TH_$!
$DEL DV_ TH_ X_$!
$E
!
$M RELA
$CAL X_=%1+%Z1*%EQ(%2,2)$FIT X_ $CA TH_(%C)=%Z1 : DV_(%C)=%DV !
: %C=%C+1 : %F=%LE(%C,%G) :%Z1=%Z1+%I$!
$E
!
$MACRO LOGODDS
$DEL DV_ LTH_ OF_ TH_$!
$PRINT 'Profile deviance of logodds for ' *n %1 ' from '!
       %a  ' to ' %z$!
$CA %G= 1 +(%Z-%A)/%I : %C=1 : %F=%G :%Z1=%A$!
$VAR %G DV_ TH_ LTH_ $!
$CA LTH_=%A+%CU(%I)-%I$!
$OFF OF_$!
$ARG LOGA %1$!
$OUT $TRA $!
$WHILE %F LOGA$!
$OUT %POC$TRA I O F W H$!
$CA TH_=%EXP(LTH_)!
$PLOT DV_ LTH_ '+'$!
```

```
$LOOK DV_ LTH_ TH_$!
$OFF$DEL DV_ LTH_ OF_ TH_$!
$E
!
$M LOGA
$CA OF_= LTH_(%C)*%1$!
$FIT $!
$CALC DV_(%C)=%DV  : %C=%C+1 : %F=%LE(%C,%G) $!
$E
!
$MACRO BINREL  !
!
!  Macro to calculate binomial relative likelihood and its
!  normal approximation.
!
!   Need to specify 5 scalars: %R ....no. of successes r
!                              %N ....no. of trials n
!                              %A ....startpoint of grid
!                              %Z ....endpoint of grid
!                              %I ....grid increment
!
!  Output: Plot 1) Relative likelihood      (+ =exact, *=normal approx.)
!               2) Log relative likelihood  (     "          "        )
!          Table of values of the following quantities:
!             P_     grid of values of p = r/n
!             R_     relative likelihood for p
!             NR_    normal approximation to R_
!             LR_    log relative likelihood for p
!             NLR_   normal approximation to LR_
!
!  Restrictions: 0 < %R < %N , 0 < %A < %Z < 1 , %I>0
!
$CA %Z1 = (%R<=0)?(%R>=%N)?(%A<=0)?(%Z>=1)?(%A>=%Z)?(%I<=0)$SW %Z1 BERR$!
$DEL R_ LR_ P_ NR_ NLR_$CA %G = 1+(%Z-%A)/%I$!
$VAR %G R_ LR_ P_ NR_ NLR_$!
$CA P_ = %A+%I*(%CU(1)-1) : %P = %R/%N !
 :  %Z3 = %R*%LOG(%P)+(%N-%R)*%LOG(1-%P)!
 :  LR_ = %R*%LOG(P_)+(%N-%R)*%LOG(1-P_)-%Z3!
 :  R_  = %EXP(LR_)!
 : NLR_ = -0.5*%N*((P_-%P)**2)/(%P*(1-%P))!
 :  NR_ = %EXP(NLR_)!
$PLOT R_ NR_ P_ '+*'$PR 'Relative likelihood (' *i %r ' out of ' *i %n !
     ')        + is exact, * is normal approx.';;$!
$PLOT LR_ NLR_ P_ '+*'$PR 'Log relative likelihood (' *i %r ' out of '!
     *i %n  ')      + is exact, * is normal approx.';;$!
$LOOK(S=1) P_ R_ NR_ LR_ NLR_$!
$DEL P_ R_ NR_ LR_ NLR_$!
$ENDM
!
$M BERR !
$PR '*** Error in macro BINREL  - one of the following restrictions'
'*** has been violated: 0 < %R < %N , 0 < %A < %Z < 1 , %I>0'$EX 2$
$ENDM
!
!
$RETURN
!
!***********************************************************************
!
$SUBFILE EXTVAL
!
!
!-----------------------------------------------------------------------
!  Author: B. J. Francis, Centre for Applied Statistics,
!                        University of Lancaster, U. K.
!  Main macros:
```

```
!       EXTVAL   Fits the extreme value distribution to survival
!                data. Survival times may be right-censored.
!       For macro EXTVAL:
!          Formal arguments:
!             %1          (obligatory) Variate containing the survival
!                         times, some of which may be right censored.
!             %2          (obligatory) Indicator or censor variate. Elements
!                         take the value 1 if uncensored and 0 if censored.
!          Macro arguments:
!             MODEL       The model formula requested (no default)
!             CONV        (optional) The convergence criterion(default .001)
!             DISP        (optional) The $DISP options used after
!                           convergence (default E)
!             CYCLE       (optional) Takes the values $CYC$ or $RECY$
!                           Determines whether $recycling is used for the
!                           iterative fitting of the EXTVAL distribution.
!                           (Default  $CYCLE$)
!          Scalar Arguments:
!             %W          Maximum number of iterations (default 15)
!  Output:
!         EXTVAL:Displays for each iteration
!                the estimate of the scale parameter,the
!                deviance and the number of degrees of freedom
!                On convergence, or after %W (15) iterations,displays
!                by default the estimates of the parameters and s.errors
!  Example of use:
!            $mac model a+b+c $endm
!            $mac disp et $endm       !  optional
!            $use EXTVAL t c$
!
!---------------------------------------------------------------------
!
!
$MAC MESS $PRI '-- Standard errors of estimates'!
' given below are underestimated'!
$$E
!
$MAC WAR1 $PRI '-- Weights not available. No weights used in fit.'$WEI $$ENDM
!
$MAC CONV .0005 $ENDM
!
$MAC DISP E $ENDM
!
$MAC CYCLE $CYCLE $ENDM
!
$MAC EXV1!
$CAL %Z6=%Z6-1: %Z6=%GT(%Z6,0)*%Z6!          update iteration counter
$OUT $TRA$CAL OFV_=%A*%1!                    update offset variate
$USE CYCLE$FIT £MODEL$OUT %Z4$TRA I O W F H$!fit model suppressing usual output
$CAL %D=%Z2 - 2*%CU(%2*%LOG(%A*%FV)-%FV)!    calculate deviance
: %B=%A/%Z7
: %F=%DF-1!                                  calculate correct DF
$PRI *5 %D %B *I %F,6$!                      and print  them out
$CAL %Z8=%CU(%1*(%FV-%2))
: %Z8=0.5*(%A-%Z1/%Z8)!                      new increment for shape parameter
: %Z8=%IF(%LT(%Z8,-5),-5,%Z8)!               should not be less than -5
: %Z9=%LE(-£CONV,%Z8)*%GE(£CONV,%Z8)!
$EXI %Z9!                                    test for convergence
$CAL %A=%A-%Z8 $$ENDM!                       update shape parameter
!
$MAC EXTVAL!
$DEL OFV_ T_!
$PR *M 20 '-- Model is ' MODEL $
$TAB THE %1 LARGEST INTO L_$CA %Z7=L_$DEL L_$
$CA T_=%1/%Z7!
 : %Z1=%CU(%2)!
```

```
 : %Z2=2*%Z1*%LOG(%Z7)!
 : %Z4=%COC !                                   store current output channel
 : %Z6=%IF(%GT(%W,0),%W,15)!                    max no. of iterations
 : %A=1 !
$YVA %2 $ERR P$CYC!
$SWI %PWF WAR1$OFF OFV_!                        switch weights off if set
$PR 'Extreme Value Fit':$
$PR : ' Deviance   shape     df'!               Print headings
 : '            parameter'!
$ARG EXV1 T_ %2$WHI %Z6 EXV1!                   use EXV1 until convergence
$USE MESS$DIS £DISP$CYC$OFF!
$CA %A=%B$DEL T_ OFV_$!
$$END
!
!
$RETURN
!
!*********************************************************************
!
$SUBFILE CLOGISTIC
!
!
!--------------------------------------------------------------------
!  Main macros:
!       LOGISTIC  Fits the LOGISTIC distribution to survival
!                 data. Survival times may be left or right-censored.
!       LLOGISTIC  Fits the LOG-LOGISTIC distribution to survival
!                 data. Survival times may be left or right-censored.
!       For macro LOGISTIC or LOG-LOGISTIC:
!               Formal arguments:
!                  %1         (obligatory) Variate containing the survival
!                             times, some of which may be censored.
!                  %2         (obligatory) Indicator or censor variate. Elements
!                             take the value 2 if left-censored, 1 if uncensored
!                             and 0 if right-censored.
!               Macro arguments:
!                  MODEL      The model formula requested (no default)
!                  DISP       (optional) The $DISP options used after
!                               convergence (default E)
!                  CONV       (optional) The convergence criterion(default .001)
!               Scalar Arguments:
!                  %W         (optional) Maximum no. of iterations(default 15)
!       Output(both macros):
!                     Displays for each iteration the estimate of the
!                     shape parameter (sigma), the deviance and the number of
!                     degrees of freedom.
!                     On convergence, or after %W (15) iterations,displays
!                     by default the estimates of the parameters and s.errors
!                     Scalar output: %d - deviance
!                                    %f - d.f.
!                                    %s - shape parameter (sigma).
!                                    %sc- sigma squared
!       Example of use:
!                  $mac MODEL a+b $endm
!                  $mac DISP et $endm      !  optional
!                  $use LOGISTIC t c$
!--------------------------------------------------------------------
!
!    Scalars used by macro:
!
!    %Z1 - No. of uncensored obns.
!    %Z2 - Initial updated estimate of shape parm.
!    %Z3 - Switch for convergence or £iterations
!    %Z4 - Current output channel
!    %Z5 - Iteration counter
!    %Z6 - Maximum number of iterations
```

```
!    %Z7 - Deviance correction for log-logistic
!    %Z8 - Increment for updated shape parameter/(at end) S squared
!    %Z9 - Switch for distribution message
!
!------------------------------------------------------------------------
!
$MAC MESS $PRI '-- Standard errors of estimates'!
' given below are underestimated'!
$$E
!
$MAC MOD1 $PRI 'LOGISTIC fit' $$ENDM
!
$MAC MOD2 $PRI 'LOG-LOGISTIC fit' $$ENDM
!
$MAC WAR1 $PRI '-- Prior weights not allowed - No weights used in fit.'$WEI $$E
!
$MAC WAR2 $PRI '-- Offset is not allowed - No offsets used in fit.'$OFF$$ENDM
!
$MAC CONV .001 $ENDM
!
$MAC DISP E $ENDM
!
$MAC LOGISTIC !
$DEL TT_$CAL TT_=%1: %Z9=1: %Z7=0$ARG LOG1 %1 %2$USE LOG1$$E
!
$MAC LLOGISTIC !
$DEL TT_$CAL TT_=%LOG(%1+%EQ(%1,0)*0.5)$CA %Z9=2:%Z7=2*%CU(%EQ(%2,1)*TT_)$!
$ARG LOG1 %1 %2$USE LOG1$$E
!
$MAC LOG2
$CAL OFS_=TT_/%S$!
$FIT £MODEL$!
$CAL P_=%FV/BD_$!
  :  %Z2=%CU((%FV-BR_)*%LP*%S)/%Z1$!
  :  %F=%DF-1$!
  :  %D=%Z7+2*%Z1*%LOG(%S)-2*%CU((BR_*%LOG(P_))+((BD_-BR_)*%LOG(1-P_)))!
$OUT %Z4$TRA I O W F H$!
$PRI *5 %D %S *I %F,6$!                      and print  them out
$EXI %Z3$!
$CAL %Z3=%GE(%Z5,%Z6)$EX %Z3$!
$OUT $TRA$
$CA %Z8=0.5*(%S-%Z2)!                       new increment for shape parameter
:   %Z3=%LE(-£CONV,%Z8)*%GE(£CONV,%Z8)!
$CAL %S=%S-%Z8:%Z5=%Z5+1 $$ENDM!                       update shape parameter
!
!
$MAC LOG1 !
$DEL WT_ BR_ BD_ OFS_ P_!
$SWI %PWF WAR1$SWI %OSF WAR2$!               switch off offset and weights
$PRI *M 20 '-- Model is ' MODEL :$!
$SWI %Z9 MOD1 MOD2$!
$CAL WT_=%EQ(%2,1)!
 :   %Z1=%CU(WT_)!
 :   %Z4=%COC !                                store current output channel
 :   %Z5=1 :%Z3=0 !
$PRI : ' Deviance   shape     df'!            Print headings
 : '             parameter'!
$OUT $TRA$!
$WEI WT_ $YVA TT_$ERR N$OFF$FIT £MODEL$!                get starting values
$CAL %S=%SQRT(3*%SC)/%PI$!
 :   %Z6=%IF(%GT(%W,0),%W,15)!                 max no. of iterations
 :   BD_=WT_+1 : BR_=%NE(%2,0) !
$ERR B BD_$YVA BR_$OFF OFS_$WEI $!
$WHI %Z5 LOG2$!
$D E$!
$CAL OFS_=(%S+1)*%LP-TT_ : %Z8=%S*%S $SCA %Z8 $FIT .$!
```

```
$USE MESS$DIS £DISP $CYC $OFF $SCA 1$!
$DEL WT_ BR_ BD_ TT_  OFS_ P_!
$$END
!
!
$RETURN
!**********************************************************************
!
$SUBFILE KAPLAN
!
$M KAPLAN !
$DEL TYP_ DO_ D1_ IND_ DT_ DS_ H_ SF_ RS_ ZA_ ZC_!
$CA TYP_=(%2==1)$TAB FOR %1 WITH TYP_ USING D1_ BY  DT_$!
$CA TYP_=(%2==0)$TAB FOR %1 WITH TYP_ USING DO_ BY  DT_$!
$CA %Z1=%CU(DT_==DT_):%Z4=%CU(%1==%1) !
$CA %Z2=%CU(D1_>0)$CA %Z9=(%Z2==0)$SWI %Z9 KPER$VAR %Z2 ZA_ ZC_$!
$VAR %Z1 IND_$CA IND_=%CU(1)-1$CA DS_=%CU(DO_+D1_)!
$CA RS_=%Z4-DS_(IND_) : H_=1-D1_/RS_ $!
:%Z3=1 :  %Z3=SF_=H_*%Z3$! find cu. prod. of H_ and store in SF_
$CA IND_=D1_>0 : IND_=IND_*%CU(IND_) : ZA_(IND_)=SF_: ZC_(IND_)=DT_$!
!$DEL TYP_ DO_ D1_ IND_ DT_ DS_ H_ SF_ RS_ $
$PR 'Kaplan-Meier Survivor function estimate'$
$PLOT (S=1) ZA_ ZC_$PR 'survivor function is stored in ZA_'$
$E
!
$M KPER !
$PR '***No uncensored death times..macro cannot continue'$!
$DEL D1_ DO_ DT_ TYP_$EX 2$ !
$E
!
$RETURN
!
!**********************************************************************
!
$SUBF CNORMAL!
!
!---------------------------------------------------------------------------
!  Author: B. J. Francis, Centre for Applied Statistics,
!                         University of Lancaster, U. K.
!  Version 1.1 GLIM 3.77 February 1987
!  Main macros:
!       NORM    Fits a normal distribution to survival data.  Survival
!               times may be right censored.
!       LNORM   Fits a log-normal distribution to survival data.  survival
!               times may be right censored.
!            Formal arguments: (Both macros)
!               %1 Value containing the survival times.
!               %2 Indicator or censor variate.  Elements take the value
!               1 if uncensored and 0 if censored.
!            Macro arguments:  (Both macros)
!               MODEL The model formula requested (obligatory)
!            Scalar arguments: (Both macros)
!               %W Maximum number of iterations (default is 15)
!  OUTPUT:      The value of the deviance and its standard error are displayed
!               after each successive fit of the model.
!  EXAMPLE OF USE:
!               $mac model G$endm
!               $use norm T C$
!               $use lnorm T C$
!---------------------------------------------------------------------------
!
!
$MAC  NORM !
$DEL T_ SR_ HZ_ AYV_$SWI %PWF NCER!
$PRI *M 20 '-- Model is ' MODEL!
$ARG NORI %1 %2$CA T_=%1:%Z1=0:%Z7=%COC$PR ' Normal Fit'$USE NORI!
```

```
$DEL AYV_ T_$$ENDM!
!
$MAC  LNORM!
$DEL T_ SR_ HZ_ AYV_$SWI %PWF NCER!
$PRI *M 20 '-- Model is ' MODEL!
$ARG NORI %1 %2$CA T_=%LOG(%1+0.5*%EQ(%1,0)):%Z1=2*%CU(T_*%2):%Z7=%COC!
$PR ' Lognormal Fit'$USE NORI!
$DEL AYV_ T_$$ENDM!
!
$MAC NORINIT!
$CAL %Z2=%CU(%2) : %Z3=%IF(%GT(%W,0),%W,15)!
$YVAR T_ $ERR N  $OU $TRA $FIT £MODEL $OU %Z7$TRA I O W F H!
$CAL %S=%SQRT(%DV/%DF)*1.1!
$PR ' Deviance   Shape      DF':'          Parameter'!
$ARG NORC T_ %2 $WHILE %Z3 NORC $USE MESS$D E$$ENDM!
!
$MAC NORC!
$CAL %Z3 =%Z3-1 : %Z3=%GT(%Z3,0)*%Z3!
 : SR_=(%1-%FV)/%S : HZ_=.3989*%EXP(-(SR_*SR_)/2)/(1-%NP(SR_))!
 :  %D=%CU(%2*SR_*SR_)-2*%CU((1-%2)*%LOG(1-%NP(SR_)))!
 :  %D=2*%Z2*(%LOG(%S)+.91894) +%D+%Z1!
$PRI *5 %D %S *-1 %DF$CA %Z4=%Z4-%D!
 : %Z4=%GE(.0001,%Z4*%Z4)$EX %Z4$CAL %Z4=%D!
$OU$TRA $CAL AYV_=%1*(%2) + (1-%2)*(%FV + %S*HZ_)!
$YVAR AYV_!
$FIT £MODEL $OU %Z7$TRA I O W F H!
$CAL %Z5=%CU(%2*((%1-%FV)**2)) : %Z6=%Z2-%CU((1-%2)*SR_*HZ_)!
:%S=%SQRT(%IF(%GT(%Z6,0),(%Z5/(%Z6+%EQ(%Z6,0))),(%S*%S)))$!
$$ENDM!
!
$MAC MESS $PRI '-- Standard errors of estimates'!
' given below are underestimated'!
$$E!
!
$MAC NCER !
$PRI '---Macro can not be used if prior weights are set.'$!
$EXI 2$$ENDM!
!
$RETURN!
!
!**********************************************************************
$SUBFILE PIECEWISE
!
!   macros to fit the piecewise exponential survival model
!            M. Green    Centre for Applied Statistics
!                        University of Lancaster, U.K.    Sept 1985
!
!     macro        parameters
!     CUTP         %1   survival time
!                  %2   censor indicator  0 - censored
!                                         1 - uncensored
!                  %3   cut points for piecewise exponential (output)
!        CUTP finds the values of the uncensored survival times
!
!     INIB         %4 - weight variate (optional)
!     INIC
!        INIB,INIC  initialise the fitting procedure for methods B,C
!        %W used to record whether a weight was set in INIB and INIC
!
!     FITB           %1   censor indicator
!     FITC
!        FITB,FITC  fit the model stored in macro MODEL using method B,C
!        %T  controls tolerance for terminating iterations
!            set at 0.00001 in INIB,INIC.
!        %N  Maximum no. of iterations set to 10
!
```

```
!     RESET        %1   cut points
!        RESET  deletes temporary variates and %1 for method C
!
!     PHAZ         %1   cut points
!        PHAZ  produces a plot of log hazard versus log survival time for
!              methods B and C only.
!
!-----------------------------------------------------------------------
!
$macro INIB !  for method B
$use INIT %1 %2 %3 $!
$cal %T=0.00001 $! tolerance
!   total W over i
$cal WJ_=0 : WJ_(BLOK)=WJ_(BLOK)+%2 $!
!   declare model
$cal %W=%a4 $!  weight set?
$units %z9 $yvar WV_ $error P $link L $weight IWT_ $!
$cal %fv=1 $arg spwb %4 $switch %W spwb $$!
$endmac
!
$macro spwb !
!   macro to expand a weight variate
!   %1   input weight variate
$cal PW_=%1 $!
$use XPCC PW_ $!
$endmac
!
$macro INIT !
$group NI_=%1 intervals * %3 * $!
$cal NI_=NI_-(%1==%3(NI_-1)) : CNI_=%cu(NI_) $!
!   expand censor variable
$cal %z9=%cu(NI_) $var %z9 TMP_ $!
$cal TMP_=0 : TMP_(CNI_)=%2 $assign %2=TMP_ $!
!   generate exposure time
$cal GAP_=%3-%3(%cu(1)-1) $var %z9 IND_ $!
$cal IND_=1 : IND_(CNI_)=1-NI_ : IND_=%cu(IND_) $!
$cal TMP_=GAP_(IND_) : TMP_(CNI_)=%1-%3(NI_-1) $!
$cal ET_=TMP_ $!
!   generate block factor
$cal %z1=0 : %z1=%if((NI_>%z1),NI_,%z1) $!
$cal IND_(CNI_)=NI_ $assign BLOK=IND_ $!
$factor BLOK %z1 $var %z1 HAZ_ WJ_ $!
$endmac
!
$macro INIC !
$cal %T=0.00001 : %N=10 $!
$cal %Z1=%cu(%1==%1) $var %Z1 NI_ D_ $!
$cal %Z1=%cu(%3==%3)+1 $var %Z1 ET_ WJ_ HAZ_ GK_ $!
$cal ET_=%3(%gl(%Z1-1,1))-%3(%gl(%Z1-1,1)-1) $!
$cal ET_(%Z1)=0 $!
$cal NI_=0 : D_=0 $!
$cal %Z3=0 : %Z2=%Z1 $arg STP2 %1 %3 $while %Z2 STP2 $!
$cal %Z3=0 : %Z2=%Z1 $arg STP3 %2 $while %Z2 STP3 $!
! declare model
$cal %W=%a4 $!  weight set?
$error P $link L $weight ITW_ $yvar WV_ $!
$cal PW_=1 $arg SPWC %4 $switch %W SPWC $!
$cal %fv=1 $use CNDI $!
$endmac
!
$macro SPWC ! set prior weight
$cal PW_=%1 $$endmac
!
$macro XPCC !
! macro to expand constant covariates
!    %1,%2,... covariates
```

```
$cal IND_=0 : IND_(CNI_-NI_+1)=1 : IND_=%cu(IND_) $!
$cal %z2=%a1+%a2+%a3+%a4+%a5+%a6+%a7+%a9 : %z1=0 $!
$arg XCCL %1 %2 %3 %4 %5 %6 %7 %8 %9 $!
$while %z2 XCCL $!
$endmac
!
$macro XCCL !
$cal %z1=%z1+1 : %z2=%z2-1 $!
$cal TMP_=%%z1(IND_) $assign %%z1=TMP_ $!
$endmac
!
$macro XPRM !
!   macro to expand repeated measurements covariates
!     %1 - ID of case
!     %2 - time of measurement
!     %3 - cut-points
!     %4,%5,...   covariates
$ca %z1=%cu(%1)$var %z1 I_$ca I_=%cu(1)$!
$FC_=(%1(I_)>%1(I_-1)) : FC_=1-FC_$!
$group IT_=%2 intervals * %3 * $!
$cal IT_=(IT_-((%2==%3(IT_-1))&(%2/=0))+CNI_(%1-1))*FC_ $!
$cal IND_=0 : IND_(IT_)=IND_(IT_)+1$!
$cal IND_=%if(%eq(BLOK,1),IND_+1,IND_)$!
$cal IND_=%cu(IND_) $delete IT_ FC_ I_$!
$cal %z2=%a4+%a5+%a6+%a7+%a8+%a9 : %z1=0 $!
$arg XRML %4 %5 %6 %7 %8 %9 $while %z2 XRML $!
$endmac
!
$macro XRML !
$cal %z1=%z1+1 : %z2=%z2-1 $!
$cal TMP_=%%z1(IND_) $assign %%z1=TMP_ $!
$endmac
!
$macro XPTR !
!   macro to generate treatment factor
!     %1 - time of start of treatment
!     %2 - cut-points
!     %3 - factor (output)
$group IT_=%1 intervals * %2 * $!
$cal IT_=IT_+CNI_(%cu(1)-1) : IT_=%if(%gt(IT_,%z9),0,IT_) $!
$factor %z9 %3 2 $cal %3=0 : %3(CNI_-NI_+1)=-1 : %3(IT_)=%3(IT_)+1 $!
$cal %3=%cu(%3) : %3=%3+2 $!
$delete IT_ $endmac
!
$macro XPTI !
!   macro to generate time variable
!     %1 - survival time
!     %2 - cut-points
!     %3 - time variable (output)
$cal IND_=1 : IND_(CNI_)=1-NI_ : IND_=%cu(IND_) $!
$cal %3=%2(IND_) : %3(CNI_)=%1 $!
$endmac
!
$macro CUTP !
!  macro to find ordered observed death times
!    %1 -  survival time
!    %2 -  censor indicator
!    %3 -  ordered death times (cut-points) (output)
$warn $tab for %1 with %2 by VAL_ using F_ $warn $!
$cal %z3=%cu(F_/=0) $var %z3 %3 $!
$cal IND_=(F_/=0) : IND_=%cu(IND_)*IND_ : %3(IND_)=VAL_ $!
$del VAL_ F_ IND_ $!
$endmac
!
$macro DEVU $warn $cal %1=2*%cu(%4*%3-%2*%log(%3)) $warn
$endmac
```

```
$macro DEVW $warn $cal %1=2*%cu(%5*(%4*%3-%2*%log(%3))) $warn
$endmac
!
$macro DEVB !
!  macro to find correct deviance
$cal FV_=HAZ_(BLOK)*%fv $!
$arg DEVU %1 %2 FV_ ET_ : DEVW %1 %2 FV_ ET_ PW_ $!
$cal %z1=%W+1 $switch %z1 DEVU DEVW $!
$endmac
!
$macro ITRB !
$cal HAZ_=0 : HAZ_(BLOK)=HAZ_(BLOK)+ET_*%fv : HAZ_=WJ_/HAZ_ $!
$cal IWT_=ET_*HAZ_(BLOK) : WV_=%1/IWT_ $switch %W PWT  $!
$fit £MODEL $!
$cal %z7=%z7+1 $!
$use DEVB %z8 %1 $!
$cal %z5=%if(%z5<%z8,%z8-%z5,%z5-%z8) : %z5=%z5/%z8 $!
$cal %z6=(%z5>%T) & (%z7/=%N) $!
$out %POC $trans 0 $print *I %z7 *6 %dv *6 %z8 $out $trans $!
$cal %z5=%z8 $!
$endmac
!
$macro PWT !
$cal IWT_=IWT_*PW_ $!
$endmac
!
$macro FITB !
$arg ITRB %1 $!
$out $trans $cal %Z5=-1 : %Z6=1 : %Z7=0 $recycle 2 $use ITRB $!
$recyc 1 $while %Z6 ITRB $!
$out %POC $trans I 0 W $!
$cal %Z8=%df-%Z1 $!
$print ; ' cycle ' *I %Z7 '      deviance '*5 %Z5 '  on '*I %Z8 ' d.f.' $!
$endmac
!
$macro STP2 !
$cal %Z3=%Z3+1 : %Z2=%Z2-1 $
$cal NI_=NI_+(%1 > %2(%Z3-1)) $!
$cal D_=%if((%1 > %2(%Z3-1)),%1-%2(%Z3-1),D_) $!
$endmac
!
$macro STP3 !
$cal %Z2=%Z2-1 : %Z3=%Z3+1 $!
$cal WJ_(%Z3)=%cu((%Z3==NI_)*%1) $!
$endmac
!
$macro CNDJ !
$cal GK_=%cu(ET_*HAZ_) $!
$cal ITW_=GK_(NI_-1)+D_*HAZ_(NI_) $!
$warn $cal WV_=%1/ITW_ : ITW_=ITW_*PW_ $warn !
$endmac
!
$macro CNDI !
$cal %Z2=%Z1 : %Z3=0 $while %Z2 STP4 $!
$endmac
!
$macro STP4 !
$cal %Z3=%Z3+1 : %Z2=%Z2-1 $!
$cal %Z4=ET_(%Z3)*%cu((%Z3 < NI_)*%fv)+%cu((%Z3==NI_)*D_*%fv) $!
$cal HAZ_(%Z3)=WJ_(%Z3)/%Z4 $!
$endmac
!
$macro ITRC !
$use CNDJ %1 $!
$fit £MODEL $!
$cal %Z7=%Z7+1 $!
```

```
$use CNDI $!
$use DEVU %z8 %1 %fv %pw $cal %z8=%z8-2*%1*%log(HAZ_(NI_)) $!
$cal %z5=%if(%z5<%z8,%z8-%z5,%z5-%z8) : %z5=%z5/%z8 $!
$cal %z6=(%z5>%T) & (%z/=%N) $!
$print 'cycle ' *I %z7 '  deviance ' *6 %z8  $!
$cal %z5=%z8 $!
$endmac
!
$macro FITC !
$out $trans $!
$cal %z5=-1 : %z6=1 : %z7=0 $use ITRC %1 $!
$recyc 1 $while %z6 ITRC $!
$out %POC $trans I O W F H $cycle $!
$cal %z8=%df-%z1 $!
$print ; ' cycle ' *I %z7 '     deviance '*5 %z5 '  on '*I %z8 ' d.f.' $!
$endmac
!
$macro PHAZ !
$var %z1 %re $!
$ass LCUT=%1,0 $!
$extr %pe $!
$warn $cal LCUT=%log(LCUT) : LHAZ=%pe(1)+%log(HAZ_) : %re=1 : %re(%z1)=0 $!
$warn !
$print '                  log hazard  v  log survival time' $!
$plot(rows=20) LHAZ LCUT '+' $!
$delete LCUT LHAZ %re $!
$endmac
!
$macro RESET !
$delete %1 NI_ D_ ET_ WJ_ HAZ_ GK_ ITW_ $!
$endmac
!
$RETURN
!
$FINISH
!\\\\\\\\\\\\\\\\\\\\\\\\\\\\\\\\\\\\\\\\\\\\\\\\\\\\\\\\\\\\\\\\\\\\\\\\\\\\\\\
```

Author Index

Subject Index